Advances in Environmental Hydraulics

Advances in Environmental Hydraulics

Topic Editors

Helena M. Ramos
Juan Antonio Rodríguez Díaz
Jorge Matos

MDPI

Basel • Beijing • Wuhan • Barcelona • Belgrade • Novi Sad • Cluj • Manchester

Topic Editors
Helena M. Ramos
Department of Civil Engineering, Architecture and Environment
University of Lisbon
Lisbon
Portugal

Juan Antonio Rodríguez Díaz
Department of Agronomy
University of Córdoba
Córdoba
Spain

Jorge Matos
CERIS, Instituto Superior Técnico
University of Lisbon
Lisboa
Portugal

Editorial Office
MDPI AG
Grosspeteranlage 5
4052 Basel, Switzerland

This is a reprint of the Topic, published open access by the journals *Energies* (ISSN 1996-1073), *Fluids* (ISSN 2311-5521), *Journal of Marine Science and Engineering* (ISSN 2077-1312), *Sustainability* (ISSN 2071-1050), and *Water* (ISSN 2073-4441), freely accessible at: https://www.mdpi.com/topics/917J1YM1K6.

For citation purposes, cite each article independently as indicated on the article page online and as indicated below:

Lastname, A.A.; Lastname, B.B. Article Title. *Journal Name* **Year**, *Volume Number*, Page Range.

ISBN 978-3-7258-6994-7 (Hbk)
ISBN 978-3-7258-6995-4 (PDF)
https://doi.org/10.3390/books978-3-7258-6995-4

Contents

About the Editors

Helena M. Ramos

Helena M. Ramos is a full-time professor at Instituto Superior Técnico (IST), University of Lisbon, and is a member of CERIS Research Center. A globally acclaimed expert in hydraulics, hydropower, hydraulic transients, and the water–energy nexus, she holds a PhD and Habilitation from IST. Ranked among Stanford's Top 2% Scientists (2021–2022) and Research.com's Top Engineering & Technology Scientists (2021), she has over 400 publications (7 books and 182 peer-reviewed journals) with an h-index of 35. She has led 32 national/international projects, supervised 50+ graduate theses, and taught 15 civil engineering courses. With editorial roles (e.g., MDPI's *Water*) and international collaborations—including a guest professorship at Nanjing University of Aeronautics and Astronautics—she excels in bridging research, education, and global impact.

Juan Antonio Rodríguez Díaz

Juan Antonio Rodríguez Díaz is a full-time professor at the University of Córdoba (Spain), specializing in agricultural water management, irrigation engineering, and the water–food–energy nexus. He holds a PhD in Water Resources Management (2004) from the University of Córdoba, where he also earned his Agricultural Engineering degree (2000). A recipient of the ICID WatSave Award for Young Researchers (2004), he completed a postdoc at Cranfield University's Centre for Water Science (UK, 2005–2007) and served as a Ramón y Cajal Researcher (2008) via Spain's Ministry of Science. With over 112 peer-reviewed journal papers, his work focuses on precision irrigation, energy use in pressurized water networks, and climate-resilient irrigation. He contributes to global best practices through EU and ICID collaborations.

Jorge Matos

Jorge Matos is an associate professor at the Department of Civil Engineering, Architecture and Georesources, Instituto Superior Técnico (IST), University of Lisbon, and a core member of CEHIDRO (Centre for Hydrosystems Research) and CERIS. Specializing in hydraulic structures, spillway hydraulics, air–water flows, physical modeling, and multi-block numerical simulations, he holds MSc and PhD degrees in Civil Engineering/Hydraulics from IST. He is a recipient of the APRH Award (2001/2002) and ASCE EWRI J.C. Stevens Award (2002) and chaired the IAHR Hydraulic Structures Committee (2005–2009), of which he served as Secretary (2003–2005). He coordinates IST's participation in KIC Innoenergy's ENTECH Master Program and contributes to the MSc in Energy Engineering and Management's Scientific Commission. His impactful research—featured in journals like *Environmental Fluid Mechanics*—and editorial roles underscore his leadership in global hydrosystems research.

water

MDPI

Editorial

Editorial: Advances in Environmental Hydraulics

Helena M. Ramos [1], Juan Antonio Rodríguez Díaz [2,*] and Jorge Matos [1]

1 Civil Engineering, Architecture and Environment Department, Civil Engineering Research and Innovation for Sustainabiity, Instituto Superior Técnico, Universidade de Lisboa, Av. Rovisco Pais, 1049-001 Lisboa, Portugal; helena.ramos@tecnico.ulisboa.pt (H.M.R.); jorge.matos@tecnico.ulisboa.pt (J.M.)
2 Department of Agronomy, University of Cordoba, Campus Rabanales, 14014 Cordoba, Spain
* Correspondence: jarodriguez@uco.es

1. Introduction

Water-related challenges represent one of the most pressing global concerns of the 21st century. Environmental hydraulics describes the study of flow behavior, hydrodynamic processes, and their environmental implications; it has emerged as a critical field for addressing sustainable water management, ecosystem preservation, and adaptive engineering solutions in the face of climate change. This Special Issue on "Advances in Environmental Hydraulics" provides an international platform for disseminating cutting-edge research and innovative engineering applications that combines theoretical foundations with practical, field-based implementations.

The scope of environmental hydraulics encompasses a diverse range of water-related problems, including sustainable management practices that optimize resource utilization, eco-hydraulic engineering solutions that restore aquatic ecosystem functions, and innovative flood management and energy nexus considerations. Modern environmental hydraulics increasingly emphasizes the integration of computational modeling with experimental validation and field observations, recognizing that complex water systems require multidisciplinary approaches that link hydrology, ecology, engineering, and socioeconomic considerations [1].

2. Context and Motivation

Contemporary water systems face unprecedented pressures from multiple directions. For instance, climate change alters precipitation patterns and hydrological cycles, demanding adaptive management strategies and improved predictive capabilities [Contribution 1]. Urbanization results in concentrated population and infrastructure in water-dependent regions, intensifying competition for the limited freshwater resources and escalating flood risks. Agricultural intensification, which depends critically on irrigation systems, consumes substantial volumes of water while simultaneously requiring a sustainable management approach to preserve water quality and ecosystem health [2]. Additionally, the nexus between water management and energy production—particularly in hydropower systems—introduces complex trade-offs between energy generation, environmental sustainability, and system efficiency.

These converging challenges necessitate integrated solutions that transcend traditional disciplinary boundaries. Environmental hydraulics provides the theoretical and practical framework for developing such solutions, incorporating the following factors:

- Advanced computational methods for flow simulation and prediction, including computational fluid dynamics (CFD), large eddy simulations (LES), and machine learning approaches;

 https://doi.org/10.3390/w18030381

- Field-based experimental investigations that validate proposed models and provide ground-truth data for understanding real-world flow behavior and hydrodynamic processes;
- Eco-hydraulic engineering solutions that restore ecological functions while maintaining water infrastructure performance;
- Innovative design approaches that enhance system efficiency, resilience, and environmental compatibility;
- Monitoring and management strategies that support adaptive water resource governance in response to climate variability.

3. Contributions Included in This Topic

This Special Issue comprises 15 peer-reviewed papers that exemplify the breadth and depth of contemporary environmental hydraulics research. The contributions span multiple dimensions of the field and are outlined in the following.

3.1. *Climate Change and Hydrological Adaptation*

The impacts of climate change on water systems represent a central challenge for environmental hydraulics. Katip and Anwar [Contribution 1] employed feed-forward neural networks to simulate climate-based hydrological changes at Doğancı Dam in Turkey, demonstrating the application of machine learning in predictive modeling under non-stationary climate conditions. Similarly, research on environmental flow assessment in the Yen River Basin (Vietnam) illustrated holistic approaches integrating hydrological, ecological, and management considerations to support adaptive water governance in the face of climate uncertainty [Contribution 2].

3.2. *Advanced Hydraulic Modeling and Simulation*

Computational approaches form the foundation of modern environmental hydraulics. There are multiple contributions advancing modeling capabilities: Liu et al. [Contribution 3] applied a hydrodynamic simulation (MIKE21) to elucidate the mechanisms governing pollutant transport in coastal waters, while numerical investigations into the optimal placement of portable fishways demonstrated the application of computational methods to enhancing ecosystem connectivity in agricultural drainage systems [Contribution 4]. Zöschg [Contribution 5] employed large eddy simulations to quantify head losses through trash racks at hydropower installations, providing actionable information for improving energy efficiency in renewable energy systems. Weijermars and Afagwu [Contribution 6] developed analytical solutions for pressure transient behavior in bounded and unbounded reservoirs, advancing fundamental theory for both groundwater and petroleum systems.

3.3. *Eco-Hydraulic Engineering and Ecosystem Restoration*

In the field of environmental hydraulics, the restoration and maintenance of aquatic ecosystem functions is increasingly emphasized. Calluaud et al. [Contribution 7] investigated flow conditions and fish passage efficiency in vertical slot fishways through detailed behavioral observations, providing empirical guidance for optimizing fish migration corridors. Research on hydraulic roughness in mountain headwater streams [Contribution 8] contributed to the understanding of natural flow–habitat relationships essential for stream restoration and ecosystem-based management. Zhang et al. [Contribution 9] characterized vegetation resistance in muddy water flows, building fundamental knowledge for modeling ecosystem–water interactions in engineered and natural systems.

 https://doi.org/10.3390/w18030381

3.4. Innovation in Hydraulic Systems Design

Contemporary hydraulic challenges demand innovative design solutions. The study on using friction groups for hydraulic parameter estimation [Contribution 10] provided practical tools for solving pressure drop, flow, and diameter estimation problems in piping and open-channel systems. Research on capillary barrier design and water storage capacity [Contribution 11] addressed subsurface water management relevant to both agricultural and environmental engineering applications. Studies on scour protection around bridge piers [Contribution 12] exemplified the application of hydraulic engineering to critical infrastructure resilience and safety.

3.5. Energy–Water Nexus

The relationship between water management and energy production remains central to sustainable development. Research on water consumption rates in cascade hydropower stations [Contribution 13] demonstrated systematic approaches to optimizing energy generation efficiency while considering water availability constraints. Investigations into flood discharge characteristics at high-altitude dam projects [Contribution 14] addressed the safety and operational challenges emerging from climate variability and topographic extremes, illustrating the complexity of integrating water security with energy production in vulnerable environments.

3.6. Nonlinear Wave Dynamics and Shallow Water Hydraulics

Fundamental advances in the understanding of wave generation and propagation in shallow water environments [Contribution 15] provided theoretical foundations that are applicable to coastal engineering, tsunami modeling, and river dynamics. These contributions bridge academic theory and practical applications in hazard mitigation and infrastructure design.

4. Overarching Themes and Future Perspectives

Several overarching themes emerge from these contributions:

Integration of modeling approaches: Effective solutions require combining computational modeling (CFD, machine learning, analytical solutions) with experimental validation and field observation. No single approach captures the full complexity of environmental hydraulic systems.

Climate adaptation and resilience: Climate change fundamentally alters hydrological patterns, necessitating adaptive management strategies informed by predictive modeling, long-term monitoring, and flexible system design.

Ecosystem–infrastructure coupling: Modern environmental hydraulics recognizes that engineered water systems and natural ecosystems are inextricably linked. Solutions must optimize functionality across both domains.

Efficiency and sustainability imperatives: Whether addressing energy generation, agricultural water use, or pollutant transport, contemporary hydraulic engineering emphasizes resource optimization and environmental protection as complementary rather than competing objectives.

Interdisciplinary knowledge integration: Environmental hydraulics problems require expertise from a variety of disciplines, including civil engineering, environmental science, ecology, economics, and data science, necessitating genuinely collaborative research approaches.

 https://doi.org/10.3390/w18030381

5. Concluding Remarks

This Special Issue demonstrates that environmental hydraulics remains a vibrant, evolving discipline that is increasingly central to global sustainability challenges. The research presented illustrates how contemporary hydraulic science combines theoretical rigor with practical applicability, sophisticated computational methods with empirical validation, and engineering innovation with ecological awareness.

The field faces continuing evolution driven by climate change impacts, urbanization pressures, agricultural transformation, and technological advancement. The future of this field must focus on emerging opportunities, such as the enhanced integration of artificial intelligence and machine learning for adaptive water management, development of nature-based solutions combining ecological and hydraulic principles, and improved connections between local hydraulic studies and global water security frameworks.

This topic represents a snapshot of current progress; future contributions will undoubtedly expand the field's capability to address increasingly complex water challenges in an uncertain environmental future.

Conflicts of Interest: The authors declare no conflicts of interest.

List of Contributions:

1. Katip, A.; Anwar, A. Simulating the impacts of climate change on the hydrology of Doğancı Dam in Bursa, Turkey, using feed-forward neural networks. *Sustainability* **2025**, *17*, 6273.
2. Tong, T.P.; Hoang, S.T.; Bui, D.Q.; Ha, N.T.; Nguyen, L.H.; Nguyen, L.M.; Tran, C.K. Applying a holistic approach to environmental flow assessment in the Yen River Basin. *Water* **2024**, *16*, 1174.
3. Liu, X.; Deng, J.; Zhang, L.; Wang, P.; Zhang, G.; Dong, X.; Sun, J. A study on chemical oxygen demand (COD) concentration distribution and its hydrodynamic mechanisms in Liaodong Bay, China. *Water* **2024**, *16*, 2135.
4. Maeda, S.; Yoshida, Y.; Yoshinari, K.; Takahashi, N. Effective placement strategies for portable fishways in agricultural drainage canals: A numerical investigation. *Sustainability* **2023**, *15*, 16283.
5. Zöschg, H. Large eddy simulations of flow past circular cylinders to determine head loss coefficients of circular bar trash racks with perpendicular inflow conditions. *Water* **2024**, *16*, 347.
6. Weijermars, R.; Afagwu, C. Pressure transient solutions for unbounded and bounded reservoirs produced and/or injected via vertical well systems with constant bottomhole pressure. *Fluids* **2024**, *9*, 199.
7. Calluaud, D.; Cornu, V.; Baran, P.; Pineau, G.; Sagnes, P.; David, L. Optimizing flow conditions and fish passage success in vertical slot fishways: Lessons from fish behavior observations. *Water* **2024**, *16*, 1718.
8. Kim, T.-H.; Lee, J.; Kim, T.; Choi, H.T.; Im, S. A framework for quantifying reach-scale hydraulic roughness in mountain headwater streams. *Water* **2024**, *16*, 647.
9. Zhang, X.; Zhu, Y.; Wu, H.; Bi, Z.; Xu, Z. Characteristics of vegetation resistance variation in muddy water flows. *Water* **2023**, *15*, 2238.
10. Brkić, D. Revised friction groups for evaluating hydraulic parameters: Pressure drop, flow, and diameter estimation. *J. Mar. Sci. Eng.* **2024**, *12*, 1663.
11. Liu, H.; Dong, J.; Liu, Q.; Geng, L.; Wang, Z.; Sun, C. Effect of median soil–particle size ratio on water storage capacity of capillary barrier. *Water* **2024**, *16*, 1774.
12. Guragain, S.; Tanaka, N. An experimental study on the effect of distance and sheltering area of a group of linearly arranged sacrificial piles on reducing local scour around a circular bridge pier under clear-water conditions. *Fluids* **2024**, *9*, 35.
13. Ai, X.; Zhang, X.; Guo, R.; Li, X.; Wang, M.; Liu, F.; Zhang, J. Study on the variation law of water consumption rate of cascade hydropower station under different conditions. *Energies* **2024**, *17*, 4966.

 https://doi.org/10.3390/w18030381

14. Liu, D.; Lian, J.; Liu, D.; Liu, F.; Ma, B.; Shi, J.; Yan, L.; Zheng, Y.; Xu, C.; Zhang, J. A monographic experimental investigation into flood discharge atomized raindrop size distributions under low ambient pressure conditions. *Water* **2025**, *17*, 1721.
15. Liu, J.; Hayatdavoodi, M.; Ertekin, R.C. A comparative study on generation and propagation of nonlinear waves in shallow waters. *J. Mar. Sci. Eng.* **2023**, *11*, 917.

References

1. UNESCO. *The United Nations World Water Development Report 2023: Partnerships and Cooperation for Water*; UNESCO Publishing: Paris, France, 2023.
2. FAO. *The State of the World'S Land and Water Resources for Food and Agriculture—Systems at Breaking Point*; FAO: Rome, Italy, 2023.

 https://doi.org/10.3390/w18030381

sustainability

MDPI

Article

Simulating the Impacts of Climate Change on the Hydrology of Doğancı Dam in Bursa, Turkey, Using Feed-Forward Neural Networks

Aslıhan Katip [1,*] and Asifa Anwar [1,2]

1 Department of Environmental Engineering, Bursa Uludağ University, Bursa 16059, Turkey
2 Division of Sustainable Development, College of Science and Engineering, Hamad Bin Khalifa University, Qatar Foundation, Doha 34110, Qatar; asan39545@hbku.edu.qa
* Correspondence: aballi@uludag.edu.tr

Abstract: Climate change continues to pose significant challenges to global water security, with dams being particularly vulnerable to hydrological cycle alterations. This study investigated the climate-based impact on the hydrology of the Doğancı dam, located in Bursa, Turkey, using feed-forward neural networks (FNNs). The modeling used meteorological parameters as inputs. The employed FNN comprised one input, hidden, and output layer. The efficacy of the models was evaluated by comparing the correlation coefficients (R), mean squared errors (MSE), and mean absolute percentage errors (MAPE). Furthermore, two training algorithms, namely Levenberg-Marquardt and resilient backpropagation, were employed to determine the algorithm that yields more accurate output predictions. The findings of the study showed that the model using air temperature, solar radiation, solar intensity, evaporation, and evapotranspiration as predictors for the water budget and water level of the Doğancı dam exhibited the lowest MSE (0.59) and MAPE (1.31%) and the highest R (0.99) compared to other models under LM training. The statistical analysis determined no significant difference ($p > 0.05$) between the Levenberg and Marquardt and resilient backpropagation training algorithms. However, a visual interpretation revealed that the Levenberg-Marquardt algorithm outperformed the resilient backpropagation, yielding lower errors, higher correlation values, and faster convergence for the models tested in this study. The novelty of this study lies in the use of certain meteorological inputs, particularly snow depth, for dam inflow forecasting, which has seldom been explored. Moreover, this study compared two widely used ANN training algorithms and applied the modeling framework to a region of strategic importance for Turkey's water security. This study highlights the effectiveness of ANN-based modeling for hydrological forecasting and determining climate-induced impacts on water bodies such as dams and reservoirs.

Keywords: climate change; Doğancı dam; feed-forward neural networks; hydrological parameters; meteorological parameters; modeling

1. Introduction

Climate change is currently one of the most significant environmental issues. The water flow in rivers and streams varies immensely with changing weather, which may cause floods during high flows and droughts [1]. Dams help store water during high flows and simultaneously allow the gradual release of water during periods of low flow [2]. In addition, the same dams act as a source of water for drinking and irrigation [3]. Hence, precise water budget predictions in dams are recommended to mitigate water management problems. In contrast, forecasting the water quantity in dams is challenging because it relies on various factors, such as streamflow parameters and climatic variations [4]. Water in a dam may also evaporate or seep into nearby rocks or soil [5]. A dam's storage capacity may be reduced by the sediments transported by streams [6]. Therefore, several conditions may impact the water quantity in dams.

In recent times, modeling has helped simulate the water budget, that is, the volume of water coming in and out of a water body and water level variations depending on previously available data [7]. For decades, water budget modelling has relied on statistical methods such as autoregressive integrated moving average and linear and non-linear regression models, but attention has shifted to artificial intelligence (AI) techniques due to their prediction accuracy [8,9]. Traditional statistical models struggle to capture the non-linear and dynamic interactions that define hydrological processes. Some established models, such as the Penman-Monteith model, rely heavily on site-specific data. When generalized or non-local data are used, these models often produce overestimations and lead to significant uncertainty in the estimation results [10]. In contrast, artificial intelligence or machine learning techniques can learn complex patterns directly from data without assuming a functional relationship, making them highly suitable for non-linear systems [11,12]. Under the umbrella of AI techniques, fuzzy neural networks, artificial neural networks (ANNs), and genetic algorithms have been utilized for hydrological modeling [13,14]. Most models rely on ANN due to its robustness and low probability of errors [15]. ANNs can approximate complex functional relationships. This makes them particularly suitable for modeling hydrological systems influenced by multiple interdependent variables [16]. Furthermore, recent developments have improved artificial neural networks to more sophisticated architectures, such as long short-term memory networks and hybrid models that integrate fuzzy logic or ensemble learning to improve accuracy, data interpretation, and generalization [17].

Research has been conducted using ANNs to predict the dam's water budget and water level. The USA's Millers Ferry dam reservoir level was estimated using an ANN [18]. In one study, autoregressive ANN models were compared to evaluate the Dez dam's monthly flow [19]. Different static and dynamic ANNs were used to predict the water budget of the Sefidroud Dam in Iran. These networks include static feedforward and non-linear autoregressive neural networks [20]. ANN applications for Sukhi reservoir water level estimation have been conducted in India [21]. Artificial neural networks are often preferred for estimating dam or reservoir-level fluctuations [22].

Similarly, ANN has been used to study the effects of climate change on water quantity within dams. Meteorological data, such as air temperature, relative humidity, wind speed, rainfall pattern, sunshine duration, and solar radiation, were used to determine evaporation from the Batu dam in Malaysia using ANN and climate-based models [23]. A long-term assessment of climate change impacts on the Norris Dam was conducted in the USA in 2017 using ANN [24]. The effect of climate variability on the Asa and Kampe dams was evaluated in Nigeria [25]. It has been shown that deep learning algorithms, especially long short-term memory models, can distinguish rainfall-runoff patterns derived from static catchment features and can be trained using a regional approach [26].

In recent years, the global scientific community has increasingly emphasized the resilience of water infrastructure, particularly dams and reservoirs, to resist the adverse impacts of climate change. The intensifying frequency of extreme weather events, including droughts and floods, underscores the urgent need for accurate forecasting tools to support climate-resilient water management strategies. In this regard, artificial neural networks have gained prominence. A related study modeled the impact of climate change on the water quality of a dam using artificial neural networks, demonstrating a high predictive accuracy [27]. Recent studies, such as Kim et al. [28], have demonstrated the effectiveness of ANN models in predicting monthly inflows to reservoirs using climate indices, enabling proactive adaptation to variable climatic conditions. Similarly, another study integrated ANN-based modeling with 30-year precipitation data to assess flood risk and operational vulnerability in Korean multipurpose dams, offering critical insights for infrastructure planning [29]. These advancements reflect a broader paradigm shift, as noted in the Intergovernmental Panel on Climate Change (IPCC) Sixth Assessment Report [30], which identifies machine learning models as key components of adaptive and anticipatory water governance.

Recent advances have expanded the capabilities of ANN-based models by incorporating deep learning architectures and hybrid approaches for improved climate impact predictions. For example, enhanced machine learning hydrological models (such as soil and water assessment tools and random forest algorithms) have been applied to the Yellow River Basin to assess the combined effects of climate change and anthropogenic activity on monthly runoff and drought [31]. While ANN is preferred for hydrological forecasting, its applications have also extended to interdisciplinary areas like regional economic integration and spatial planning under climate variability, as demonstrated in the Yangtze River Economic Belt study by Chen et al. [32]. These examples demonstrate the growing academic and practical relevance of ANN and related AI models for climate resilience and water systems. Despite these advantages, ANN-based models also face challenges, such as the risk of overfitting and dependency on high-quality input data [33]. These ongoing challenges have motivated the continuous refinement of model design, training algorithm selection, and input variable optimization to enhance the predictive performance and reliability of water resource applications.

The annual precipitation will decrease by 157 mm (25%) between 2020 and 2050, compared to 1970 to 2000 in Turkey [34]. These statistics show that Turkey's risk of experiencing drought is higher, eventually leading to water scarcity. Hence, proper water management is needed, and predicting water quantity is crucial for this purpose. In Bursa, long-term meteorological data from 1984 to 2013 show that the highest and lowest average monthly temperatures recorded were 25.1 °C and 5.6 °C, with the mean annual temperature between 14 °C and 16 °C. This shows a significant fluctuation in air temperature over the years. Likewise, the simulations performed in a study by Katip [34] using the standardized precipitation index were indicative of meteorological and agricultural droughts in Bursa. This implies growing vulnerability to climate extremes in the region.

This research used artificial neural networks to determine the relationship between water quantity, i.e., water level, water inflow, and outflow fluctuations in the Doğancı dam in Bursa, Turkey, with changing climatic parameters relevant to meteorological drought. The water budget of the Doğancı dam decreased from 92% to 72% compared to that in 2011 [35]. The mentioned dam is one of the primary sources of meeting the water needs in Bursa, a city in Turkey with a population of 3,056,120 [36]. The basic aim of the study was to determine the accuracy of ANN in modeling the water quantity or hydrology-related data of the Doğancı dam and to analyze which meteorological parameters are the best fit for the models. These meteorological parameters were representative of climate data. Long-term variations in meteorological parameters are associated with climate change.

ANN, an accurate forecasting tool, has been used to predict the water budgets and water levels of dams in Turkey. The daily flow to Turkey's Ermenek hydroelectric dam reservoir was simulated using deep recurrent neural network models [37]. Considering climatic factors, the effects of rainfall and temperature changes on Tahtalı and Gördes dam flows were evaluated for 2010–2099 [38]. The flow of the Namazgah dam was also predicted using ANN with air temperature, vapor pressure, and precipitation as input data [39]. The efficiency of ANN relies on the behavior between the chosen inputs and outputs and the understanding of the operation of neural networks [40].

Prior studies have often relied on single-variable models, such as precipitation or temperature alone. In contrast, this study systematically compared multiple combinations of meteorological inputs to identify the most predictive parameters using a robust feed-forward neural network architecture. The comparative evaluation of two ANN training algorithms (Levenberg-Marquardt and resilient backpropagation) provides a methodological contribution to improving hydrological forecasting performance. Thus, beyond its local applicability, this study contributes to the advancement of ANN-based modeling frameworks for climate-resilient water management. Thus, tackling the widespread global issue of climate-related factors that pose a threat to water security worldwide. By enhancing the accuracy of hydrological forecasting under climate variability, the proposed modeling framework can support decision-makers in developing robust water management plans and adapting operational strategies to ensure long-term water security. For these reasons, this study is important for making an international scientific contribution to sustainable water management.

2. Materials and Methods

In the current study, ANNs were employed to simulate the hydrology (volume, water level, and incoming and outgoing water flow rates) of the Doğancı Dam using various combinations of meteorological parameters as inputs. The Doğancı dam is an earth-fill dam that started operating in 1983, with a net volume of 2,520,000 m^3, a height of 65 m above the riverbank, a normal lake area of 1.55 km^2, and a maximum lake volume of 43.3 hm^3. Each year, it delivers roughly 125 hm^3 of drinking water [41]. It is one of the major dams for the drinking water supply in Bursa, along with the Nilufer dam. The mentioned dams receive their water supply primarily from Mount Uludağ. The Doğancı Dam is located in Bursa, which lies on the northwestern slopes of Mount Uludağ and in the southeast near the Marmara Sea, Turkey [42]. The main districts of Bursa are Osmangazi, Yildrim, Nilufer, Görukle, Mustafakemalpaşa, Gursu, and Kestel (Figure 1). Bursa experiences Mediterranean-style summers from June to September, which are hot and dry, preceded by chilly, wet winters. Bursa is one of Turkey's major industrial and agricultural cities [43].

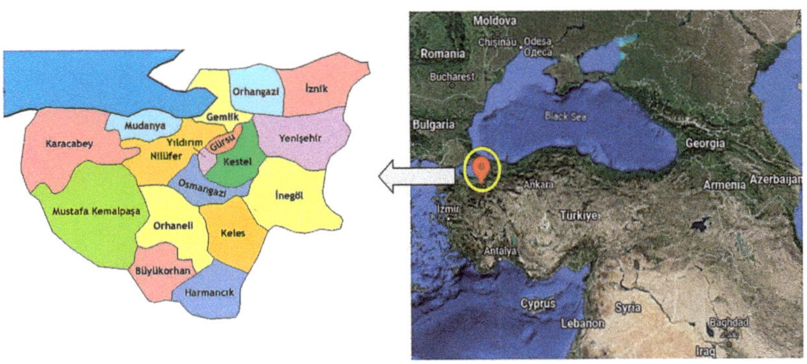

Figure 1. Bursa city map (**left**) and its location in Turkey (**right**) (acquired from Google Maps on 8 June 2025).

2.1. *Artificial Neural Networks and Their Requisites*

ANN prediction models are proven to be helpful, accurate, and easy-to-use forecasting tools, especially for many purposes in which processes are complex [44]. ANNs offer several benefits for water modeling, including: (i) modeling does not require a collection of integrated algorithms, enabling a quicker and more adaptable modeling approach; (ii) non-linear relationships can be dealt with carefully with minimal difficulty [45]; and (iii) individual expertise can be implemented into the model's design [46,47]. The goal of ANNs is to swiftly provide findings by exploring the vast parallel network of processing units while being able to cater to the loss and failure of particular networks, which defines the robustness of the ANN [48]. Figure 2 highlights the important steps in ANN modeling.

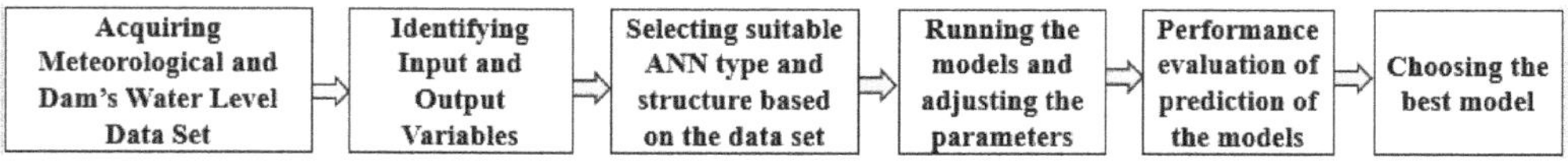

Figure 2. Steps performed using artificial neural networks.

2.2. Data Attainment

The first step in the analysis was the data collection. Monthly meteorological data were obtained from the Meteorological Department in Bursa. These data were representative of all districts within Bursa from 1999 to 2019. Conversely, the hydrological data for the Doğancı dam from 1999 to 2021 were obtained from Bursa's Water and Sewage Authority (BUSKI). The inflow and outflow data used for the Doğancı Dam were obtained as aggregated values, representing the total flow at the dam from the upstream reservoirs or other similar sources. The meteorological data comprised the monthly average values of relative humidity, vapor pressure, air temperature, wind speed, total daily global solar radiation, solar intensity, precipitation, evaporation, evapotranspiration, and snow depth. The hydrological data comprised the monthly average volume, incoming and outgoing water flow rates, and water level of the Doğancı dam. As mentioned, monthly data of the parameters were used in this study due to the completeness of long-term meteorological and hydrological records in this format. Additionally, the use of monthly data supports the long-term trend analysis presented in this study and helps to smooth short-term variability, making it suitable for water resource planning.

The models were trained and tested for data associated with 1999–2019, as the values were complete for both sets. The meteorological data obtained over a long-term period of 20 years were used as indicators of climate change [49]. The statistical values for the parameters selected for this research are listed in Table 1.

Table 1. Statistical values of meteorological and hydrological parameters (1999–2019).

Parameters	Average ± Standard Deviation	Range (Minimum–Maximum)
Meteorological Parameters of Bursa		
Monthly Average Vapor Pressure (hPa)	12.09 ± 4.59	5.01–22.52
Monthly Average Air Temperature (°C)	15.24 ± 7.23	2.18–27.64
Monthly Average Relative Humidity (%)	69.56 ± 7.56	50.13–87.86
Monthly Average Wind Speed (m/s)	2.06 ± 0.46	0.77–3.30
Monthly Average Total Precipitation (mm)	1.94 ± 2.20	0.00–26.70
Monthly Average Total Daily Global Solar Radiation (kWh/m^2)	3.38 ± 2.09	0.05–7.16
Monthly Average Total Daily Solar Intensity (cal/cm^2)	323.12 ± 161.83	53.13–644.72
Monthly Average Total Evapotranspiration (mm)	4.60 ± 4.05	0.09–30.15
Monthly Average Total Evaporation (mm)	5.26 ± 2.55	1.00–13.30
Monthly Average Total Snow Depth (cm)	0.98 ± 3.18	0.00–30.75

Table 1. *Cont.*

Parameters	Average ± Standard Deviation	Range (Minimum–Maximum)
Hydrological Parameters of Doğancı Dam		
Monthly Average Volume (hm^3)	29.80 ± 7.88	8.92–41.35
Monthly Average Incoming Water Flow Rate (m^3/day)	497,783.67 ± 530,709.90	1555.20–2,959,638.17
Monthly Average Outgoing Water Flow Rate (m^3/day)	239,507.67 ± 57,012.68	44,250.00–398,533.68
Monthly Average Water Level (m)	325.40 ± 6.05	305.38–333.35

The statistical analysis showed that the average flow rate of incoming water to the dam was higher over all the examined years than the outgoing flow rate. The inflow to the dam increases as the snowfall on Mount Uludağ melts after winter. However, climate change and an increasing population may alter these parameters.

2.3. The Selection of the ANN Model

A multi-layer feed-forward neural network was chosen for this study because it has yielded successful results in previous studies for forecasting the water quantity of dams, reservoirs, and lakes [22,24,50]. The typical architecture of a feedforward neural network is shown in Figure 3. In the given network structure, x and y represent the inputs and outputs, respectively, and h denotes the hidden nodes.

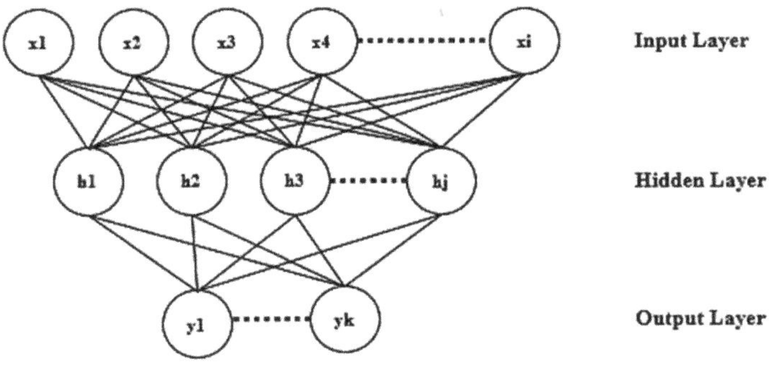

Figure 3. Structure of a typical feed-forward neural network.

The input layer stores the provided data, the hidden layer processes the data using mathematical equations and functions, and the output layer provides the final output based on the set targets. Each layer has its own set of neurons. These neurons are linked to every neuron or node in the layer ahead. The output of the network is defined as

$$Output = f(\theta i) \tag{1}$$

where, $f(\theta i)$ is the transfer function. θi is the potential of the ith neuron in the input layer, which can be given by the following formula:

$$\theta i = \sum_{i=1}^{R} w_i x_i + \beta_i \tag{2}$$

In Equation (2), β_i is the bias, and w_i is the weight coefficient. x_i is the input parameter. The weight coefficient detects whether and how much of a signal from a certain neuron should be transferred to the next layer. R is the total number of input neurons [51].

In this study, meteorological parameters were selected as inputs for the feed-forward neural network. The water quantity parameters, i.e., the volume, water level, inflows, and outflows of the Doğancı Dam, were introduced as outputs.

2.4. Training of a Feed-Forward Neural Network

A feed-forward neural network operates in the training and prediction modes. In the training mode, the weights start with arbitrary values and are adjusted iteratively. Epochs are used to describe each repetition of the entire training set. The network modifies the values in each epoch to minimize the error [51].

Two types of training algorithms were used in the current analysis due to their successful application in previous similar studies. The first was resilient backpropagation [23,52], and the second was the Levenberg-Marquardt algorithm [53,54]. The resilient backpropagation (RProp) algorithms are extremely quick and accurate, especially compared to conjugate gradient methods. Rprop algorithms are quite rigorous regarding the selection of their hyperparameters and are only reliant on the sign of the integrals of the decision variables rather than their amount, making them appropriate for applications where the gradient is quantitatively projected. Some resilient backpropagation algorithms use weight-backtracking. The weight-backtracking algorithm differs because it considers the progress of the partial derivatives and aggregates inaccuracy [55].

The gradient descent and Gauss-Newton methods are two numerical optimization strategies integrated into the Levenberg-Marquardt (LM) algorithm. The gradient descent method reduces the sum of the squared errors by updating the variables along the steepest-descent direction, while the Gauss-Newton method minimizes the total squared errors by presuming that the least-squares function is quadratic in the parameters and calculating the smallest value of the quadratic. If the parameters are far from their optimal value, the Levenberg-Marquardt technique mimics a gradient-descent method, whereas when the parameters are close to their desired value, it appears like the Gauss-Newton method [56].

2.5. Performance Evaluation

A feed-forward neural network provides accurate predictions if the model has a high correlation coefficient R and low mean squared error (MSE) and mean absolute percentage error (MAPE). The correlation coefficient (R) shows how well the predicted output values correlate with the target outputs and provides an idea of the prediction efficiency of the model. The R-values can vary between -1 and 1, with 1 indicating a positive correlation [57]. In contrast, MSE is the error that is equated by using the following formula:

$$\mathrm{MSE} = \frac{1}{M}\sum_{j=1}^{M}(y_k - \hat{y}_k)^2 \quad (3)$$

In Equation (3), M is the total number of outputs, y_k represents the observed output value, and $\hat{y}_k$ is the predicted value of the same parameter. The lower the error value, the better the prediction [58].

MAPE is another metric used for regression models. It is calculated by comparing each predicted value with its corresponding actual output value in terms of relative errors. Thus, unbiased statistics are provided to evaluate the predictive performance of a model [59]. The formula for evaluating the mean absolute percentage error is as follows:

$$\mathrm{MAPE} = \frac{1}{M}\sum_{j=1}^{M}\left|\left(\frac{y_k - \hat{y}_k}{y_k}\right)\right| \times 100 \quad (4)$$

2.6. ANN Application

As discussed, feed-forward neural networks were applied to examine the meteorological parameters that can impact the hydrological data of the Doğancı dam. The feedforward neural network was structured as one input, hidden, and output layer [18,21,50]. The artificial neural network toolbox in MATLAB version R2022a [60] was employed for the study. The typical structure of a feedforward neural network displayed on the MATLAB interface is shown in Figure 4. The monthly readings of the parameters considered from 1999 to 2019 were analyzed. Therefore, the dataset comprised approximately 252 values for each parameter, with very few missing values. These missing records were addressed by interpolating the adjacent data points. Linear interpolation was chosen because it is an appropriate technique for handling time-dependent datasets [61]. Next, 70% of the dataset was used for training, and 30% for testing and validation, i.e., 176 values of each variable were used for training, and the remaining were used for testing and validation [23,62]. Prior to model training, the data were normalized using the following equation:

$$D = \frac{D_i - D_{min}}{D_{max} - D_{min}} \tag{5}$$

where D represents a dimensionless normalized value. In contrast, D_i is the normalized value for the ith measurement in the data, and D_{max} and D_{min} are the maximum and minimum normalized scores of all the training and testing data taken.

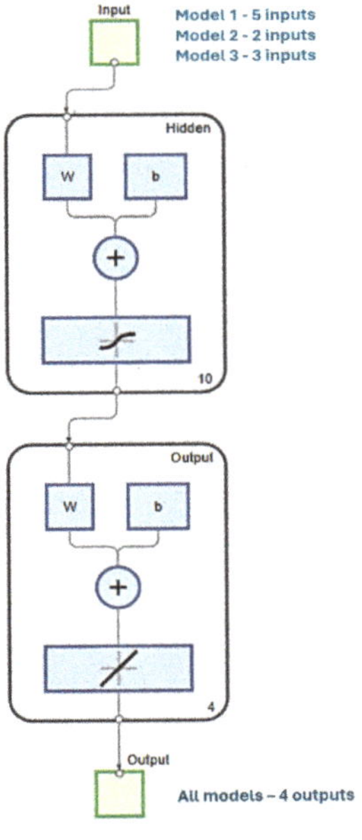

Figure 4. The structure of a feed-forward neural network, as acquired from MATLAB.

A sigmoid activation function was used in the hidden layer, and a linear function was applied to the output layer. The epochs were set to 1000 [21,63]. The number of nodes in the hidden layer was determined using a hit-and-trial method. Through trials between one and fifty hidden nodes, the best results were acquired when the network had ten nodes in the hidden layer, while it failed to give any improvement in results if more than ten nodes were present. The MSE was set to target 0. Two training algorithms (RProp and LM) were also tested to determine which algorithm was best suited for the prediction.

Both algorithms have associated advantages and disadvantages; however, the goal was to ascertain the most accurate algorithm for the data provided.

Three different models were trained and tested. In the first model, the monthly average data of air temperature, evaporation, evapotranspiration, total daily solar radiation, and solar intensity were used as input data, and the monthly average data of the Doğancı Dam's water level, volume, and incoming and outgoing water flow rates were targeted as outputs. The outputs were maintained the same for each model. The only difference was the set of input parameters. For the second model, the monthly average wind speed and vapor pressure were used as inputs. For the last model, the monthly average precipitation, snow depth, and relative humidity were selected as input parameters. Therefore, the aim of this study was to explore which meteorological parameters can accurately predict the hydrological parameters of the Durance dam, so that such an analysis can aid in predicting the future water quantity. The R, MSE, and MAPE values were compared to determine the best model based on the best predictive fit.

3. Results and Discussion

The results obtained from the detailed analysis are presented in Table 2. This can be portrayed by having an insight into the R, MSE, and MAPE values for all the models that model 1 gave the best fit. Considering both algorithms, Model 1 had a high R-value (0.99) for the entire dataset. In addition, it had the lowest MSE (0.59 in the case of LM and 46.8 using RProp) compared to the other models, which had considerably larger errors in the validation stage. In other words, all the models had the same output parameters, but the analysis showed that the model using air temperature, global solar radiation, solar intensity, evaporation, and evapotranspiration as input meteorological parameters defined and correlated best with the change in the hydrological data of the Doğancı dam. Hence, it can be interpreted that a change in the meteorological parameters of Model 1 can significantly impact the dam's water volume, flow rates, and level. In previous studies, it has been observed that temperature and evaporation parameters (used in Model 1) had a greater effect on precipitation-flow changes and hydrological droughts in water resources [64,65].

The magnitude of errors in Models 2 and 3 was quite high (ranging between 2.4×10^{10} to 7.4×10^{10} overall), and R-values were also lower in comparison to Model 1 (approximately 0.82 to 0.83 in the case of Model 2 and 0.68 to 0.70 for Model 3). The inputs for Model 2 were wind speed and vapor pressure. Likewise, precipitation, snow depth, and relative humidity were the inputs for Model 3.

Model 2 indicated a good correlation between the observed and predicted values. However, considering the error, which was reasonably high (regardless of the type of training algorithm used), it may not be the preferred model for future prediction of the hydrology of the Doğancı dam. Model 3 showed a similar trend, with a lower correlation (up to 0.70) than Model 2 for the observed and predicted values of the hydrological parameters of the dam.

The mean absolute percentage error presented in Table 2 also offers a clear comparison of the predictive accuracy of Model 1. The aforementioned model trained using the Levenberg-Marquardt algorithm yielded the lowest MAPE value of 1.31%, indicating excellent predictive and generalization capabilities. In contrast, the same model trained with the resilient backpropagation algorithm yielded a higher MAPE of 13.5%, although it was still within an acceptable range. Model 2 showed moderate performance, with MAPE values ranging between 37% and 38%. Model 3 had the highest MAPE, reaching 184.27%, which indicated a substantial deviation between the predicted and observed output values. MAPE values over 100% exhibited nonlinear and weak correlations between the input and output parameters, leading to reduced predictive power. These limitations reduce the model's ability to generalize effectively and highlight the importance of input selection. Therefore, this discussion further strengthens the reliability of Model 1, making it the optimal choice for forecasting the hydrological parameters of the Doğancı dam.

Table 2. Results from Feed-forward Neural Network Application to Meteorological and Hydrological Data for Doğancı Dam.

Model	Inputs	Outputs	ANN Layout	Checks R Training	Testing	Validation	Whole Dataset	MSE	MAPE (%)
1	Monthly average: • air temperature (°C) • total daily global solar radiation (kWh/m²) • total daily solar • intensity (cal/cm²) • evaporation (mm) • evapotranspiration (mm)	Monthly average: • volume (hm³) • incoming water flow rate (m³/day) • outgoing water flow rate (m³/day) • water level (m)	5-10-4	RProp: 0.9908 LM: 1	RProp: 1 LM: 1	RProp: 1 LM: 1	RProp: 0.99913 LM: 0.99995	RProp: 46.8353 LM: 0.59257	RProp: 13.5 LM: 1.31
2	Monthly average: • wind speed (m/s) • vapor pressure (hPa)	Monthly average: • volume (hm³) • incoming water flow rate (m³/day) • outgoing water flow rate (m³/day) • water level (m)	2-10-4	RProp: 0.83633 LM: 0.89876	RProp: 0.79397 LM: 0.62345	RProp: 0.85413 LM: 0.93957	RProp: 0.82412 LM: 0.83057	RProp: 4.0×10^{10} LM: 2.4×10^{10}	RProp: 38.88 LM: 37.3
3	Monthly average: • precipitation (mm) • total snow depth (cm) • relative humidity (%)	Monthly average: • volume (hm³) • incoming water flow rate (m³/day) • outgoing water flow rate (m³/day) • water level (m)	3-10-4	RProp: 0.69929 LM: 0.69352	RProp: 0.70283 LM: 0.7626	RProp: 0.60448 LM: 0.65079	RProp: 0.6858 LM: 0.69542	RProp: 7.4×10^{10} LM: 6.7×10^{10}	RProp: 152.52 LM: 184.27

The acquired mean squared error and coefficient values from Model 1 were compared with those obtained from previous studies (Table 3). When considering the errors, the MSE obtained using the LM training algorithm in Model 1 was 0.59, which was higher than that of Studies 1 and 6, as listed in Table 3. However, it was still lower than similar studies on the Norris, Doroozdan, Asa, Kampe, and Batu dams, which still show its acceptability for predictions. The error for Model 1 using the Rprop training algorithm was 46.84, higher than all the enlisted studies except Study 3 (Doroozdan dam). Because the errors must be as low as possible, it was suggested that the LM training algorithm performed better in our study.

However, when comparing the correlation coefficients, R in Model 1 was approximated to be 0.99 for the entire dataset. This R-value was closest to those of studies 4, 5, and 6 and higher than those of the other relevant studies listed in Table 3. When discussing the training algorithms, LM and Rprop showed almost the same high correlation between the observed and predicted outcomes.

In addition to other inputs, precipitation is the preferred meteorological parameter when simulating water availability in dams (as perceived from previous studies). However, in our study, Model 1, which performed better than the other tested models, did not include precipitation as an input. Therefore, this study reveals that precipitation is not the sole parameter for such determinations. Rather, other parameters, such as air temperature, solar intensity or radiation, evaporation, or evapotranspiration, can also provide valuable outcomes regarding the water volume, flow rate, and/or level of the dam. The closest example of such a case is a study by Benzaghta, et al. [23] on the Batu dam.

Table 3. References to past studies that used meteorological parameters as inputs to predict the water quantity parameters of a dam.

Study No.	Dam (Location)	Inputs (Meteorological Parameters)	Outputs (Hydrological Parameters)	Errors	Correlation Coefficient	Reference
1	Yalova Gökçe (Turkey)	Annual precipitation	Dam's water level	0.11	0.87	[53]
2	Norris (America)	Annual precipitation and air temperature	Inflow volume to the dam and water level	11	0.81	[24]
3	Doroozdan (Iran)	Monthly precipitation	Inflow volume to the dam	34.60×10^6	0.64	[62]
4	Asa and Kampe (Nigeria)	Annual air temperature, precipitation, and evapotranspiration	Inflow rate to the dam	-	0.99	[25]
5	Batu (Malaysia)	Daily air temperature, wind speed, relative humidity, precipitation, sunshine duration, and solar radiation	Evaporation from the dam	1.22	0.96	[23]
6	Yarseli (Turkey)	Daily precipitation	Dam's water level	0.135	0.99	[54]

Other modeling approaches have also been applied for dam inflow prediction, yielding varying levels of performance. For instance, Awan and Bae [66] utilized an adaptive neuro-fuzzy inference system to forecast long-term monthly inflows to major dams in South Korea using diverse meteorological input variables. Their optimal model achieved a correlation coefficient of 0.97. In a separate study on Egypt's Aswan Dam, a forecast-based adaptive reservoir operation framework was implemented to evaluate the dam's long-term resilience to climate change scenarios. This approach, which employed monthly precipitation and temperature data, achieved a maximum correlation coefficient of 0.92 [67]. Another model, the variation analog model, was modified by Amnatsan et al. [68] for the standardized inflow management of the Sirikit dam in Thailand during low-flow and high-flow periods. Their model produced a minimum root mean square error of 115.55 and a peak R-value of 0.98. In comparison, the present research demonstrates competitive predictive performance, further validating the reliability and robustness of the proposed ANN-based model.

The regression plots are shown in Figure 5. The line labeled Y = T represents the ideal case, where the predicted output exactly matches the actual output value. Hence, there was perfect agreement between the model prediction and actual data. The closer the regression fit line to Y = T, the better the predictions. In Model 1, the regression line lay almost exactly on the Y = T line, demonstrating that it fitted the target (water volume, flow rates, and water level of the dam) very well. In contrast, the plots for Models 2 and 3 exhibited significant deviations from the Y = T line, indicating underfitting and poor predictive alignment. The data are dispersed and have many outliers in Cleavage models 2 and 3, which shows that the data are heteroscedastic. This term indicates a significant variance in the errors for Models 2 and 3, which causes the data points on the regression graphs to move away from the Y = T line. This again revealed that Models 2 and 3 may not provide accurate predictions as Model 1.

The second objective was to identify the best training algorithm among the resilient backpropagation and Levenberg-Marquardt algorithms. Figure 6 shows the graphs of the MSE trend for the training, validation, and test sets across epochs. In all cases, the LM-trained models exhibited lower errors than those trained with Rprop. It can also be noted that the LM algorithm yielded the best performance for the models in terms of convergence with a lower number of epochs. Hence, stabilizing earlier indicates faster learning and better generalization. Model 1, which outperformed the other models, demonstrated high fluctuations in the error curves. Therefore, it indicates higher validation losses and lower predictive accuracy. Overall, the LM algorithm outperformed Rprop in all three models in terms of convergence speed, error minimization, and predictive reliability.

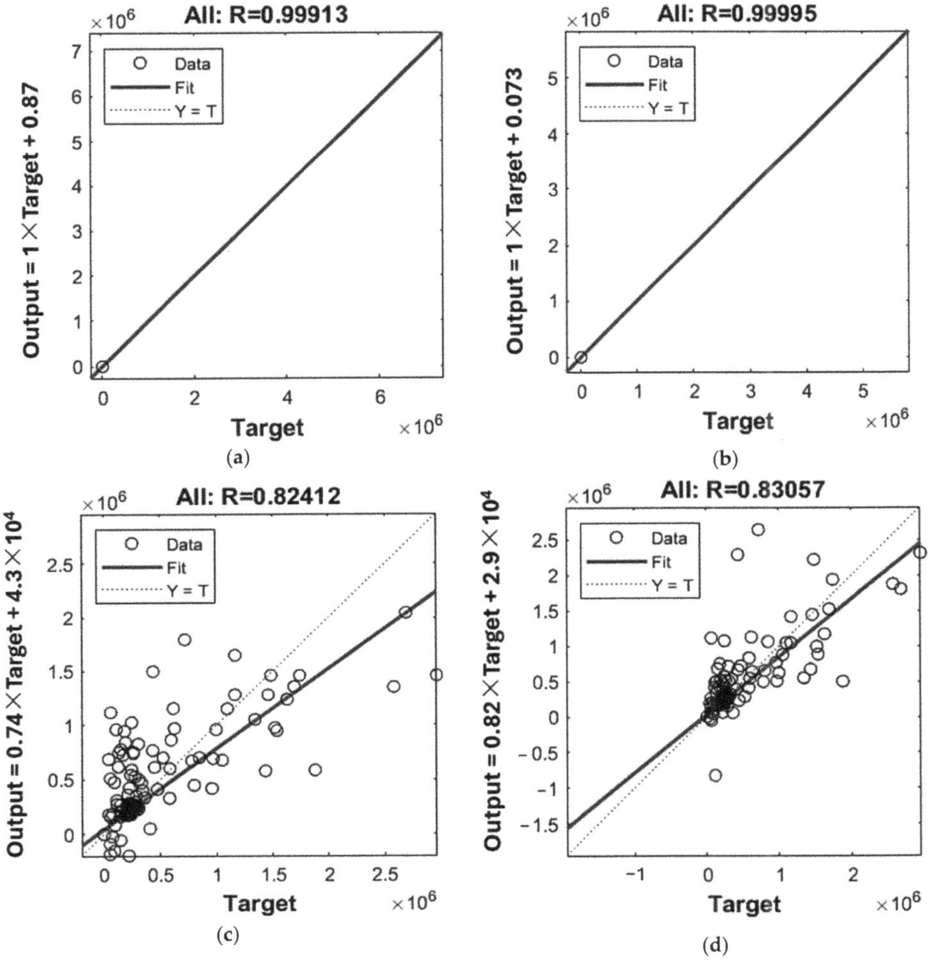

Figure 5. *Cont.*

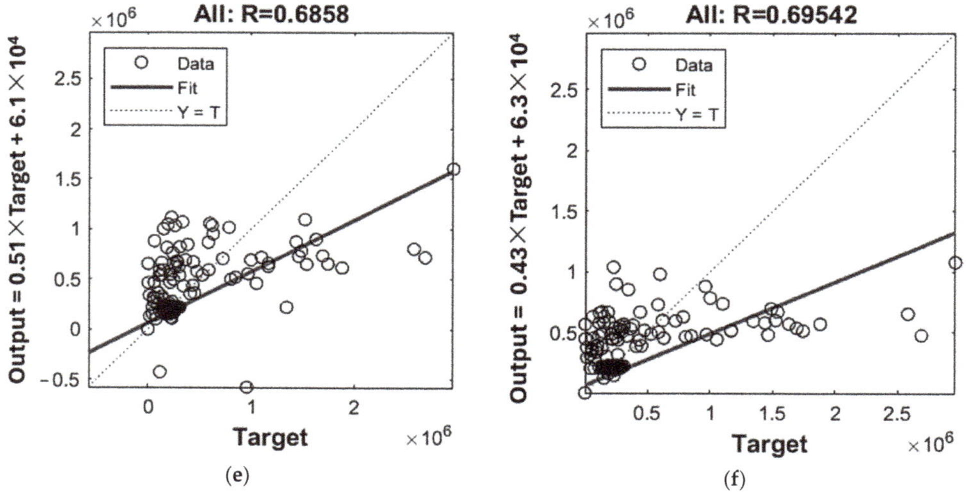

Figure 5. Regression plots for the prediction results of the tested models using the Rprop ((**a**)—model 1, (**c**)—model 2, and (**e**)—model 3) and LM ((**b**)—model 1, (**d**)—model 2, and (**f**)—model 3) training algorithms.

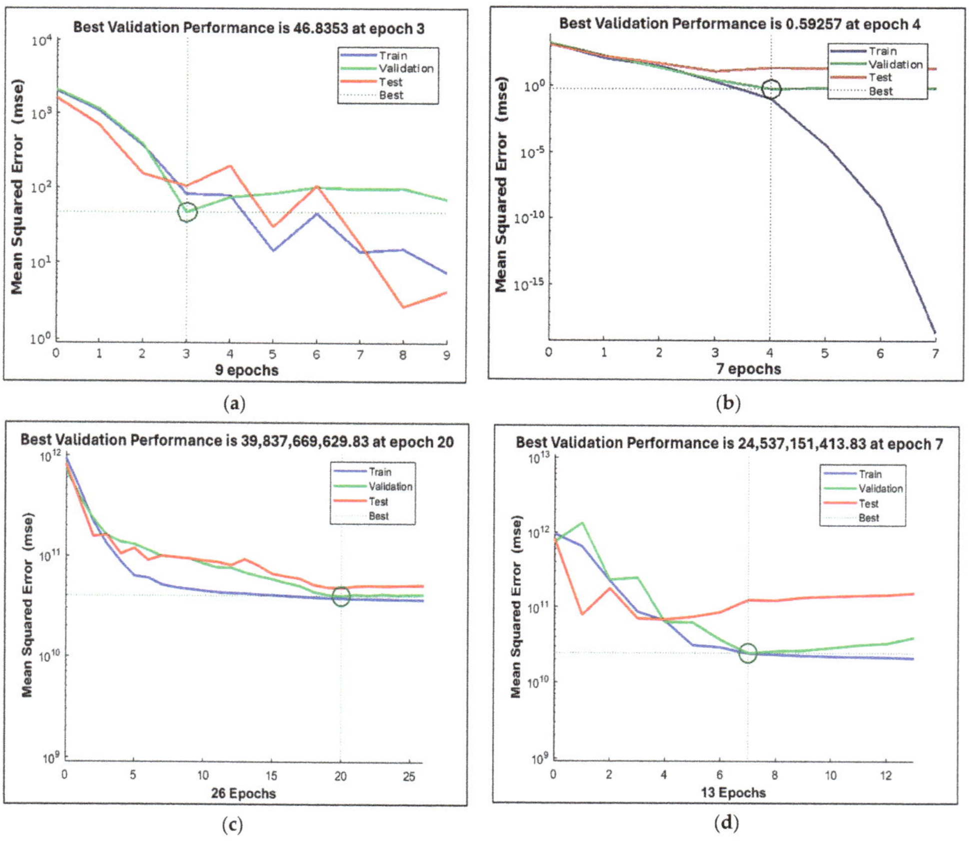

Figure 6. *Cont.*

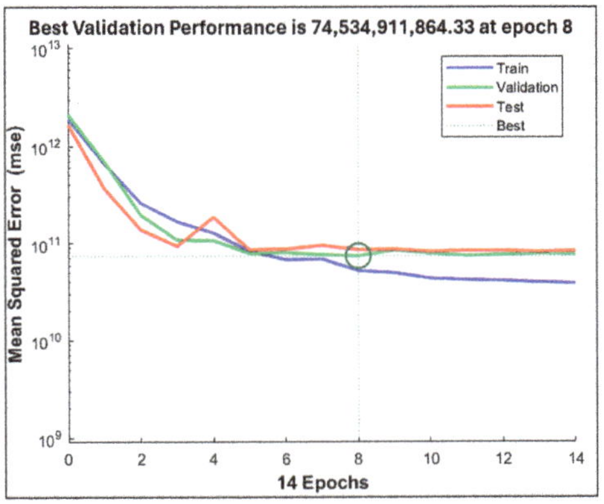

(**e**)

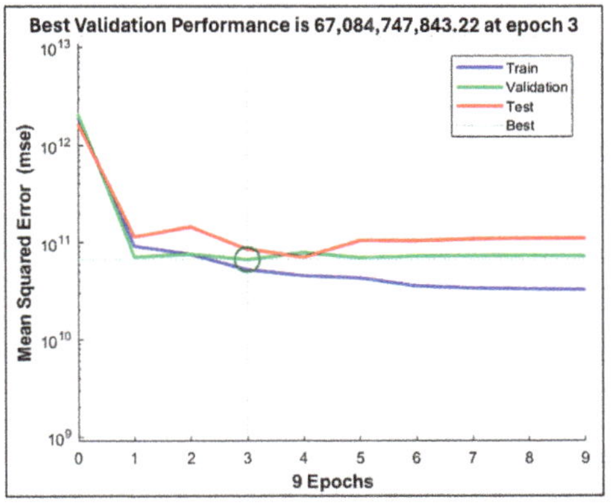

(**f**)

Figure 6. Graphs depicting the validation performance of the models using Rprop ((**a**)—model 1, (**c**)—model 2, and (**e**)—model 3) and LM ((**b**)—model 1, (**d**)—model 2, and (**f**)—model 3) training algorithms.

In addition, the gradient graphs in the training state showed similar outcomes (Figure 7). The gradient is a vector of partial derivatives that indicates the extent to which the error changes with respect to the weight in the networks and adjusts the weights to minimize the error. If the gradient is large, the model is far from the optimum, and vice versa. The Rprop algorithm displayed a slower and less smooth gradient reduction. Hence, this indicates that the Rprop training algorithm struggled to efficiently minimize the loss function for complex input patterns or non-linear relationships. This may contribute to suboptimal prediction accuracy and higher error metrics (as seen in the MAPE and MSE values). The LM algorithm reflected comparatively lower gradient values across all models, further confirming its suitability for training artificial neural networks in hydrological applications.

Resilient backpropagation has also been documented in the previous literature to have slow training speeds and instability over time, which can lead to oscillatory behavior during learning and susceptibility to getting stuck in local minima. In contrast, convergence is faster with the Levenberg–Marquardt algorithm, which offers a more efficient and stable approach for training artificial neural networks [69,70].

Analysis of variance (ANOVA) was performed on JASP (version 0.16.3) [71] for the correlation, mean squared error, and mean absolute percentage error values obtained from the neural network using the two training algorithms for all the tested models. As shown in Table 4(a), for a comparison between the R values for the examined algorithms, the p-value is 0.965 (as $p > 0.05$ with a 95% confidence level), which indicates that the difference was not significant between the R values provided for different models. The effect size (η^2) was also minimal (5.541×10^{-4}), indicating that the difference between the two training algorithms was negligible. The same was true for the MSE and MAPE values for both algorithms, where the p-values were 0.807 and 0.936, and the effect sizes were 0.017 and 0.002, respectively. Hence, both training algorithms performed the same with our data and can be applied when using feedforward neural networks to determine the climatic effects on the hydrology of a dam, as statistically proven.

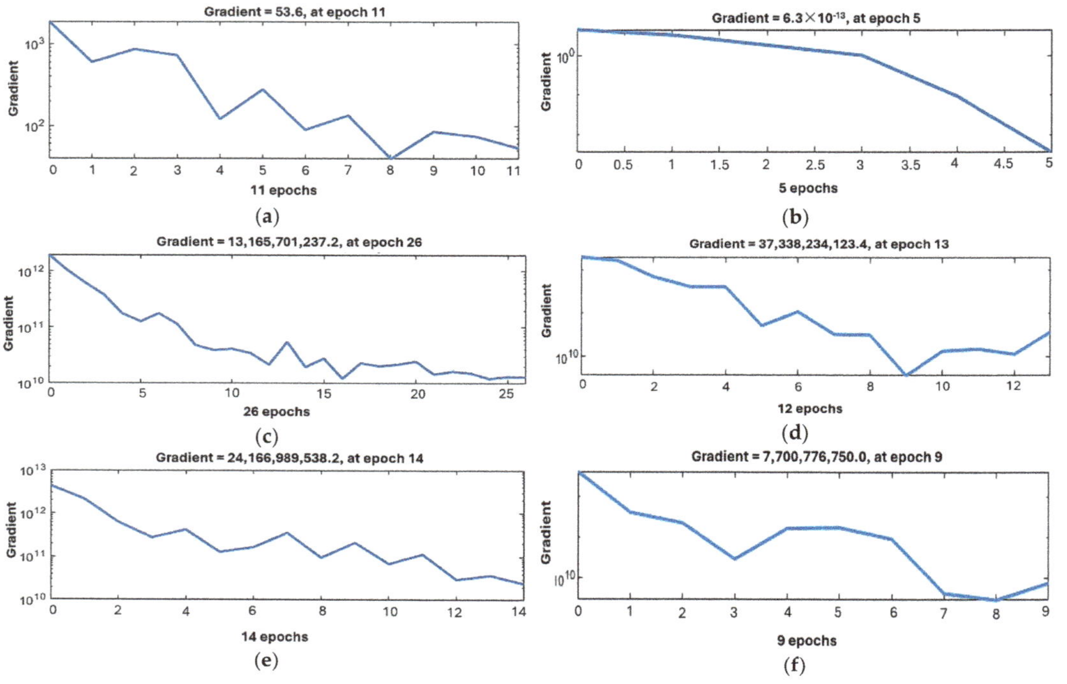

Figure 7. Training state of the models using Rprop ((**a**)—model 1, (**c**)—model 2, and (**e**)—model 3) and LM ((**b**)—model 1, (**d**)—model 2, and (**f**)—model 3) training algorithms.

Table 4. ANOVA results for (a) R, (b) MSE, and (c) MAPE values of the models when comparing algorithms.

Cases	Sum of Squares	df	Mean Square	F	p	η^2
Algorithms	5.298×10^{-5}	1	5.298×10^{-5}	0.002	0.965	5.541×10^{-4}
Residuals	0.096	4	0.024			
			(a)			
Algorithms	8.627×10^{19}	1	8.627×10^{19}	0.068	0.807	0.017
Residuals	5.086×10^{21}	4	1.272×10^{21}			
			(b)			
Algorithms	54.361	1	54.361	0.007	0.936	0.002
Residuals	29,756.237	4	7439.059			
			(c)			

However, visually interpreting the results and based on the discussion on the performance and training state of the models, a slight edge can be given to the Levenberg-Marquardt algorithm, where correlation values for all the studied models were slightly higher and error values lower compared to the resilient backpropagation training algorithm in most of the cases. The mean squared error values of Model 1 with the Levenberg-Marquardt training algorithm were also well related to previous studies in similar fields.

4. Conclusions

Climate change has adversely affected the environment. However, its impact on water resources is alarming, especially on the water budget of water bodies. This study investigated one such impact. Artificial neural networks were used to develop and predict the influence of meteorological changes on the hydrological characteristics of the Doğancı dam, which included volume, inflows, outflows, and the water level of the Doğancı dam. The results concluded that Model 1 performed the best, i.e., air temperature, solar radiation, solar intensity, evaporation, and evapotranspiration best predicted the water volume, flow rates, and water level of the Doğancı dam. Model 1 had the least mean squared and mean absolute percentage error, i.e., 0.59 and 1.31%, respectively, and the highest R (0.99) for the whole dataset among the other defined models tested using the Levenberg-Marquardt training algorithm. The analysis was based on the correlation values and mean squared errors. In addition, it was observed that using resilient backpropagation or Levenberg-Marquardt training algorithms does not statistically affect the analytical results. However, visually comparing the R, MSE, and MAPE values, graphical interpretations, and convergence patterns, the Levenberg-Marquardt algorithm performs slightly better than the resilient backpropagation. Hence, this study describes the appropriateness and efficiency of artificial neural networks for predictions in hydrology and the meteorological parameters that can most precisely simulate the water availability of dams. The results of this study offer practical value for government bodies, water resource planners, and climate adaptation policymakers. Accurate prediction of dam hydrology under changing climates can inform proactive reservoir management, drought preparedness, and long-term infrastructure planning.

However, certain limitations of this study must be acknowledged. The models were trained and tested on data from previous years without incorporating future climate projections, which may constrain their applicability for long-term planning. Additionally, the model's performance is contingent on the quality and completeness of the input data, which may compromise prediction reliability. Future research should focus on integrating downscaled climate model outputs based on future scenarios and testing their generalizability across different hydrological systems. Additionally, future studies may explore other established algorithms, such as evolutionary algorithms, for training and comparing ANN models.

Author Contributions: A.K. Conceptualization, formal analysis, investigation, methodology, resources, software, supervision, validation, writing—review and editing. A.A. Formal analysis, investigation, methodology, software, validation, visualization, writing, and original draft. All authors have read and agreed to the published version of the manuscript.

Funding: This research received no external funding.

Institutional Review Board Statement: Not applicable.

Informed Consent Statement: Not applicable.

Data Availability Statement: The datasets used in this study can be accessed from Asifa (2024). Asifa44/Meteorological-and-Hydrological-Data-for-the-Research-: Data used for research (v1.0). Zenodo. https://doi.org/10.5281/zenodo.13839069 (under Creative Commons Attribution 4.0 International Rights).

Acknowledgments: The authors would like to thank Bursa Water and Sewerage Administration (Bursa Su ve Kanalizasyon İdaresi-BUSKİ) and Bursa's Meteorological Department for their assistance in providing the datasets for this study. The authors would also acknowledge the State Hydraulic Works (Devlet Su İşleri-DSİ), which provided resources for this research.

Conflicts of Interest: The authors declare that they have no competing financial interests or personal relationships that could have influenced the work reported in this paper.

References

1. Cooper, S.D.; Lake, P.S.; Sabater, S.; Melack, J.M.; Sabo, J.L. The effects of land use changes on streams and rivers in mediterranean climates. *Hydrobiologia* **2013**, *719*, 383–425. [CrossRef]
2. Kondolf, G.M.; Gao, Y.; Annandale, G.W.; Morris, G.L.; Jiang, E.; Zhang, J.; Cao, Y.; Carling, P.; Fu, K.; Guo, Q.; et al. Sustainable sediment management in reservoirs and regulated rivers: Experiences from five continents. *Earth's Future* **2014**, *2*, 256–280. [CrossRef]
3. Hogeboom, R.J.; Knook, L.; Hoekstra, A.Y. The blue water footprint of the world's artificial reservoirs for hydroelectricity, irrigation, residential and industrial water supply, flood protection, fishing and recreation. *Adv. Water Resour.* **2018**, *113*, 285–294. [CrossRef]
4. Chamoglou, M.; Papadimitriou, T.; Kagalou, I. Key-descriptors for the functioning of a Mediterranean reservoir: The case of the new Lake Karla-Greece. *Environ. Process.* **2014**, *1*, 127–135. [CrossRef]
5. Mahmood, R.; Pielke Sr, R.A.; Hubbard, K.G.; Niyogi, D.; Dirmeyer, P.A.; McAlpine, C.; Carleton, A.M.; Hale, R.; Gameda, S.; Beltrán-Przekurat, A.; et al. Land cover changes and their biogeophysical effects on climate. *Int. J. Climatol.* **2014**, *34*, 929–953. [CrossRef]
6. Casserly, C.M.; Turner, J.N.; O'sullivan, J.J.; Bruen, M.; Bullock, C.; Atkinson, S.; Kelly-Quinn, M. Impact of low-head dams on bedload transport rates in coarse-bedded streams. *Sci. Total Environ.* **2020**, *716*, 136908. [CrossRef]
7. Zhang, J.; Zhu, Y.; Zhang, X.; Ye, M.; Yang, J. Developing a Long Short-Term Memory (LSTM) based model for predicting water table depth in agricultural areas. *J. Hydrol.* **2018**, *561*, 918–929. [CrossRef]
8. Awad, M.; Zaid-Alkelani, M. Prediction of water demand using artificial neural networks models and statistical model. *Int. J. Intell. Syst. Appl.* **2019**, *11*, 40. [CrossRef]
9. Elshorbagy, A.; Simonovic, S.; Panu, U. Performance evaluation of artificial neural networks for runoff prediction. *J. Hydrol. Eng.* **2000**, *5*, 424–427. [CrossRef]
10. McJannet, D.; Cook, F.; Burn, S. Comparison of techniques for estimating evaporation from an irrigation water storage. *Water Resour. Res.* **2013**, *49*, 1415–1428. [CrossRef]
11. Ouali, D.; Chebana, F.; Ouarda, T.B. Fully nonlinear statistical and machine-learning approaches for hydrological frequency estimation at ungauged sites. *J. Adv. Model. Earth Syst.* **2017**, *9*, 1292–1306. [CrossRef]
12. Yaseen, Z.M. A new benchmark on machine learning methodologies for hydrological processes modelling: A comprehensive review for limitations and future research directions. *Knowl.-Based Eng. Sci.* **2023**, *4*, 65–103. [CrossRef]
13. Nourani, V.; Molajou, A.; Najafi, H.; Danandeh Mehr, A. Emotional ANN (EANN): A new generation of neural networks for hydrological modeling in IoT. In *Artificial Intelligence in IoT*; Springer: Cham, Switzerland; Geneva, Switzerland, 2019; pp. 45–61.
14. Tabbussum, R.; Dar, A.Q. Performance evaluation of artificial intelligence paradigms—Artificial neural networks, fuzzy logic, and adaptive neuro-fuzzy inference system for flood prediction. *Environ. Sci. Pollut. Res.* **2021**, *28*, 25265–25282. [CrossRef]
15. Yaseen, Z.M.; Naganna, S.R.; Sa'adi, Z.; Samui, P.; Ghorbani, M.A.; Salih, S.Q.; Shahid, S. Hourly river flow forecasting: Application of emotional neural network versus multiple machine learning paradigms. *Water Resour. Manag.* **2020**, *34*, 1075–1091. [CrossRef]
16. Oyebode, O.; Stretch, D. Neural network modeling of hydrological systems: A review of implementation techniques. *Nat. Resour. Model.* **2019**, *32*, e12189. [CrossRef]
17. Huang, Y. Advances in artificial neural networks–methodological development and application. *Algorithms* **2009**, *2*, 973–1007. [CrossRef]
18. Üneş, F.; Demirci, M.; Kişi, Ö. Prediction of millers ferry dam reservoir level in USA using artificial neural network. *Period. Polytech. Civ. Eng.* **2015**, *59*, 309–318. [CrossRef]
19. Valipour, M.; Banihabib, M.E.; Behbahani, S.M.R. Comparison of the ARMA, ARIMA, and the autoregressive artificial neural network models in forecasting the monthly inflow of Dez dam reservoir. *J. Hydrol.* **2013**, *476*, 433–441. [CrossRef]
20. Hadiyan, P.P.; Moeini, R.; Ehsanzadeh, E. Application of static and dynamic artificial neural networks for forecasting inflow discharges, case study: Sefidroud Dam reservoir. *Sustain. Comput. Inform. Syst.* **2020**, *27*, 100401. [CrossRef]
21. Rani, S.; Parekh, F. Application of artificial neural network (ANN) for reservoir water level forecasting. *Int. J. Sci. Res.* **2014**, *3*, 1077–1082.
22. Üneş, F.; Demirci, M.; Başar, B.; Kaya, Y.Z.; Varçın, H. Estimating dam reservoir level fluctuations using data-driven techniques. *Pol. J. Environ. Stud.* **2019**, *28*, 3451–3462. [CrossRef] [PubMed]
23. Benzaghta, M.A.; Mohammed, T.A.; Ghazali, A.H.; Soom, M.A.M. Prediction of evaporation in tropical climate using artificial neural network and climate based models. *Sci. Res. Essays* **2012**, *7*, 3133–3148.

24. Rungee, J.; Kim, U. Long-term assessment of climate change impacts on Tennessee Valley Authority reservoir operations: Norris Dam. *Water* **2017**, *9*, 649. [CrossRef]
25. Salami, A.W.; Ibrahim, H.; Sojobi, A.O. Evaluation of impact of climate variability on water resources and yield capacity of selected reservoirs in the north central Nigeria. *Environ. Eng. Res.* **2015**, *20*, 290–297. [CrossRef]
26. Ley, A.; Bormann, H.; Casper, M. Climate change impact assessment on a German lowland river using long short-term memory and conceptual hydrological models. *J. Hydrol. Reg. Stud.* **2025**, *59*, 102426. [CrossRef]
27. Katip, A.; Anwar, A. Modeling the Influence of Climate Change on the Water Quality of Doğancı Dam in Bursa, Turkey, Using Artificial Neural Networks. *Water* **2025**, *17*, 728. [CrossRef]
28. Kim, T.; Shin, J.-Y.; Kim, H.; Kim, S.; Heo, J.-H. The use of large-scale climate indices in monthly reservoir inflow forecasting and its application on time series and artificial intelligence models. *Water* **2019**, *11*, 374. [CrossRef]
29. Lee, J. A Study on the Improvement of Flood Season in Korea Considering the 21st Century Observations. Master's Thesis, Department of Civil and Environmental Engineering, College of Engineering, Seoul National University, Seoul, Republic of Korea, 2022. Available online: https://dcollection.snu.ac.kr/common/orgView/000000173218 (accessed on 26 February 2025).
30. IPCC. *Climate Change 2022: Mitigation of Climate Change*; Cambridge University Press: Cambridge, UK, 2022.
31. Wang, L.; Li, Y.; Biswas, A.; Zhao, Y.; Niu, B.; Siddique, K.H. Assessing climate change and human impacts on runoff and hydrological droughts in the Yellow River Basin using a machine learning-enhanced hydrological modeling approach. *J. Environ. Manag.* **2025**, *380*, 125091. [CrossRef]
32. Chen, D.; Hu, W.; Li, Y.; Zhang, C.; Lu, X.; Cheng, H. Exploring the temporal and spatial effects of city size on regional economic integration: Evidence from the Yangtze River Economic Belt in China. *Land Use Policy* **2023**, *132*, 106770. [CrossRef]
33. Xu, L.; Pombo, N.; Pires, I.M.; Garcia, N.M. Is Overfitting in a Neural Network a Reliable Model for the Recognition of Activities of Daily Living? In Proceedings of the 5th EAI International Conference on Smart Objects and Technologies for Social Good, Valencia, Spain, 25–27 September 2019; pp. 223–226.
34. Pachauri, R.K.; Pachauri, R.K.; Allen, M.R.; Barros, V.R.; Broome, J.; Cramer, W.; Christ, R.; Church, J.A.; Clarke, L.; Dahe, Q.; et al. *Contribution of Working Groups I, II and III to the Fifth Assessment Report of the Intergovernmental Panel on Climate Change in Climate Change 2014: Synthesis Report*; IPCC: Geneva, Switzerland, 2014.
35. BUSKI. Baraj Doluluk Oranları. Available online: https://www.buski.gov.tr/baraj-detay (accessed on 7 January 2025).
36. Koç, A.A.; Codron, J.M.; Tekelioğlu, Y.; Lemeilleur, S.; Tozanli, S.; Aksoy, Ş.; Bignebat, C.; Demirer, R.; Mencet, N. Restructuring of Agrifood Chains in Turkey. In *Regoverning Markets Agrifood Sector Studies (A)*; IIED: London, UK, 2007.
37. Apaydin, H.; Feizi, H.; Sattari, M.T.; Colak, M.S.; Shamshirband, S.; Chau, K.-W. Comparative analysis of recurrent neural network architectures for reservoir inflow forecasting. *Water* **2020**, *12*, 1500. [CrossRef]
38. Okkan, U. Aylık Yağış ve Sıcaklık Değişimlerinin İzmir Içme Suyu Havzalarının Akımlarına Etkileri. Master's Thesis, Fen Bilimleri Enstitüsü, Dokuz Eylül Üniversitesi, İzmir, Turkey, 2009.
39. Raşit, A. Namazgâh Barajında Meteorolojik Veriler Kullanılarak Yapay Sinir Ağları ile Akışın Tahmin Edilmesi. Master's Thesis, Fen Bilimleri Enstitüsü, Kocaeli Üniversitesi, İzmit, Turkey, 2019. Available online: http://dspace.kocaeli.edu.tr:8080/xmlui/handle/11493/16790 (accessed on 12 December 2024).
40. Mata, J. Interpretation of concrete dam behaviour with artificial neural network and multiple linear regression models. *Eng. Struct.* **2011**, *33*, 903–910. [CrossRef]
41. Ozsoy, G.; Aksoy, E.; Karaata, E. Estimating soil loss of Doganci Dam watershed, northwest Turkey and lifetime analyze of Doganci Dam using multi-year remotely sensed data and GIS techniques. *Soil-Water J.* **2013**, *2*, 927–934.
42. Stotz, C.L. The Bursa region of Turkey. *Geogr. Rev.* **1939**, *29*, 81–100. [CrossRef]
43. Erünal, E. Examining age structure and estimating mortality rates in Ottoman Bursa using mid-nineteenth-century population registers. *Middle East. Stud.* **2020**, *57*, 179–196. [CrossRef]
44. Noor, R.M.; Ahmad, Z.; Don, M.M.; Uzir, M. Modelling and control of different types of polymerization processes using neural networks technique: A review. *Can. J. Chem. Eng.* **2010**, *88*, 1065–1084. [CrossRef]
45. Tabari, H.; Talaee, P.H. Reconstruction of river water quality missing data using artificial neural networks. *Water Qual. Res. J. Can.* **2015**, *50*, 326–335. [CrossRef]
46. O'Reilly, G.; Bezuidenhout, C.; Bezuidenhout, J. Artificial neural networks: Applications in the drinking water sector. *Water Supply* **2018**, *18*, 1869–1887. [CrossRef]
47. Najah, A.; El-Shafie, A.; Karim, O.A.; El-Shafie, A.H. Application of artificial neural networks for water quality prediction. *Neural Comput. Appl.* **2013**, *22*, 187–201. [CrossRef]
48. Kang, B.; Kim, Y.D.; Lee, J.M.; Kim, S.J. Hydro-environmental runoff projection under GCM scenario downscaled by artificial neural network in the Namgang Dam watershed, Korea. *KSCE J. Civ. Eng.* **2015**, *19*, 434–445. [CrossRef]
49. Kruczyk, M. Long series of GNSS integrated precipitable water as a climate change indicator. *Rep. Geod. Geoinform.* **2015**, *99*, 1–18. [CrossRef]

50. El-Mahdy, M.E.-S.; El-Abd, W.A.; Morsi, F.I. Forecasting lake evaporation under a changing climate with an integrated artificial neural network model: A case study Lake Nasser, Egypt. *J. Afr. Earth Sci.* **2021**, *179*, 104191. [CrossRef]
51. Svozil, D.; Kvasnicka, V.; Pospichal, J. Introduction to multi-layer feed-forward neural networks. *Chemom. Intell. Lab. Syst.* **1997**, *39*, 43–62. [CrossRef]
52. Mohamad, R.N.M.R.; Ishak, W.H.W. Forecasting the flood stage of a reservoir based on the changes in upstream rainfall pattern. *J. Technol. Oper. Manag.* **2019**, *14*, 46–52.
53. Damla, Y.; Temiz, T.; Keskin, E. Yapay sinir ağı kullanılarak su seviyesinin tahmin edilmesi: Yalova Gökçe barajı örneği. *Kırklareli Üniversitesi Mühendislik Fen Bilim. Derg.* **2020**, *6*, 32–49. [CrossRef]
54. Çalım, M.M. Estimation of Dam Reservoir Level with Artificial Neural Network Method. Master's Thesis, Mustafa Kemal Üniversitesi, Antakya, Turkey, 2008. Available online: https://tez.yok.gov.tr/UlusalTezMerkezi/tezDetay.jsp?id=tPtS_heSMyicaKWhD0vd6w&no=hXyjBcHCUojyy4SSrKgk7A (accessed on 16 January 2025).
55. Igel, C.; Toussaint, M.; Weishui, W. Rprop using the natural gradient. *Trends Appl. Constr. Approx.* **2005**, *151*, 259–272. [CrossRef]
56. Gavin, H.P. The Levenberg-Marquardt Algorithm for Nonlinear Least Squares Curve-Fitting Problems. Master's Thesis, Department of Civil and Environmental Engineering, Duke University, Durham, CA, USA, 2019.
57. Taylor, R. Interpretation of the correlation coefficient: A basic review. *J. Diagn. Med. Sonogr.* **1990**, *6*, 35–39. [CrossRef]
58. Wu, S.; McAuley, K.; Harris, T. Selection of simplified models: I. Analysis of model-selection criteria using mean-squared error. *Can. J. Chem. Eng.* **2011**, *89*, 148–158. [CrossRef]
59. Chicco, D.; Warrens, M.J.; Jurman, G. The coefficient of determination R-squared is more informative than SMAPE, MAE, MAPE, MSE and RMSE in regression analysis evaluation. *Peerj Comput. Sci.* **2021**, *7*, e623. [CrossRef]
60. Mathworks. Deep Learning Toolbox-Design, Train, Analyze, and Simulate Deep Learning Networks. In *Deep Learning Toolbox*; The Mathworks, Inc.: Natick, MA, USA, 2022.
61. Wahir, N.; Nor, M.; Rusiman, M.; Gopal, K. Treatment of outliers via interpolation method with neural network forecast performances. *J. Phys. Conf. Ser.* **2017**, *995*, 012025. [CrossRef]
62. Rezaeianzadeh, M.; Stein, A.; Cox, J.P. Drought forecasting using Markov chain model and artificial neural networks. *Water Resour. Manag.* **2016**, *30*, 2245–2259. [CrossRef]
63. Üneş, F.; Varçın, H.; Dindar, K.K. Yapay sinir ağları yaklaşımı ile tahtaköprü barajındaki aylık buharlaşma tahmini. *Eng. Sci.* **2011**, *6*, 114–125.
64. Khosravi, K.; Farooque, A.A.; Karbasi, M.; Ali, M.; Heddam, S.; Faghfouri, A.; Abolfathi, S. Enhanced water quality prediction model using advanced Hybridized resampling alternating tree-based and deep learning algorithms. *Environ. Sci. Pollut. Res.* **2025**, *32*, 6405–6424. [CrossRef] [PubMed]
65. Khosravi, K.; Farooque, A.A.; Naghibi, A.; Heddam, S.; Sharafati, A.; Hatamiafkoueieh, J.; Abolfathi, S. Enhancing pan evaporation predictions: Accuracy and uncertainty in hybrid machine learning models. *Ecol. Inform.* **2025**, *85*, 102933. [CrossRef]
66. Awan, J.A.; Bae, D.-H. Improving ANFIS based model for long-term dam inflow prediction by incorporating monthly rainfall forecasts. *Water Resour. Manag.* **2014**, *28*, 1185–1199. [CrossRef]
67. Eldardiry, H.; Hossain, F. The value of long-term streamflow forecasts in adaptive reservoir operation: The case of the High Aswan Dam in the transboundary Nile River basin. *J. Hydrometeorol.* **2021**, *22*, 1099–1115. [CrossRef]
68. Amnatsan, S.; Yoshikawa, S.; Kanae, S. Improved forecasting of extreme monthly reservoir inflow using an analogue-based forecasting method: A case study of the Sirikit Dam in Thailand. *Water* **2018**, *10*, 1614. [CrossRef]
69. Beale, M.H.; Hagan, M.T.; Demuth, H.B. Neural network toolbox. In *User's Guide, MathWorks*; The Mathworks, Inc.: Natick, MA, USA, 2010; Volume 2, pp. 77–81.
70. Adeloye, A.; De Munari, A. Artificial neural network based generalized storage–yield–reliability models using the Levenberg–Marquardt algorithm. *J. Hydrol.* **2006**, *326*, 215–230. [CrossRef]
71. Goss-Sampson, M. *Statistical analysis in JASP: A Guide for Students*; JASP: Amsterdam, The Netherlands, 2019.

water

MDPI

Article

Applying a Holistic Approach to Environmental Flow Assessment in the Yen River Basin

Tuan Phuc Tong [1], Son Thanh Hoang [1], Dung Quang Bui [1], Ngoc Trong Ha [2], Linh Ha Nguyen [3], Lan Minh Nguyen [4] and Chau Kim Tran [5,*]

1 Institute of Geography, Vietnam Academy of Science and Technology, 18 Hoang Quoc Viet, Cau Giay, Hanoi 100000, Vietnam; ol.tuantp.thna@gmail.com (T.P.T.); htson@ig.vast.vn (S.T.H.); quangdung259@gmail.com (D.Q.B.)

2 Department of Environment, Ministry of Natural Resources and Environment, 10 Ton That Thuyet, Cau Giay, Hanoi 100000, Vietnam; hatrongngoc@gmail.com

3 Environment Faculty, Hanoi University of Natural Resources and Environment, 41 A Phu Dien, North-Tu Liem, Hanoi 100000, Vietnam; nhlinh@hunre.edu.vn

4 Center for Water Resources Quality and Protection, 93/95 Vu Xuan Thieu, Long Bien, Hanoi 100000, Vietnam; minhlanks@gmail.com

5 Water Engineering Faculty, Thuyloi University, 175 Tay Son, Dong Da, Hanoi 100000, Vietnam

* Correspondence: kimchau_hwru@tlu.edu.vn; Tel.: +84-912871247

Abstract: Environmental flow assessment is an essential tool in water resource management. This study employs a holistic approach to evaluate the environmental flow in the Yen Basin, Thanh Hoa, Vietnam. Based on information gathered from a field survey, the Yen River system is divided into five reaches, and environmental objectives and ecological assets are identified in each reach. Hydrological and hydraulic mathematical models are applied to simulate the flow regime in the river, demonstrating their potential to assess environmental flow, especially in basins with limited data. The detailed results from the mathematical model facilitate selecting environmental flow components to address specific objectives for each river reach. By analyzing and selecting the flow regime, this study aims to ensure environmental protection while also considering basin development requirements, laying the groundwork for defining prescribed flow regimes in basin water management.

Keywords: environmental flow; Yen River; dry season baseflows; low season pulses; flood season baseflows; flood season pulses; bankfull flows; overbank flows

1. Introduction

Water resource management around the world has been under a lot of strain recently due to the world's rapid socio-economic development and the growing need to preserve ecosystems. When it comes to ungauged basins, these difficulties increase greatly. The entire ecosystem may suffer from decreased river flows. An approach to reducing this issue will, therefore, be to quantify environmental flows on river systems. Environmental flows describe the quantity, timing, and quality of water flows required to sustain freshwater and estuarine ecosystems and human livelihoods and well-being that depend on these ecosystems [1]. In the last two decades, environmental flows have been increasingly recognized as a vital contributor to the continuing provision of environmental goods and services upon which people's lives and livelihoods depend [2]. Assessing the impact of human activities on the hydrological regime can be used to develop management strategies for maintaining or restoring environmental flows in rivers and streams [3].

In Vietnam, environmental flow is becoming an increasingly pressing issue and is essential for the planning of water resources, the establishment of reservoir operating procedures, and the planning of water allocation in river basins. In 2017, the Ministry of Natural Resources and Environment issued Circular 64/2017/TT-BTNMT [4] to regulate the determination of the minimum flow on rivers and streams downstream of reservoirs and dams. However, this method provides a single minimum flow value that is applicable throughout the year, which is intended to protect only the essential ecosystem functions necessary for the normal development of aquatic ecosystems. Consequently, a more comprehensive approach is necessary to ensure the sustainable development of the ecosystem, which requires environmental flows that protect a wider range of ecological functions and goals. This must be a flow regime rather than a single flow value.

The management of water resources in specific basins requires the utilization of various approaches for evaluating environmental flows. These methods can be classified into four categories, namely Hydrological Index Methodologies, Hydraulic Rating Methodologies, Habitat Simulation Methodologies, and Holistic methodologies [5,6]. Hydrological Index Methodologies often assume that the ecosystem will suffer if the flow rate decreases below a critical value. The critical value can be determined based on historical flow to estimate the appropriate rate. The popular methods that belong to this category are the Montana Method [7], Range of Variability Approach [8], and Indicators of Hydrologic Alteration (HIA) [9]. The second category is Hydraulic Rating Methodologies, which use hydraulic characteristics such as river width, average depth, flow velocity, wetted parameter, and hydraulic radius to define the environmental flow. The widely accepted method for this category is the Wetted Perimeter Method [10]. Another category is Habitat Simulation Methodologies, which is a more complex way to compute environmental flow. This approach develops a relationship between the stream flow and the habitat requirements of aquatic organisms in a certain living period. The method most representative of this classification is the Instream Flow Incremental Methodology [11]. Currently, Holistic Methodologies have gained prominence as a means of evaluating environmental flows. For most rivers today, there is a conflict between environmental protection and water demands. Finding a solution that balances opposing goals is necessary. These methodologies involve a comprehensive assessment of river systems at a more detailed scale, considering all significant abiotic and biotic components that constitute the ecosystem. This ensures that the full spectrum of flows and their temporal and spatial variability are considered. In addition, the methodologies also take into account issues related to water use needs on the river. This is the outstanding advantage of these methodologies compared to the methodologies mentioned above. For example, Gippel et al. (2009) [3] provided a generic approach for environmental flow assessment. Kinga et al. (1998) [12] developed the Building Block Methodology (BBM) to estimate the environmental flow in Victoria, Australia. As the methods become more complex, the time, resources, and capabilities required to conduct the assessment also increase, but so does the level of confidence in the ability of the flow regime to support system values. Furthermore, holistic methods often extend beyond ecosystem values to include social values and ecosystem goods and services associated with specific flow regimes. Globally, these methods are increasingly being recognized as the standard approach for assessing the environmental flow of rivers. Despite various detailed assessment methods developed in different regions and countries, many of these holistic methods share common characteristics, which are to combine several hydrological, hydraulic, and habitat analysis tools that exist in simpler environmental flow methods.

Most studies on environmental flows focus on freshwater, with a scarcity of studies in estuary areas [3]. In this area, the flow is influenced by both river currents and tides, resulting in an exceedingly intricate hydraulic regime. The direction of flow can switch between towards the sea (positive flow) and towards the river (negative flow). This presents a significant obstacle when attempting to establish the environmental flow. In addition to the complexity of the hydraulic regime, seawater intrusion is an important factor in the spatial and temporal distribution of an estuary ecosystem [13]. Therefore, researching how to assess environmental flows in estuaries is urgently required to fill the existing gap.

The purpose of this research is to determine the environmental flow required for sustainable development in the Yen River basin. To accomplish this, the research consists of three phases, specifically: (1) develop hydrological and hydraulic models to simulate hydrological and water quality data for data limitations in the basin, (2) divide the river system into functionally similar segments using a comprehensive approach, and (3) identify the environmental flow components for each segment. The main research outcome contributes to a better understanding of the ecohydrological contexts of the Yen River. It intends not only to protect a greater range of ecological assets but also to be a part of basin planning processes.

2. Materials and Methods

2.1. Research Area

The research area is the Yen River basin in Thanh Hoa province, Vietnam. This basin spans a total of 1996 square kilometers and is characterized by a distribution of mountain forests covering 45.2% of the area, while plains and semi-mountainous regions account for 49.5% of the total. According to Vietnamese standard TCVN 13615 [14], the primary stream within this basin is the Yen River, which originates in the high mountains of Binh Yen commune and flows out to the sea. The Yen River is accompanied by eight tributaries, and the network of rivers and streams within the basin is illustrated in Figure 1 and Table 1.

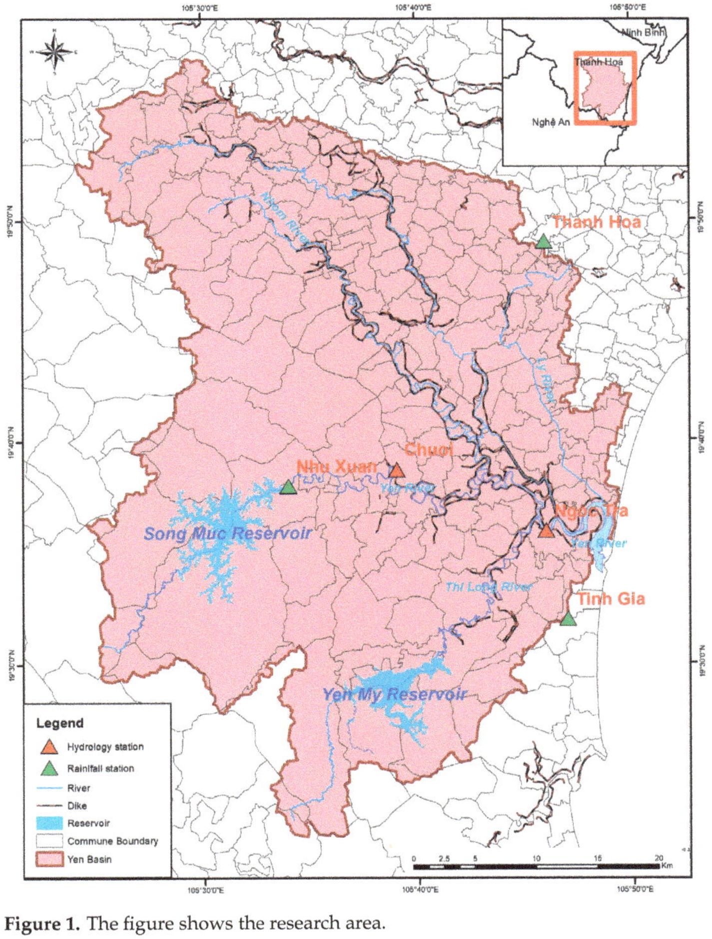

Figure 1. The figure shows the research area.

Table 1. The table shows a summary of collected and surveyed data.

ID	Data	Source	Example
1	Hydrometeorological and salinity data	Hydrometeorological stations	Figure A1
2	Operation data	Reservoir owners	Figure A2
3	River cross-section	Field survey to collect	Figure A3
4	Bed material	Field survey to sample bed material	Figure A4
5	Ecological information	Collect from both documents and questionnaires	Figure A5
6	Water use information	Collect from both documents and questionnaires	

The largest tributary in this river system is Thi Long River. Figure 1 presents the river network in the basin. A significant issue facing the Yen River basin is salt intrusion. During the dry season, which lasts from December to June, the area is heavily impacted by tidal fluctuations and the encroachment of salt intrusion. This is due to a decline in rainfall, resulting in a reduced flow and water level of the Yen River, which is unable to overcome the tidal pressure. Consequently, the tides and salinity follow the river, penetrating deep into the mainland.

In the study basin, there are two main reservoirs, Song Muc and Yen My, which were constructed in the late 1980s (Figure 1). Their respective capacities amount to 322.65 and 124.51 million cubic meters. These reservoirs have a crucial role in supplying water during the dry season and regulating floods in the flood season. In recent years, data have been collected concerning the operation of these reservoirs, which is valuable information for research purposes when calibrating hydrological models.

The network of hydrometeorological stations in the basin is somewhat limited (Figure 1). There are only two hydrological stations in the basin, Chuoi and Ngoc Tra, which measure only the water levels and rainfall. Additionally, there are some rainfall stations that provide long-duration and good-quality rain data, ensuring the reliability of research calculations.

2.2. Methodology

A flow regime can be thought of as comprising various flow components, which are specified in detail during the planning process. These components are typically characterized by their magnitude, frequency, duration, and timing (season) of occurrence, as well as their role in supporting critical ecological and physical processes within the river corridor. Commonly recognized flow components include baseflows in the dry season, low season pulses, baseflows in the flood season, flood season pulses, bankfull, and overbank flows. The specification of the desired flow regime in terms of flow components allows a river manager to implement recommendations by directing controlled flows to a river or by preventing inappropriate diversion of flows from a river.

The timing, frequency, and duration of flows are related to the life cycle needs of biota, while the hydraulic conditions may encourage or discourage certain behaviors or even exceed the capacity of some organisms to respond. Aquatic biota exists within a tolerable range of flow velocity and depth. The hydraulic parameters of a river system are subject to significant variation in relation to the discharge of water, and this has a corresponding impact on the availability of certain types of micro-habitat and locations, such as backwaters, tributaries, wetlands, floodplains, and estuaries. The specific hydraulic conditions experienced by a given reach are influenced by the physical characteristics of the river, including its bed material and in-stream large wood or vegetation, as well as the pattern of flows and the supply of sediment. The physical form of the river is shaped by these factors, as well as by the resistance of the channel boundary to change. In rivers where water quality is a limiting factor for biota, it is important to consider the variation in water quality parameters in relation to flow. In estuaries, this is particularly relevant, as the location and characteristics of salinity gradients can be strongly influenced by inflows from the river. To assess the environmental flow in Yen River, this research follows the steps in Figure 2 below. This method is based on the method proposed in [3].

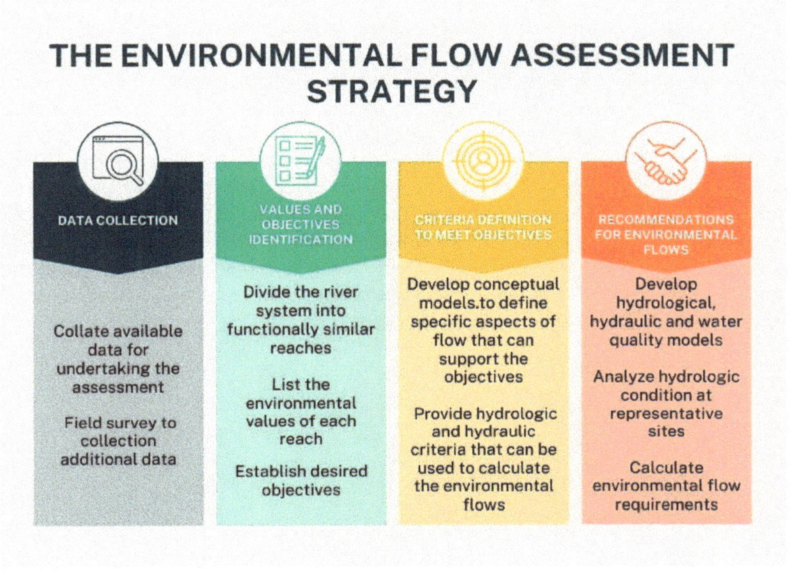

Figure 2. The figure shows the environmental flow assessment strategy.

The first step of this research is the process of collecting data and conducting field surveys in the research area. During this process, rainfall data from all of the stations shown in Figure 1 were collected. The water level data were gathered from both hydrological stations, Chuoi and Ngoc Tra, while salinity data were only available at Ngoc Tra station. These stations belong to the Vietnamese hydrometeorological station system. The data at these stations ensures reliability. The reservoir operation data were also gathered for analysis in the subsequent steps. These data included inflow and release discharge measured by reservoir owners. This is useful information when validating hydrological models. The researchers conducted the survey to collect 37 cross-sections along two main rivers, the Yen River and Thi Long River, from the downstream of the Yen My and Song Muc reservoirs to the sea. These data are used as the input for hydraulic models. In addition, 5 sediment data types were sampled to define the grain gradation curve. The study also conducted interviews with local people through the use of survey questionnaires. In total, 28 survey questionnaires were performed in this study. In each questionnaire, information about the current status of river segments, such as information relating to floods, drought, and ecological assets, was provided. Through surveys, this information also helped to determine the desired objectives (water demand, bank erosion protection, navigation, . . .) for each river segment. Additionally, related documents were collected by the authors, including annual reports of water use planning, annual reports of dike systems in the study area, the operating procedures of the two reservoirs, and the current ecosystem reports of Ben En National Park. These data were also collected from local authorities such as the Thanh Hoa Department of Agriculture and Rural Development, Song Chu Irrigation Exploitation Company, and Ben En National Forest Management Board. Therefore, the information from these documents is official and highly reliable. The collected and surveyed data are summarized in Table 1 below. Examples of the data are presented in the Appendix A in Figures A1–A5. To simplify the river system, this study divided it into several homogeneous river sections and a representative location of each section. At the end of this step, the stake-holder consultation process was used to set the objective concerning the environmental conditions of each section.

The purpose of comprehensive environmental flow assessments is the identification of environmental objectives relating to river basins. The assessment of environmental flow also defines the flow regime required to meet those objectives (e.g., to sustain important assets in the desired condition). This research was conducted using conceptual models with the aim of determining specific criteria for each river section to meet the proposed requirements. These models can show the relationships between the flow components and the objectives identified in previous steps. Moreover, this study also identified measurable criteria for each of the above-mentioned environmental objectives and flow components. These quantitative criteria can be flow, water level, and salinity for each specific river section.

Within an estuary, the flow of the river is heavily influenced by tidal fluctuations. Due to the scarcity of flow measurement data in the study area, a methodology that integrates hydrological, hydraulic, and water quality models was employed in this study. In this study, the Mike NAM model [15] was applied to convert rainfall to runoff. The structure of the Mike NAM model is presented in Figure 3 below.

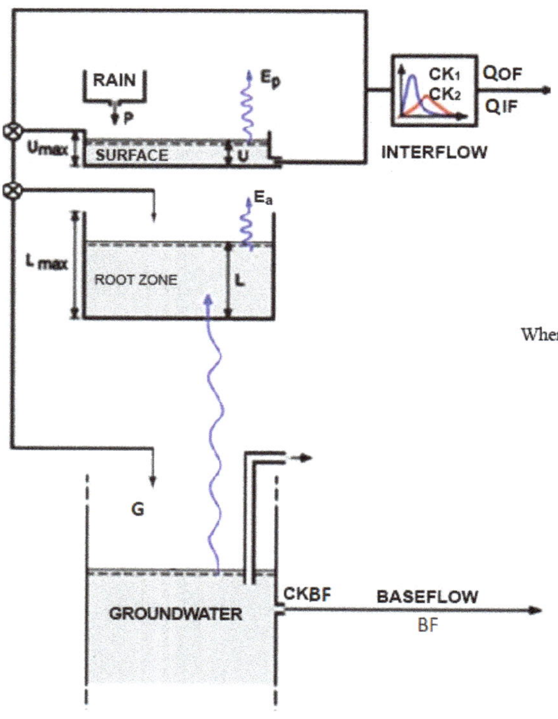

Where P is precipitation (mm);
Ep is potential evapotranspiration (mm);
Ea is actual evapotranspiration (mm);
Umax is the maximum water content in surface storage (mm);
Lmax is the maximum water content in root zone storage (mm);
PN is excess water (mm);
L is moisture content in root zone storage (mm);
U is moisture content in surface storage (mm);
CK1 and CK2 are time constant for routing overland flow (h);
G is infiltrating water (mm);
QIF is interflow; QOF is overland flow (m3/s)
CKBF is time constant for routing baseflow (h);
BF is baseflow (m3/s)

Figure 3. The figure shows the structure of the NAM model (source [16]).

The Yen River basin was subdivided into several sub-basins to meet the requirements of the lump parameter model. This approach has been employed in numerous prior studies [17–19]. The hydrological model (Mike NAN) outputted the boundaries of the hydraulic model. In this study, a Mike 11 hydraulic model was used for the study area, as illustrated in Figure 4. The foundation of the hydrodynamic module 'Saint Venant' equations includes continuity equations and momentum equations. The advection–dispersion equation presents the conservation of mass of a substance in a solution in an advection–dispersion model.

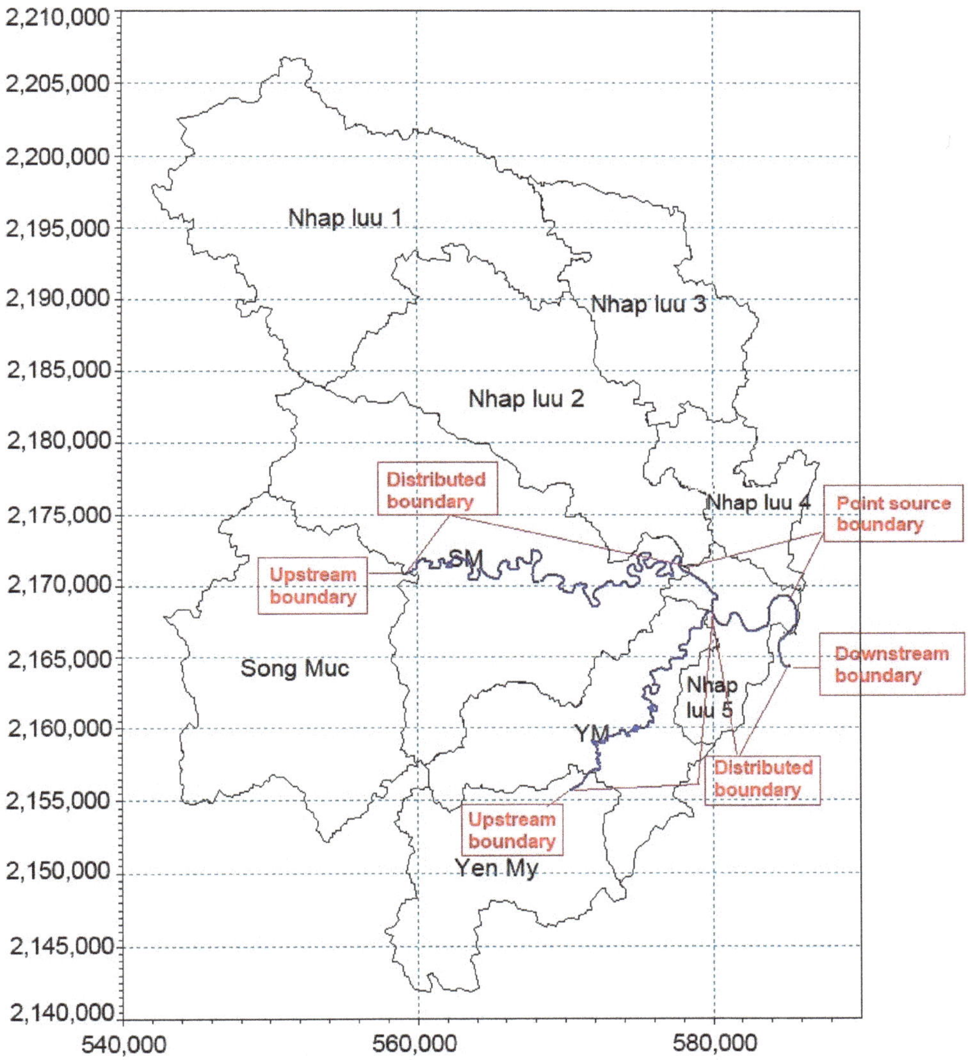

Figure 4. The figure shows the Yen River hydraulic model.

$$\text{Continuity equation}: \frac{\partial Q}{\partial x} + \frac{\partial A}{\partial t} = q \tag{1}$$

$$\text{Momentum equation}: \frac{\partial Q}{\partial t} + \frac{\partial\left(\frac{\alpha Q^2}{A}\right)}{\partial x} + gA\frac{\partial h}{\partial x} + \frac{gQ|Q|}{C^2AR} = 0 \tag{2}$$

$$\text{Advection–dispersion equation}: \frac{\partial AC}{\partial t} + \frac{\partial QC}{\partial x} + \frac{\partial}{\partial x}\left(AD\frac{\partial C}{\partial x}\right) = -AKC + C_2\mathbf{q} \tag{3}$$

where Q—discharge (m^3/s);
A—flow area (m^2);

q—lateral inflow (m^2/s);
h—stage above datum (m);
C—Chezy resistance coefficient ($m^{1/2}/s$);
R—hydraulic or resistance radius (m);
α—momentum distribution coefficient (-);
C—concentration (PSU for salinity);
D—dispersion coefficient (m^2/s);
$\mathbf{K}$—linear decay coefficient (h^{-1});
$\mathbf{C_2}$—source/sink concentration (PSU for salinity);
x—space coordinate (m/s);
t—time coordinate (s).

The model has two upstream boundaries, including the released discharge from the Song Muc and Yen My reservoirs (the sub-basins Song Muc and Yen My, respectively). The downstream boundary of the model was based on tide data. At the downstream boundary, the salinity is assumed to be equal to 32 PSU, according to the result of [20]. The model comprised five contribution boundaries. Due to missing cross-section data, the contribution discharges of the Nhon River and Ly River (Figure 1) are derived from the hydrological model's calculation results from the sub-basins of Nhap Luu 1, Nhap Luu 2, Nhap Luu 3, and Nhap Luu 4, respectively. They are considered to be the point source boundaries in the model and are shown in Figure 4. The contribution discharge from the SM, YM, and Nhap Luu 5 sub-basins contribute to river systems along the river. This leads to the distribution boundaries being used to simulate them (Figure 4). Then, the daily flow of the river system is determined in order to carry out the environmental flow assessment. At each representative location, the flow regime is determined through a number of factors, such as flow duration curve, hydrograph, etc. Based on these results, the frequency and recurrence period of the flow components that meet the established criteria can be determined. However, the setting of environmental flows is a socio-economic as well as scientific process [21]. While natural sciences can provide information on the implications of a changing flow regime, it is the socio-political decisions that determine which ecosystem services should be protected and, therefore, which environmental flows should be provided. This may involve adjusting or trading off environmental objectives against other uses and making decisions on which flow components will be protected or provided to maintain specific ecosystem goods and services. The goal of this step is to perform rigorous scientific analysis in order to identify an appropriate flow regime for both environmental maintenance and social development.

3. Results

3.1. River Segments and Objectives

From the collected information, the study has an overall view of the current state of the basin. This information establishes a common understanding of the intended outcomes for the research area. The river system can be divided into five river reaches (Figure 5). However, each section of the river will have different goals and orders. The first reach is located upstream of the Song Muc reservoir, which includes Ben En National Park. This is an area with a rich and diverse ecosystem, comprising 1417 species of higher plants and 1536 species of animals. The area plays a significant role in preserving ecosystem diversity and abundance in the region. This reach is not affected by saltwater intrusion due to the role of the Song Muc reservoir. The second reach is upstream of the Yen My reservoir, which is not affected by saltwater intrusion, similar to the first segment. This reach has less developed ecosystems compared with the previous reach, with the river's main purpose being to provide water for domestic and agricultural demands. The downstream rivers of the reservoirs are all affected by saltwater intrusion, but the level of impact is not the same. According to local people, high concentrations of salt water can pervade inland from 16.5 to 23.2 km. This means that in reaches 3 and 4, from the reservoirs to the junction of Yen and Thi Long rivers, saltwater intrusion can occur in the dry season. As a result, the most important objective is to maintain suitable salinity conditions for domestic and agricultural demands. The reach from the junction of the Yen and Thi Long rivers to the sea is dominated by the tide (reach 5). The salinity of this river reach is very high in the dry season. In addition, in this segment, it is important to pay attention to navigate and maintain estuary openings. In conclusion, the objectives in the research area can be grouped into five categories, including hydrology, biology, water quality, connectivity, and geomorphology.

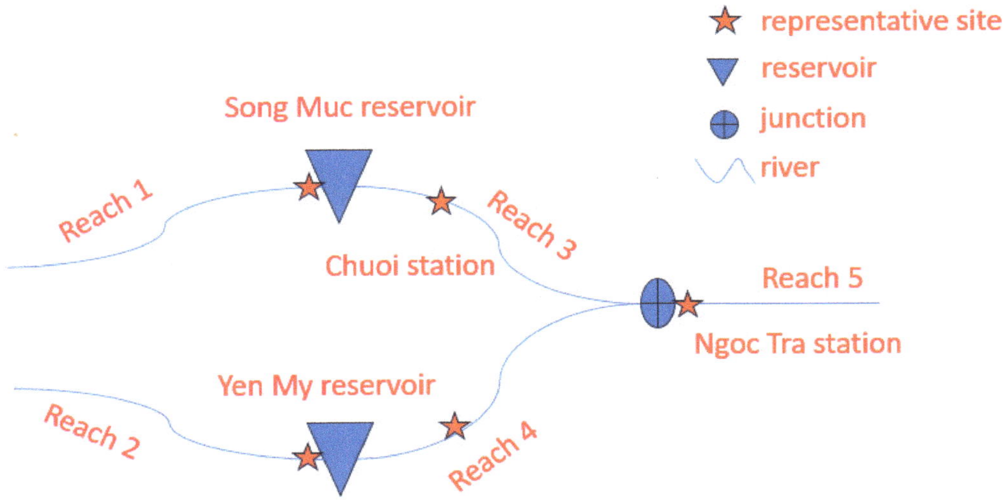

Figure 5. The figure shows river segments and representative locations.

Table 2 shows information about the river segments as well as their representative sites. For reach 1 and reach 2, the representative sites are located at the reservoir inlet. Reach 3 and reach 5 have hydrological stations, so their hydrological station locations are selected to be representative sites for each section. The representative position of reach 4 is the cross-section at Trai Ron. Establishing desired objectives for each section of a river involves defining desired outcomes. The detailed objectives of each river are listed in Table 3.

Table 2. The river segments in the Yen River system.

ID	Name	River	Description	Representative Site
1	Reach 1	Yen	Upstream to Song Muc reservoir	Song Muc reservoir
2	Reach 2	Thi Long	Upstream to Yen My reservoir	Yen My reservoir
3	Reach 3	Yen	Song Muc reservoir to junction	Chuoi station
4	Reach 4	Thi Long	Yen My reservoir to junction	
5	Reach 5	Yen	Junction to the sea	Ngoc Tra station

3.2. Conceptual Model

The flow regime in stream systems contributes to the physical processes and ecosystem functions required to maintain reaches and assure the normal development of aquatic ecosystems. There are many such physical processes and related ecosystem functions. To present a current understanding of these relationships, conceptual models are often an attractive option due to their simplicity. According to [22], there are several forms of conceptual model. They can be text descriptions, process diagrams, or mathematical relationships. In this research, a text description is selected to identify the specific aspects of flow that can support the objectives of each reach. The results are summarized in Table 3.

Dry season baseflows are low-level persistent flows during the dry season. This flow component maintains sufficient width and adequate depth at riffles and pools. It helps to ensure essential habitats for fish and other instream fauna. There is a minimum flow in the channel to maintain instream vegetation and provide water for soils and banks to support riparian vegetation. This flow component purges invasive, introduced species from aquatic and riparian communities and provides drinking water for terrestrial animals. This is especially important for reach 1, where ecological environmental requirements are the top priority. This flow value is usually taken from the smallest monthly flow to the average of the three smallest monthly flows [4]. The value of the dry season baseflow needs to be maintained during the dry season to ensure environmental values.

Dry season pulses have larger flows than dry season baseflows. In a natural system, there are increased flows that result from localized rainfall events. This flow component provides the wetting of instream features, including benches and bars. These hydraulic criteria determine the value of dry season pulses. Based on cross-sectional survey data, this study determined the corresponding flow to flood benches and bars in each reach. Additionally, this flow contributes to the scouring of fine sediment from riffle zones, provides water in pools to break up developing stratification, aids the localized movement of fish between habitats, supports fish spawning, and helps to limit saline intrusions. Dry season pulses should occur at least once every 2 months to prevent water quality issues.

During the flood season, the baseflows of a river are at their lowest, providing only minimal flooding. However, these flows have the advantage of providing a larger habitat for aquatic species. Additionally, they help restore normal water quality by removing pollutants and preventing the migration and colonization of riparian vegetation into the channel. Flood season baseflow occurs with the onset of the flood season and the demand during the flood season.

Flood season pulses are typically larger and longer than dry season pulses and are in line with the natural seasonality of the river. These pulses help move sediment through the river. A high-velocity flow helps to clean out fine sediment from shallow riffle habitats, which determine the size of stream bed substrates (sand, gravel, and cobble). The particle size of the bed material can be sampled in field surveys. The criteria for flood season pulses are a flow value that mobilizes d_{50} of the bed material. These flows flush away waste products and pollutants, which improves the quality of instream habitats. They also allow better connectivity between habitats. According to the experiences of local people, it is necessary to have two to three heavy rainfall events each year.

Bankfull flows are the maximum discharge that a stream channel can carry without overflowing onto the floodplain. At this point, the water level reaches the top of the channel, which is known as the bankfull stage. According to flood frequency analysis, rivers typically reach their bankfull state once every 1.5–2 years [23]. These flows prevent riparian vegetation from encroaching into the channel and inundating these features within the channel. Bankfull flows input nutrients into the river system, which supports the production of food for aquatic species. They also play an important role in sediment movement within the river and maintain the shape and form of the channel, which is very important for navigation purposes. In addition, they help to clean sediment that has accumulated in important aquatic habitats.

Overbank flows inundate floodplains, which is important for providing lateral connectivity with the channel and floodplain. Overbank flow thresholds are usually determined by the maximum flow in a period of 2 years (small floods) to 100 years (large floods) [24]. These flow events will deposit sediments onto the floodplain, supporting the growth of vegetation (both natural and agricultural). Additionally, these flows trigger a new phase in the life cycle of the ecosystem and introduce nutrients into the river system, which supports the production of food for fish. In addition, overbank flows are the main water supply sources for reservoirs. This is also a requirement to maintain the estuary opening in reach 5.

Table 3. The table shows the flow components relevant to the objectives.

Flow Component	Objective	Categories	Reach	Hydraulic Criteria
Dry season baseflows	Maintain sufficient width of adequate depth at riffles and pool	Hydrology	1, 2	The flow is in range from the smallest monthly flow to the average of 3 smallest months flow
	Ensure the essential habitat for the fish and other instream fauna	Biology	All	
	Maintain instream vegetation	Biology	All	
	Support riparian vegetation	Connectivity	1, 2	
	Purge invasive, introduced species from aquatic and riparian communities	Biology	All	
	Provide drinking water for terrestrial animals	Connectivity	All	
Dry season pulses	Provide for the wetting of instream features, including benches and bars	Hydrology	1, 2	The flow has the water level at bar level
	Contribute to scouring of fine sediment	Geomorphology	3, 4, 5	
	Prevent water quality issues	Water quality	3, 4, 5	
	Support fish spawning	Biology	All	
	Limit saline intrusions	Water quality	3, 4, 5	
Flood season baseflows	Provide larger habitats for aquatic species	Biology	All	The flow occurs at onset of flood season
	Restore normal water quality	Water quality	All	
	Preventing the migration and colonization of riparian vegetation into the channel	Connectivity	All	
Flood season pulses	Determine the size of stream bed substrates	Geomorphology	All	The flow has a velocity that can mobilize d_{50} of the bed material
	Allowing for better connectivity between habitats	Connectivity	All	
	Flush away waste products and pollutants	Water quality	All	
Bankfull flows	Prevent riparian vegetation from encroaching into the channel	Connectivity	All	The flood discharge occurs once every 1.5 ÷ 2 years
	Support the production of food for aquatic species	Connectivity	All	
	Maintaining the shape and form of the channel	Geomorphology	All	
Overbank flows	provide lateral connectivity	Connectivity	All	The flood discharge has P = 1 ÷ 50%
	Deposit sediments onto the floodplain	Geomorphology	1, 2	
	Trigger a new phase in the life cycle	Biology	All	
	Provide water	Hydrology	1, 2	

Notes: Q: discharge; Z: water level; D: water depth; V: velocity.

3.3. Hydrological and Hydraulic Model

In order to ensure the accuracy of the calculation results, hydrological and hydraulic models need to be calibrated and validated with the observed data to ensure accuracy. There is no hydrological station measuring discharge in the basin. However, Song Muc reservoir station has measured the inflow to the reservoir in recent years. Therefore, the study uses data from 1 January 2021 to 31 December 2023 to calibrate and validate the hydrological model. The results of the calculated and observed flow are shown in Figure 6. It can be seen that the calculated hydrograph is consistent with the observed data. Table 4 quantifies this concordance through the NASH index [25]. The Nash index can be estimated using Equation (4). According to [26], this result is at a good level. Thus, the set of parameters of the hydrological model well reflects the current status of the Song Muc reservoir basin. The validated parameters are applied to the sub-basins in the model described in Figure 4 to estimate the flow in a long-term period in the whole Yen River basin.

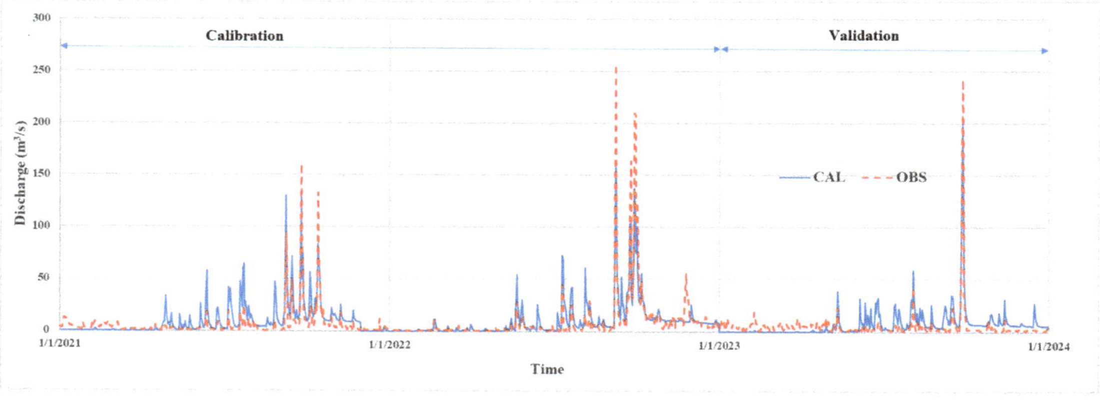

Figure 6. The figure shows the calculated and observed results of the hydrological model.

Table 4. The table shows the Nash index of the hydrological and hydraulic models.

ID	Mode	Variable	Location	Nash Index	
				Calibration	Validation
1	Hydrology	Discharge	Song Muc reservoir	0.72	0.71
2	Hydraulic	Water level	Ngoc Tra station	0.98	0.98
3	Hydraulic	Salinity	Ngoc Tra station	---	---

$$NASH = 1 - \frac{\sum_{t-1}^{T}\left(Q_0^t - Q_m^t\right)^2}{\sum_{t-1}^{T}\left(Q_0^t - \overline{Q}_0\right)^2} \quad (4)$$

where $\overline{Q}_0$ is the mean of observed discharges (m^3/s);

Q_m^t is the calculated discharges at time t (m^3/s);

Q_0^t is the observed discharges at time t (m^3/s).

Similar to the hydrological model, the hydraulic model was calibrated and validated using the observed water level data from Ngoc Tra station. The data from 1 January 2013 to 31 December 2017 were used to calibrate the model, while the period data from 1 January 2018 to 31 December 2022 were used to validate the model. The model results are shown in Figure 7 and Table 4. The results also show that the model simulates the hydraulic regime of the Yen River basin well. For the seawater intrusion issue, the result is not as good as the hydraulic regime. However, the simulated salinity value is also consistent with the observed data. This means that this model can be used to determine the daily flow of the river system to serve the next stage of the environmental flow assessment.

The study simulates long-term flow regimes in the study area using validated hydrological and hydraulic models. Figure 8 shows the daily flow at representative locations. Based on these results, the annual module of flow in the Yen River region is about 64 $L/s.km^2$. According to the criteria for seasonal classification [27], the flood season is a continuous period in which the average monthly flow is greater than the average annual flow with a frequency of occurrence greater than 50%. The flow regime in the Yen River basin is divided into two distinct seasons. The flood season is 5 months, from July to November. The total flood volume is 48% to 61% of the total annual volume. In the flood season, the Yen basin is faced with two to three tropical depressions or storms each year, which causes major floods in the basin. The largest flood peak usually appears in July or August. The dry season in the basin lasts from December to June of the following year. During this time, there is little rainfall, leading to the small discharge in the river and the deep inland saltwater intrusion. The average salinity in the dry season at Ngoc Tra station is about 23 PSU.

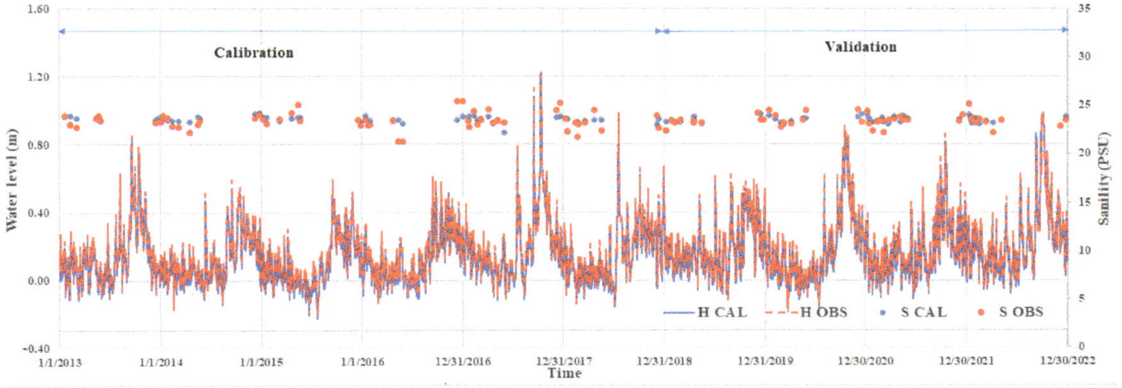

Figure 7. The figure shows the calculated and observed results for the hydraulic model.

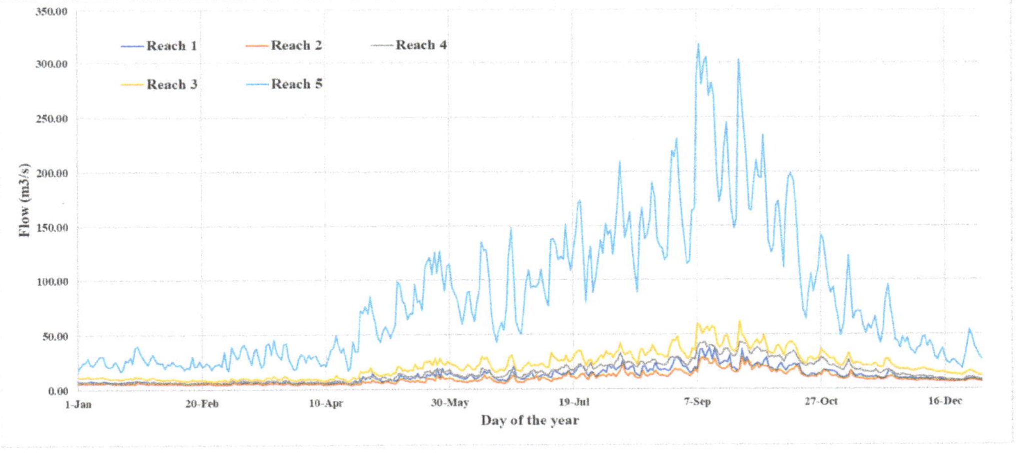

Figure 8. The figure shows a hydrograph of average daily flows at representative sites.

3.4. Environmental Flow Assessment

The assessment of environmental flow requirements forms the basis of decision making. According to Vietnam's Law on Water Resources (2012) [28], basin development needs to consider the water use demands of sectors and localities and environmental protection. It means that the processes used to identify environmental objectives for the river basin, such as important environmental assets or processes to be sustained, are not enough. Establishing desired objectives for each section of a river involves defining desired outcomes; however, achieving these goals is a complex undertaking [29]. The river not only has environmental protection purposes but also caters to other water users. Moreover, a river basin is a highly interconnected system in which activities in upstream areas can have an impact on downstream and vice versa. It is necessary to consider the entire system rather than limiting activities to each individual river section. For example, upstream reservoir operation can affect downstream flooding. Additionally, objectives may sometimes conflict with each other. For instance, preventing seawater intrusion during the dry season can conflict with the need to maintain water levels in the reservoir for the water supply. This can lead to trade-offs, and negotiations are often necessary to reconcile these requirements with the needs of other users. This may involve adjusting or trading off environmental objectives against other uses. In this domain of research, the majority of the reaches in the basin, with the exception of reach 1, which remains unaffected by human influence, serve to provide water for agricultural purposes. This calls for satisfying both quantity and quality requirements. With regard to the quantity requirements, the river water level during the water intake period must ensure sufficient depth for pumps to operate. As for the quality requirements, the salinity in the river should not surpass 4PSU since it will impair crop productivity. According to the findings of the surveys conducted on the pumps situated along the river, the minimum depth required to fulfill this condition must be greater than 0.5 m. For river sections 3 and 5, special attention needs to be given to the water transportation needs. For a 20-ton vessel to enter the river mouth, the primary channel width must be wider than 20 m, and the minimum depth needs to range between 1 and 1.2 m. All of these criteria must be taken into account when defining the environmental flow.

An assessment of environmental flows should include the identification of environmental objectives for the river basin and the determination of the flow regime required to achieve those objectives. Based on the hydraulic criteria shown in Table 3, the flow regime was analyzed and selected to not only meet environmental protection requirements but also pay attention to other issues. The study recommends the dry season baseflow value equal to the average flow of the 3 driest months in a long-term period. This is the highest value intended to ensure maximum environmental protection requirements while still being consistent with the promulgated legal documents [4]. This value also helps to ensure environmental protection and reduce pressure on reservoirs when having to release water in the dry season. The flood season baseflow value is determined as an average of monthly flows in July, the onset of the flood season. During this period, due to the abundant amount of water, water usage requirements can be fully adapted. The concentrated flow serves environmental purposes. Selecting the average value for the long term in the basin can ensure the stability of the ecosystem. Based on the hydrological statistics calculated from a modeled flow time series, this study defines the magnitude, frequencies, and durations of events (pulses, bankfulls, and overbanks). The magnitude of pulse events is selected by the 20% exceedance daily flow rate, which was identified by examining the duration of the flow curves for the two sites in the flood and dry seasons. After selecting this value, it is checked to ensure ecological requirements, such as the wetting bar in the dry season and mobilizing coarse material in the flood season. The frequencies corresponding to bankfull and overbank flows are 2% and 1.5% exceedance of the maximum flow rate, respectively. Detailed results are given in Table 5.

Table 5. The table shows the flow component summary of the Yen River system.

Flow Component		Dry Season Baseflow	Dry Season Pulse	Flood Season Baseflow	Flood Season Pulse	Bankfull	Overbank Flow
Flow rate (m^3/s)	Reach 1	4.03	8.92	14.5	21.7	133	158
	Reach 2	3.19	6.83	10.6	17.3	95	113
	Reach 3	4.16	17.6	26.2	39.5	187	213
	Reach 4	5.45	11.7	18.1	31.3	113	129
	Reach 5	15.3	54.5	117	192	1057	1205
Frequency (per year)		Continuous	4	Continuous	2	0.67	0.5
Duration (days)		Dry season	1	Flood season	2	3	3

4. Discussion

The main outcome of the present study is the insight into the holistic approach to environmental flow assessment. In this research, a flow regime with seven components is quantitatively determined for five river segments of the Yen River basin. When comparing our results with those of older studies, it must be pointed out some improvements have been made in both the dry and flood seasons. Traditional methods, such as Montana the Method [7] or the Wetted Perimeter Method [10], only define critical values to establish a minimum condition for the preservation of the ecosystem. This seems inappropriate since these methods only consider baseflows. Events such as floods now also play an important role in ecosystems. Flood management does not pose a problem for the study area. However, it is a particularly serious problem that has occurred in many areas. Typical examples in Vietnam are the Red River or Mekong River basins, where the upstream reservoirs can store the majority of the flood volume. To maintain a healthy ecosystem, it is also necessary to specify the flood events that are necessary in terms of magnitude, frequency, and duration. In the dry season, our results go beyond previous reports by Gippel et al. (2009) [3]. In this study, we not only combine information on ecological and other assets associated with the river system but also integrate water use needs and legal documents related to the river. Environmental flow components are determined holistically. In addition to ensuring environmental protection, the values given also take into account other water requirements in the catchment area and are consistent with existing legal documents. This ensures a balance of values between environmental protection and the development needs of the catchment area. The regime for environmental flows is now specifically defined using this approach. It facilitates the management of water resources in the basin.

Although some quality aspects are covered, this study focuses primarily on the amount of water that needs to be managed in a catchment to support healthy rivers. This is also a limitation of this study. To resolve this issue, additional observations related to water quality issues need to be conducted to provide input data for modeling. Additionally, this research has other limitations. There are no suitable ecological data from the Yen River to validate the flow regime found. This leads to a weak alignment of the environmental goals with the hydraulic criteria. The proposed hydraulic criteria are subjective and sometimes do not meet all environmental objectives. However, it is still a complex problem that cannot be solved under current research conditions. Ignoring the disadvantages mentioned, there are still many advantages of the approach that need to be mentioned. This research approach is very suitable for limited data sets and for low-budget projects. Field surveying to find out the unique characteristics of each range is essential. However, this content is completely feasible in any basin. If this is not possible, cross-sectional surveys can be replaced by cross-sections processed from digital elevation models. There are now many free DEMs available. These methods are not recommended by the authors because the accuracy is not high. The rainfall input of hydrological models can be easily collected from the rainfall station system. In cases where the collection is impossible, satellite rain sources can be alternative data sources.

Based on the analysis presented, it is evident that the holistic approach employed in environmental flow assessment is viable despite certain constraints. The methods employed seek to strike a balance between safeguarding the environment and fulfilling development demands. This approach can be applied to environmental flow assessments for all other catchments. By integrating mathematical models that simulate the flow regime in the catchment area and the process of field survey and data collection to evaluate river status, the methods are effective. Mathematical models are highly versatile tools that offer detailed information about the hydraulic condition of river basins. Furthermore, the additional data gathered from the investigation process, as requested, are diverse and dependent on the characteristics of the basins. At present, there is no specific standard that governs these additional data. Therefore, future research should focus on enhancing the framework to address these limitations.

5. Conclusions

In this study, a holistic environmental flow assessment is applied to the Yen River, Thanh Hoa, Vietnam. This approach achieves an appropriate balance between allowing for robust analysis whilst reducing complexity and cost as much as possible. The information investigated and collected is not too complicated and detailed. The mathematical model provides fast, detailed results with feasible input data that can be collected in most regions in Vietnam. This approach can be applied even in regions with limited data availability.

The assessment result is a whole flow regime, not a single minimum flow value applied throughout the year. This regime includes the magnitude, frequency, and duration of six flow components (dry season baseflow, dry season pulse, flood season baseflow, flood season pulse, bankfull, and overbank). This study was conducted to determine the environmental flow of five reaches in the research basin. This environmental flow ensures that specific, distinctive environmental objectives of each reach are met. In addition, it is consistent with the existing legal and policy requirements in Vietnam and considers water use demands. This issue provides a greater level of confidence in the protection of ecological assets and supporting basin development.

Author Contributions: Conceptualization, C.K.T. and T.P.T.; methodology, C.K.T.; validation, S.T.H. and N.T.H.; formal analysis, N.T.H.; investigation, S.T.H., D.Q.B. and T.P.T.; resources, N.T.H. and L.M.N.; data curation, S.T.H.; writing—original draft preparation, L.H.N.; writing—review and editing, C.K.T.; visualization, L.H.N.; funding acquisition, T.P.T. All authors have read and agreed to the published version of the manuscript.

Funding: This research was supported by the authorized project for Vietnam Academy of Science and Technology, namely "Investigate and evaluate the ecogeomorphologycal conditions of Ben En National Park, propose sustainable solutions for national parks in Vietnam", grant number UQĐTCB.04/23-24.

Data Availability Statement: The data presented in this study belongs to the project "Investigate and evaluate the ecogeomorphologycal conditions of Ben En National Park, propose sustainable solutions for national parks in Vietnam". The data are not publicly available due to commercial restrictions.

Conflicts of Interest: The authors declare no conflicts of interest.

Appendix A

River: Yen　　Year: 2022　　Station: Ngoc Tra

Daily water level

Datum: Standard.　　Unit: cm

Day \ Month	I	II	III	IV	V	VI	VII	VIII	IX	X	XI	XII
1	15	25	6	19	12	20	13	14	9	95	49	39
2	18	35	−2	27	15	21	16	12	16	96	63	45
3	27	20	−4	41	22	11	10	9	19	99	64	42
4	31	26	3	34	18	7	2	13	29	92	38	32
5	34	26	6	23	12	5	10	18	31	92	30	16
6	32	29	7	8	9	0	4	21	37	92	36	25
7	32	30	9	1	4	−2	−1	23	42	92	32	25
8	39	24	16	4	9	−6	6	28	68	87	21	30
9	27	19	8	16	10	0	13	40	83	81	22	29
10	18	17	7	19	15	3	21	43	79	66	22	32
11	2	16	8	4	8	3	35	54	64	58	20	23
12	10	15	11	8	0	11	64	41	47	55	17	31
13	18	21	10	6	-3	9	60	36	34	57	19	35
14	15	−2	4	4	6	8	60	29	28	58	22	39
15	11	14	−4	9	16	12	56	23	28	67	24	38
16	18	17	1	27	10	24	49	21	33	52	34	37
17	17	20	2	37	24	29	40	19	29	41	29	28
18	2	25	1	23	25	26	27	19	31	50	31	32
19	19	37	5	12	28	18	24	20	34	76	20	34
20	25	17	17	8	29	13	24	26	29	61	27	23
21	26	31	23	8	28	4	32	32	33	51	28	10
22	33	35	18	7	44	4	25	28	50	42	23	4
23	31	12	10	6	49	3	23	23	56	23	20	12
24	18	19	11	9	39	6	16	16	72	23	26	24
25	13	19	17	9	27	17	7	11	83	51	42	33
26	18	17	13	2	23	9	8	20	74	48	46	30
27	16	11	15	4	21	3	13	4	61	40	50	35
28	15	10	6	6	24	-2	12	8	96	36	53	40
29	33		23	4	21	5	11	3	90	45	47	24
30	37		22	2	14	9	14	2	96	55	34	13
31	34		17		14		12	4		54		15

Figure A1. Water level data at Ngoc Tra station in 2022.

Reservoir	Time	Date	Basin	Water level (m)	Inflow (m^3/s)	Released discharge (m^3/s)
Song Muc	1	8/25/2019	CP	29.78	8	6
Song Muc	7	8/25/2019	CP	29.79	10	0
Song Muc	13	8/25/2019	CP	29.79	8.22	5.9
Song Muc	19	8/25/2019	CP	29.79	10.2	6
Song Muc	1	8/27/2019	CP	29.78	8	6
Song Muc	7	8/27/2019	CP	29.79	10	0
Song Muc	13	8/27/2019	CP	29.79	8.22	5.9
Song Muc	19	8/27/2019	CP	29.79	10.2	6
Song Muc	19	8/30/2019	CP	29.79	11.76	4.5
Song Muc	19	9/1/2019	CP	30.33	10.5	0.4
Song Muc	1	9/5/2019	CP	30.62	7	4.5
Song Muc	7	9/5/2019	CP	30.63	9	4.5
Song Muc	13	9/5/2019	CP	30.63	8	4.5
Song Muc	19	9/5/2019	CP	30.64	7	4.5
Song Muc	1	9/6/2019	CP	30.65	7	4
Song Muc	7	9/6/2019	CP	30.66	9	3
Song Muc	13	9/6/2019	CP	30.66	9	3
Song Muc	19	9/6/2019	CP	30.67	7	0
Song Muc	1	9/7/2019	CP	30.67	7	3.9
Song Muc	7	9/7/2019	CP	30.67	9	3
Song Muc	13	9/7/2019	CP	30.68	8	3
Song Muc	19	9/7/2019	CP	30.68	6	0
Song Muc	1	9/8/2019	CP	30.68	8	3.9
Song Muc	7	9/8/2019	CP	30.68	8	3
Song Muc	13	9/8/2019	CP	30.68	8	3
Song Muc	19	9/8/2019	CP	30.69	6	0

Figure A2. The figure shows an example of Song Muc reservoir operation data.

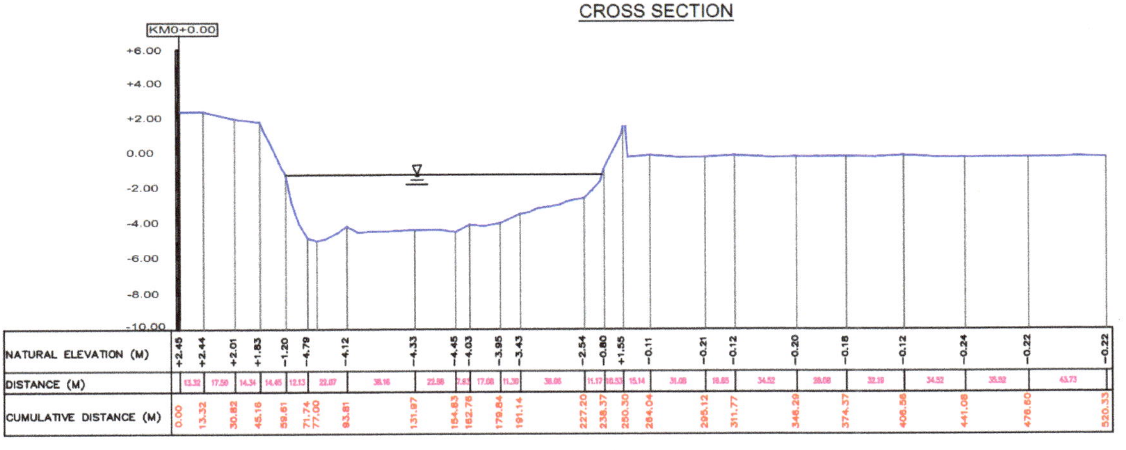

Figure A3. The figure shows cross-section data at Ngoc Tra station.

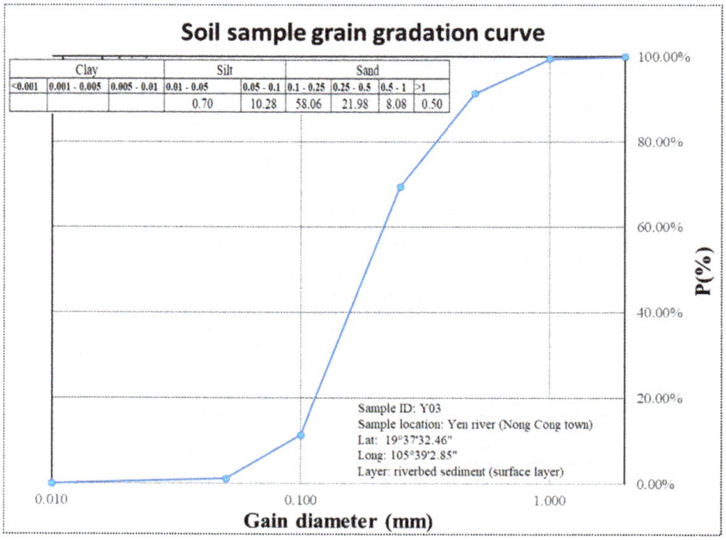

Figure A4. The figure shows the soil sample grain gradation curve.

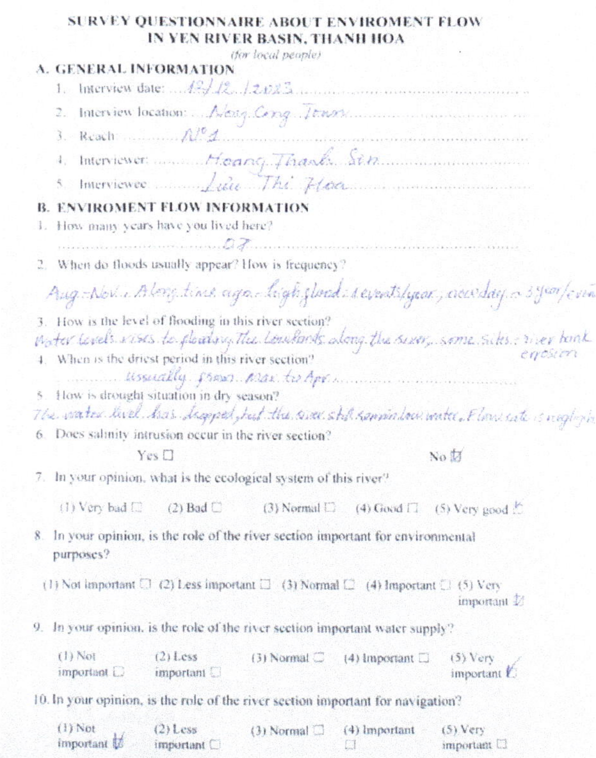

SURVEY QUESTIONNAIRE ABOUT ENVIROMENT FLOW
IN YEN RIVER BASIN, THANH HOA
(for local people)

A. GENERAL INFORMATION

1. Interview date: 19/12/2023
2. Interview location: Nong Cong Town
3. Reach: N°1
4. Interviewer: Hoang Thanh Son
5. Interviewee: Luu Thi Hoa

B. ENVIROMENT FLOW INFORMATION

1. How many years have you lived here?
   07
2. When do floods usually appear? How is frequency?
   Aug–Nov. A long time ago, high flood: 1 events/year, nowday ~ 3 year/even
3. How is the level of flooding in this river section?
   Water levels rises to flooding the lowlands along the river, some sites: river bank erosion
4. When is the driest period in this river section?
   usually from Mar. to Apr.
5. How is drought situation in dry season?
   The water level has dropped, but the river still remain low water. Flow rate is negligib
6. Does salinity intrusion occur in the river section?
   Yes ☐ No ☑
7. In your opinion, what is the ecological system of this river?
   (1) Very bad ☐ (2) Bad ☐ (3) Normal ☐ (4) Good ☐ (5) Very good ☑
8. In your opinion, is the role of the river section important for environmental purposes?
   (1) Not important ☐ (2) Less important ☐ (3) Normal ☐ (4) Important ☐ (5) Very important ☑
9. In your opinion, is the role of the river section important water supply?
   (1) Not important ☐ (2) Less important ☐ (3) Normal ☐ (4) Important ☐ (5) Very important ☑
10. In your opinion, is the role of the river section important for navigation?
   (1) Not important ☑ (2) Less important ☐ (3) Normal ☐ (4) Important ☐ (5) Very important ☐

Figure A5. *Cont.*

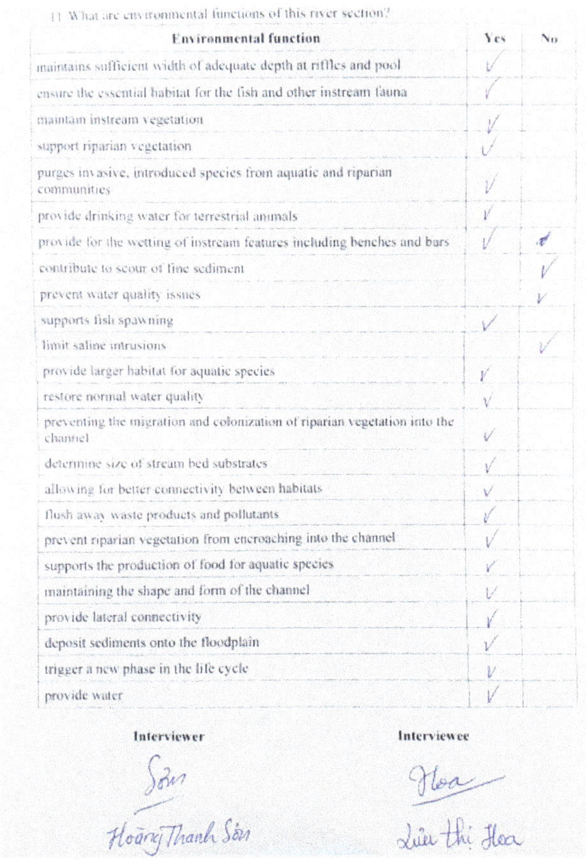

11. What are environmental functions of this river section?

Environmental function	Yes	No
maintains sufficient width of adequate depth at riffles and pool	✓	
ensure the essential habitat for the fish and other instream fauna	✓	
maintain instream vegetation	✓	
support riparian vegetation	✓	
purges invasive, introduced species from aquatic and riparian communities	✓	
provide drinking water for terrestrial animals	✓	
provide for the wetting of instream features including benches and bars	✓	✓
contribute to scour of fine sediment		✓
prevent water quality issues		✓
supports fish spawning	✓	
limit saline intrusions		✓
provide larger habitat for aquatic species	✓	
restore normal water quality	✓	
preventing the migration and colonization of riparian vegetation into the channel	✓	
determine size of stream bed substrates	✓	
allowing for better connectivity between habitats	✓	
flush away waste products and pollutants	✓	
prevent riparian vegetation from encroaching into the channel	✓	
supports the production of food for aquatic species	✓	
maintaining the shape and form of the channel	✓	
provide lateral connectivity	✓	
deposit sediments onto the floodplain	✓	
trigger a new phase in the life cycle	✓	
provide water	✓	

Interviewer Sơn Hoàng Thanh Sơn

Interviewee Hoa Lưu thị Hoa

Figure A5. The figure shows an example questionnaire.

References

1. The Brisbane Declaration. In Proceedings of the 10th International Riversymposium and International Environmental Flows Conference, Brisbane, Australia, 3–6 September 2007; pp. 1–5. Available online: http://www.conservationgateway.org/Documents/Brisbane-Declaration-English.pdf (accessed on 23 November 2023).
2. Chan, K.M.A.; Satterfield, T.; Goldstein, J. Rethinking ecosystem services to better address and navigate cultural values. *Ecol. Econ.* **2012**, *74*, 8–18. [CrossRef]
3. Gippel, C.J.; Bond, N.R.; James, C.; Wang, X. An Asset-based, Holistic, Environmental Flows Assessment Approach. *Int. J. Water Resour. Dev.* **2009**, *25*, 301–330. [CrossRef]
4. Vietnamese Ministry of Natural Resources and Environment. Circular no 64/2017/TT-BTNMT Regulations on Determining Minimum Flows in Rivers, Streams and Downstream of Reservoirs and Dams. Ha Noi, Dec. 2017. Available online: https://thuvienphapluat.vn/van-ban/Tai-nguyen-Moi-truong/Thong-tu-64-2017-TT-BTNMT-quy-dinh-xac-dinh-dong-chay-toi-thieu-tren-song-suoi-ha-luu-cac-ho-chua-372413.aspx (accessed on 21 November 2023). (In Vietnamese)
5. Poff, N.L.R.; Tharme, R.E.; Arthington, A.H. Evolution of Environmental Flows Assessment Science, Principles, and Methodologies. In *Water for the Environment: From Policy and Science to Implementation and Managemen*; Elsevier Inc.: Amsterdam, The Netherlands, 2017; pp. 203–236. [CrossRef]
6. Wu, X.; Lenon, G.P.; Dong, Y. The calculation of riverine ecological instream flows and runoff profit-loss analysis in a coal mining area of northern China. *River Res. Appl.* **2020**, *36*, 760–768. [CrossRef]
7. Tennant, D.L. Instream Flow Regimens for Fish, Wildlife, Recreation and Related Environmental Resources. *Fisheries* **1976**, *1*, 6–10. [CrossRef]
8. Richter, B.D.; Baumgartner, J.V.; Powell, J.; Braun, D.P. A Method for Assessing Hydrologic Alteration within Ecosystems. *Conserv. Biol.* **1996**, *10*, 1163–1174. [CrossRef]

9. The Nature Conservancy. *Indicators of Hydrologic Alteration: Version. 7.1 User's Manual*; The Nature Conservancy: Arlington County, VA, USA, 2009.
10. Leathe, S.A.; Nelson, F.A. *A Literature Evaluation of Montana's Wetted Perimeter Inflection Point Method for Deriving Instream Flow Recommendations*; Department of Fish, Wildlife, and Parks: Helena, MT, USA, 1986.
11. Stalnaker, C.; Lamb, B.L.; Henriksen, J.; Bovee, K.; Bartholow, J. *The Instream Flow Incremental Methodology: A Primer for IFIM*; National Biological Service: Washington, DC, USA, 1995.
12. Kinga, J.; Louwb, D. Instream flow assessments for regulated rivers in South Africa using the Building Block Methodology. *Aquat. Ecosyst. Health Manag.* **1998**, *1*, 109–124. [CrossRef]
13. Williams, J.; Jenkins, G.P.; Hindell, J.S.; Swearer, S.E. Linking environmental flows with the distribution of black bream Acanthopagrus butcheri eggs, larvae and prey in a drought affected estuary. *Mar. Ecol. Prog. Ser.* **2013**, *483*, 273–287. [CrossRef]
14. Vietnamese Ministry of Natural Resources and Environment. Decision No. 341/QD-BTNMT on Promulgating the List of Rivers within Provinces. Ha Noi, Mar. 2012. Available online: https://thuvienphapluat.vn/van-ban/Linh-vuc-khac/Quyet-dinh-341-QD-BTNMT-Danh-muc-luu-vuc-song-noi-tinh-137249.aspx (accessed on 21 November 2023). (In Vietnamese).
15. DHI. *Mike 11 User Manual*; 2020; Available online: https://www.mikepoweredbydhi.com/products/mike-11 (accessed on 21 November 2023).
16. Willmott, C.J. On the Validation of Models. *Phys. Geogr.* **1981**, *2*, 184–194. [CrossRef]
17. Tran, C.K.; Do, D.A. Application of Integrated Hydro-Hydraulic Model to Assess the Impact of Infrastructure on Flood Drainage Issue —Case Study for The Downstream of Kone-Ha Thanh River Basin. *J. Water Environ. Technol.* **2022**, *20*, 45–58. [CrossRef]
18. Tran, C.K.; Nguyen, T.T.; Do, D.A.; Nguyen, L.H. Developing the Regression Equations to Determine the Bankfull Discharge from the Basin Characteristics. *Arch. Hydroeng. Environ. Mech.* **2021**, *68*, 103–117. [CrossRef]
19. Tran, C.K.; Nguyen, T.T.; Nguyen, T.T. Primarily Results of a Real-Time Flash Flood Warning System in Vietnam. *Civ. Eng. J.* **2021**, *7*, 747–762. [CrossRef]
20. Niu, M.; Kang, J.; Chen, Z. The Climatic Characteristics of Surface Salinity in the South China Sea and the Adjacent Northwest Pacific Ocean. *IOP Conf. Ser. Earth Environ. Sci.* **2018**, *171*, 012028. [CrossRef]
21. Tickner, D.; Kaushal, N.; Speed, R.; Tharme, R. Editorial: Implementing Environmental Flows: Lessons for Policy and Practice. *Front. Environ. Sci.* **2020**, *8*, 106. [CrossRef]
22. Argent, R.M.; Sojda, R.S.; Guipponi, C.; McIntosh, B.; Voinov, A.A.; Maier, H.R. Best practices for conceptual modelling in environmental planning and management. *Environ. Model. Softw.* **2016**, *80*, 113–121. [CrossRef]
23. Leopold, L.B. *A View of the River*; Harvard University Press: Cambridge, MA, USA, 1994.
24. Bertola, M.; Viglione, A.; Vorogushyn, S.; Lun, D.; Merz, B.; Blöschl, G. Do small and large floods have the same drivers of change? A regional attribution analysis in Europe. *Hydrol. Earth Syst. Sci.* **2021**, *25*, 1347–1364. [CrossRef]
25. Nash, J.E.; Sutcliffe, J.V. Ireland's water budget—Model validation and a greenhouse experiment. *J. Hydrol.* **1970**, *10*, 282–290. [CrossRef]
26. Mccuen, R.H.; Knight, Z.; Cutter, A.G. Evaluation of the Nash-Sutcliffe Efficiency Index. *J. Hydrol. Eng.* **2006**, *11*, 597–602. [CrossRef]
27. Thuyloi University. Vietnamese Standard TCVN 13615: 2022 Calculation of Design Hydrological Elements. Ha Noi, 2022. Available online: https://kttnn.tlu.edu.vn/Portals/0/BM_Thuyvan/TCVNTV_2022.pdf (accessed on 27 February 2024).
28. Vietnam National Assembly. Law on Water Resources 17/2012/QH13. Ha Noi, 2012. Available online: https://thuvienphapluat.vn/van-ban/Tai-nguyen-Moi-truong/Luat-tai-nguyen-nuoc-2012-142767.aspx (accessed on 21 November 2023). (In Vietnamese)
29. Gawne, B.; Capon, S.J.; Hale, J.; Brooks, S.S.; Campbell, C.; Stewardson, M.J.; Grace, M.R.; Stoffels, R.J.; Guarino, F.; Everingham, P. Different conceptualizations of river basins to inform management of environmental flows. *Front. Environ. Sci.* **2018**, *6*, 111. [CrossRef]

water

MDPI

Article

A Study on Chemical Oxygen Demand (COD) Concentration Distribution and Its Hydrodynamic Mechanisms in Liaodong Bay, China

Xincang Liu [1,2], Jiahui Deng [1,2], Lianjie Zhang [1,2], Ping Wang [1,2,*], Guangshuai Zhang [1,2], Xiangke Dong [1,2] and Jiawen Sun [1,2]

1 National Marine Environmental Monitoring Center, Dalian 116023, China; xcliu@nmemc.org.cn (X.L.); djhdut@163.com (J.D.); ljzhang@nmemc.org.cn (L.Z.); gszhang@nmemc.org.cn (G.Z.); xkdong@nmemc.org.cn (X.D.); jwsun@nmemc.org.cn (J.S.)

2 State Environmental Protection Key Laboratory of Marine Ecological Environment Remediation, Dalian 116023, China

* Correspondence: wangping@nmemc.org.cn

Abstract: In order to reveal the impact of hydrodynamic conditions on the transport and diffusion of pollutants in Liaodong Bay in China, this article uses MIKE21 to establish a numerical model to simulate the hydrodynamic mechanisms of tidal currents and residual currents in Liaodong Bay. The model has been calibrated using observation data from 10 stations, and the simulation results of the tidal currents, Euler residual currents, Lagrangian residual currents, and particle tracking in Liaodong Bay have been calculated. Subsequently, a comparative analysis is conducted based on the abovementioned data and measured data, exploring the impact of hydrodynamic conditions on the transport and diffusion of COD in Liaodong Bay. The research results in this article indicate that high concentration COD areas are mainly concentrated in the coastal areas around the estuary of the Liao River and the Daliao River, and river input is the main source of COD in Liaodong Bay. The Euler residual circulation can form COD enrichment in some areas, which is significantly higher than the background concentration, and the large-scale transportation of COD after entering Liaodong Bay is determined by the Lagrangian residual current. The particle tracking results in the estuarine area can effectively characterize the actual transportation of pollutants. The results of the Lagrangian residual flow and particle tracking in the bay indicate that river pollutants are mainly transported to the west bank after entering Liaodong Bay. The distribution of a COD concentration of 1.5 mg/L confirms this finding. The research findings presented in this paper offer valuable insights into the spatial distribution and transportation mechanisms of pollutants. These results hold significant implications for pollution prevention and mitigation strategies in comparable bay environments.

Keywords: COD; Liaodong Bay; Eulerian residual current; Lagrangian residual current; particle tracking; MIKE21

1. Introduction

Bays are formed by the concave part of the coastline, and the twists and turns in the coastline lead to the widespread existence of bays at the sea–land boundary. Due to the shallow water depth, the low disturbance from wind and waves, and calm water bodies, bays have become an important base for humans to engage in marine-related economic activities and have helped to develop tourism. At the same time, high-intensity human activities continue to damage the ecological environment of bays, and the accelerating industrialization process and gradually increasing population have led to an increasing number of pollutants being generated [1]. These pollutants are discharged into the sea through rivers and sewage outlets, leading to the deterioration of seawater quality year by year [2]. Marine disasters, such as red tides and algal blooms [3–6], occur frequently.

This issue is particularly prominent in developing countries, such as China, especially in semi-enclosed bays, such as Liaodong Bay [6,7], which have shallow water depths and poor water exchange capacity.

The water quality (WQ) of the sea is the core component of the ecological environment quality in bays. If the water quality in the bay is severely polluted all year round, human activities that rely on the bay will be affected. Therefore, seeking effective solutions is crucial. On the one hand, practical actions should be taken, including the formulation of national water quality criteria [8,9] to constrain and limit the amount of major pollutants [10], seawater quality monitoring should be carried out [11], and targeted protection and remediation efforts should take place [12]. On the other hand, theoretical research should be conducted on the hydrodynamic and material transportation [13] in bays, in order to understand the transportation process of pollutants in seawater and the pollution mechanism of bays, guiding the development of governance measures.

Liaodong Bay is located in the northeastern part of the Bohai Sea in China. With an average water depth of 22 m and a total area of 30,597 km^2 [14,15], it is the northernmost semi-enclosed bay in China in terms of latitude [16]. Furthermore, it represents one of the most significant estuarine ecosystems in northern China [17] (Figure 1). Economically developed cities, such as Dalian, Yingkou, Panjin, Jinzhou, and Huludao, are distributed around Liaodong Bay [18]. The water in the bay has been severely polluted [19,20], and the complex diffusion of high-intensity pollution emissions driven by the tidal current field in the bay has made it very difficult to solve the water quality problem. Many studies have been conducted on the distribution of nutrients, the river input, and the nitrogen–phosphorus ratio (N/P) impact on eutrophication in Liaodong Bay and the Bohai Sea area [21,22], as well as the distribution and sources of heavy metals in sediment in key areas [23,24], the spatial distribution of pollutants, and the environmental risks [25,26]. However, the understanding of how the transport of pollutants (COD) in the bay is affected by hydrodynamic forces is still lacking.

This paper establishes a numerical model of the entire Bohai Sea, including Liaodong Bay; the model was validated by measured data from 10 tidal current stations. On the basis of the calculated tidal current (TC) field, the Euler residual current (ERC), Lagrange residual current (LRC), and particle tracking (PT) were calculated. Then, the distribution of COD in the bay during the same period was analyzed in comparison to the above calculated results, and the hydrodynamic mechanism of the COD concentration distribution in Liaodong Bay was explained. The conclusion of this article is of great significance for the prevention and control of pollution in bays mainly affected by river inputs, such as the formulation of water quality standards and criteria for various regions of the bay, and the preparation of marine environment protection plans.

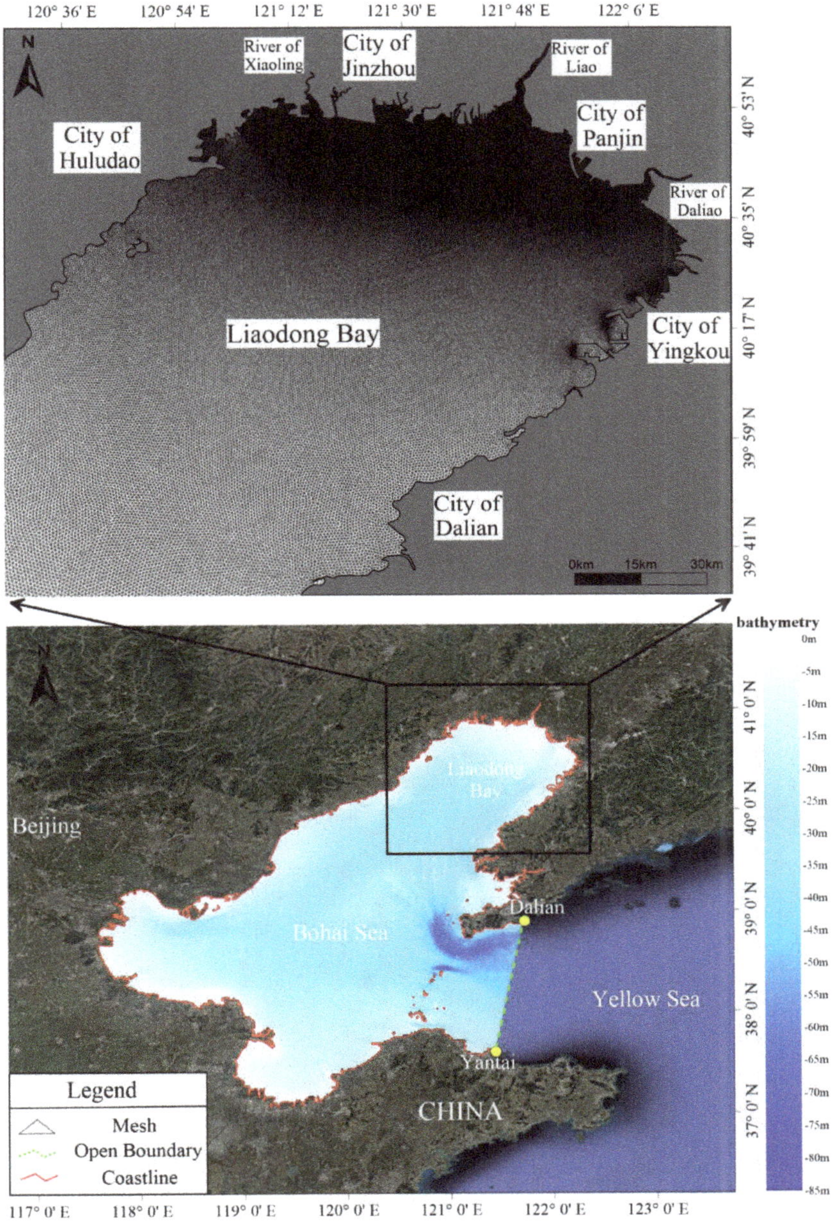

Figure 1. Study area and computational grid.

2. Numerical Model and Monitoring Data

2.1. Numerical Model

In this paper, the two-dimensional numerical model MIKE21, developed by the Danish Hydraulic Institute, is used for modeling. Its user-friendly interface and flexible unstructured grid system mean that it is widely used in numerical simulations of rivers, oceans, and lakes. The model covers the entire Bohai Sea, with coordinates ranging from 117.53 E to 122.34 E and 37.02 N to 41.09 N. The model uses a triangular grid to dissect the computational domain and reduces the grid scale of Liaodong Bay. The side length of the smallest grid is about 50 m. The open boundary of the model, which is dictated by the water-level boundary, is the straight line connecting the Yantai and Dalian harbors. The water-level data are obtained from the Tide Table published by the National Marine Information Center and are inputted into the model as the open boundary after the datum adjustment. The water-depth data used here are based on the digitized nautical chart published by the Navigation Guarantee Department of the Chinese Navy Headquarters and the digital nautical chart of the DHI C-MAP. Liaodong Bay is mainly affected by periodic tidal currents, so this paper mainly studies the distribution characteristics of COD under tidal currents. In order to make the conclusion universal, factors with strong regional or irregular spatiotemporal features, like temperature, salinity, and meteorological variables, are not considered here.

2.2. Eulerian and Lagrangian Residual Currents

The ERC is the average of the tidal velocity field, which can be defined as [27]:

$$v_E = \frac{1}{T}\int_0^T \vec{v}\,\mathrm{d}t \tag{1}$$

In this paper, the calculation is performed according to the following equation:

$$v_E = \frac{1}{N}\sum\nolimits_{i=1}^{N} \vec{v} \tag{2}$$

where v_E is the ERC on the current grid; $N = \frac{T}{\Delta t}$; N is the number of computation time steps; and T is the computation time, totaling 750 h, which can encompass the COD observation period and is an integer multiple of the tidal cycle. Δt is the computation time step and $\vec{v}$ is the TC on the grid.

Stokes drift (SD) represents the net mass transport resulting from the asymmetry between flood and ebb tidal currents, and can be defined as [27]:

$$v_S = \frac{1}{T \cdot D}\int_0^T \vec{v}\cdot\xi\,\mathrm{d}t \tag{3}$$

In this paper, the calculation is performed according to the following equation:

$$v_S = \frac{1}{N \cdot D}\sum\nolimits_{i=1}^{N} \vec{v}\cdot\xi \tag{4}$$

where D is the distance from the mean sea level to the sea floor and ξ is the water surface elevation above the mean sea level.

The LRC is the average of the tidal flux field, which can be expressed as the vector sum of the ERC and SD [27], so the LRC v_L on the grid can be expressed as:

$$v_L = \frac{1}{N \cdot D}\sum\nolimits_{i=1}^{N} \vec{v}\cdot\xi + \frac{1}{N}\sum\nolimits_{i=1}^{N} \vec{v} \tag{5}$$

2.3. Particle Tracking Model

The particle tracking (PT) model calculates the transportation of dissolved, suspended, and sedimented substances discharged or accidently spilled into lakes, estuaries, coastal areas, and at open sea. The PT model can be used to calculate the transportation direction and distance of COD in Liaodong Bay under tidal currents, which is helpful to understand the influence of hydrodynamic mechanisms on the COD concentration distribution, and the equation can be expressed as [28]:

$$\begin{cases} \mathrm{d}x(t) = U(x,y,t)\mathrm{d}t + \mathrm{d}x'(t) \\ \mathrm{d}y(t) = V(x,y,t)\mathrm{d}t + \mathrm{d}y'(t) \end{cases} \tag{6}$$

where dx and dy are the displacement in the x and y directions, respectively. Moreover, $dx'(t)$ and $dy'(t)$ are the random walk displacements due to horizontal turbulent diffusion. The zonal (U) and meridional (V) components of the current velocity $\vec{v}$ are derived from numerical simulations. When particles reach open boundaries, they exit the computational domain and cease to participate in subsequent calculations. Upon reaching the coastline, particles are reflected back into the computational domain. The direction and velocity of this reflection are computed based on the incident angle of the particles and the prevailing flow field conditions, ensuring that the post-reflection movement adheres to physical principles.

2.4. Shoreline, Water Depth, and COD Data

The shoreline data were derived from remote sensing images, extracted according to the mean spring tides high tide line, and appropriately adjusted according to the accuracy of the model grid. The bathymetry data were obtained from the nautical charts published by the Department of Navigation Protection of the Chinese People's Liberation Army and the DHI C-MAP. The COD data were from the WQ data released by the National Seawater Quality Monitoring Information Public System of the Ministry of Ecology and Environment of the People's Republic of China, monitored in the autumn of 2021 (October and November) by the National Marine Environment Monitoring Center.

3. Results

3.1. COD Concentration Monitoring Results

The overall COD concentration in Liaodong Bay in autumn of 2021 shows a higher concentration in the estuarine region and gradually decreases toward the sea. It can be seen from Figure 2 that the highest COD concentration is 4.9 mg/L in the area between the Liao River and Daliao River, which is 5~6 times higher than the concentration of COD in the mouth of the bay and the center of Liaodong Bay. This shows that Liao River and Daliao River in Liaodong Bay are the main pollution sources in terms of COD in Liaodong Bay. The COD concentration decreases with the distance from the mouth of the rivers, but the contour plot of the COD concentration shows that there are also COD areas with a higher concentration far from the estuary (regions in the figure with COD concentrations above 3.5 mg/L) in the north side of Dalian and west side of Yingkou. This appears anomalous against the background of COD degradation over time.

3.2. MIKE21 Model Verification

In this paper, the model is calibrated using current data measured at 10 stations, as shown in Figure 3. These data, including velocity and flow direction, are from the observation results of Tianjin Survey and Design Institute for Water Transport Engineering Co., Ltd., in Liaodong Bay from 18 September to 19 September 2021. The sea current observations were carried out using an acoustic Doppler current profiler, with the water column divided into six vertical layers. The equipment underwent a self-check before the observations. The model started at 00:00 on 15 September 2021. The initial conditions set the water level and velocity field to zero. The time step for the open boundary water-level conditions is one hour, which includes the seasonal variations in the water level. The comparisons between the model-calculated data and the measured data for all stations are shown in Figure 4. The validation results indicate that the model-calculated values and the measured values of flow velocity and direction are relatively consistent across all stations. The range of the average flow velocity error is between 0.01 and 0.11 m/s, and the average flow direction deviation is between 0.16° and 19.76°, with the error values within a reasonable range. The discrepancies between simulated and observed tidal current values may be attributed to observational errors, as well as non-linear effects, such as topography and bottom friction. From the comparison results, it can be seen that the numerical model can effectively represent the tidal current field in Liaodong Bay and can be used as a basic model for further research work.

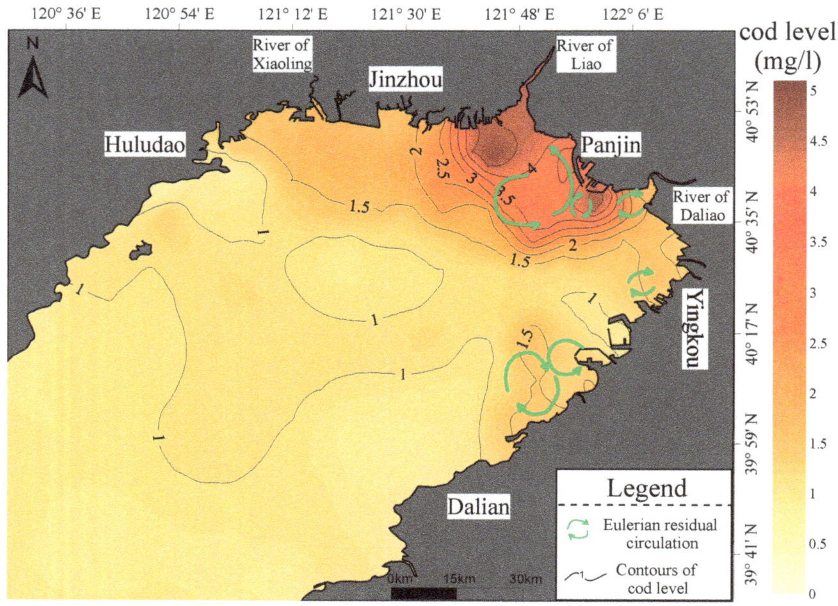

Figure 2. Distribution of COD concentration in Liaodong Bay in autumn 2021.

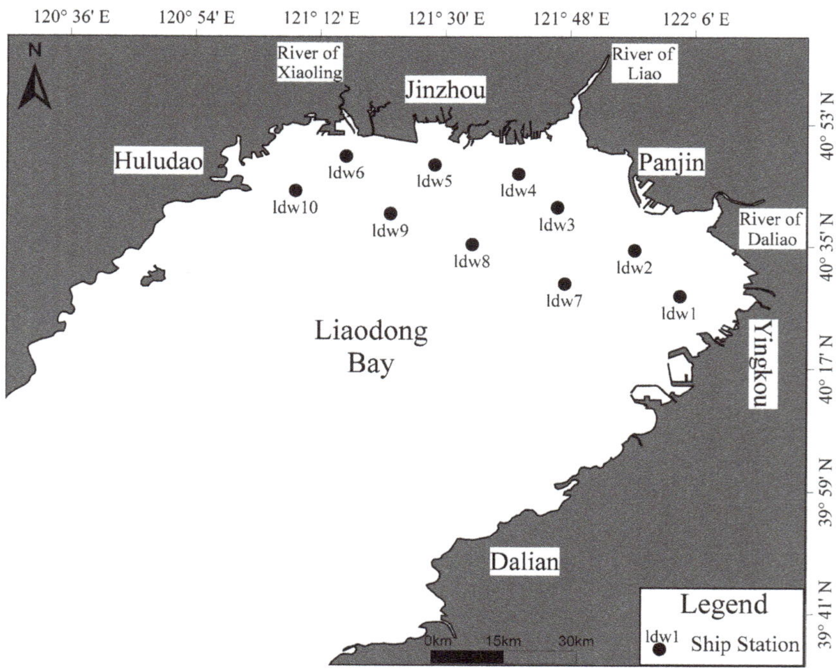

Figure 3. Current observation location.

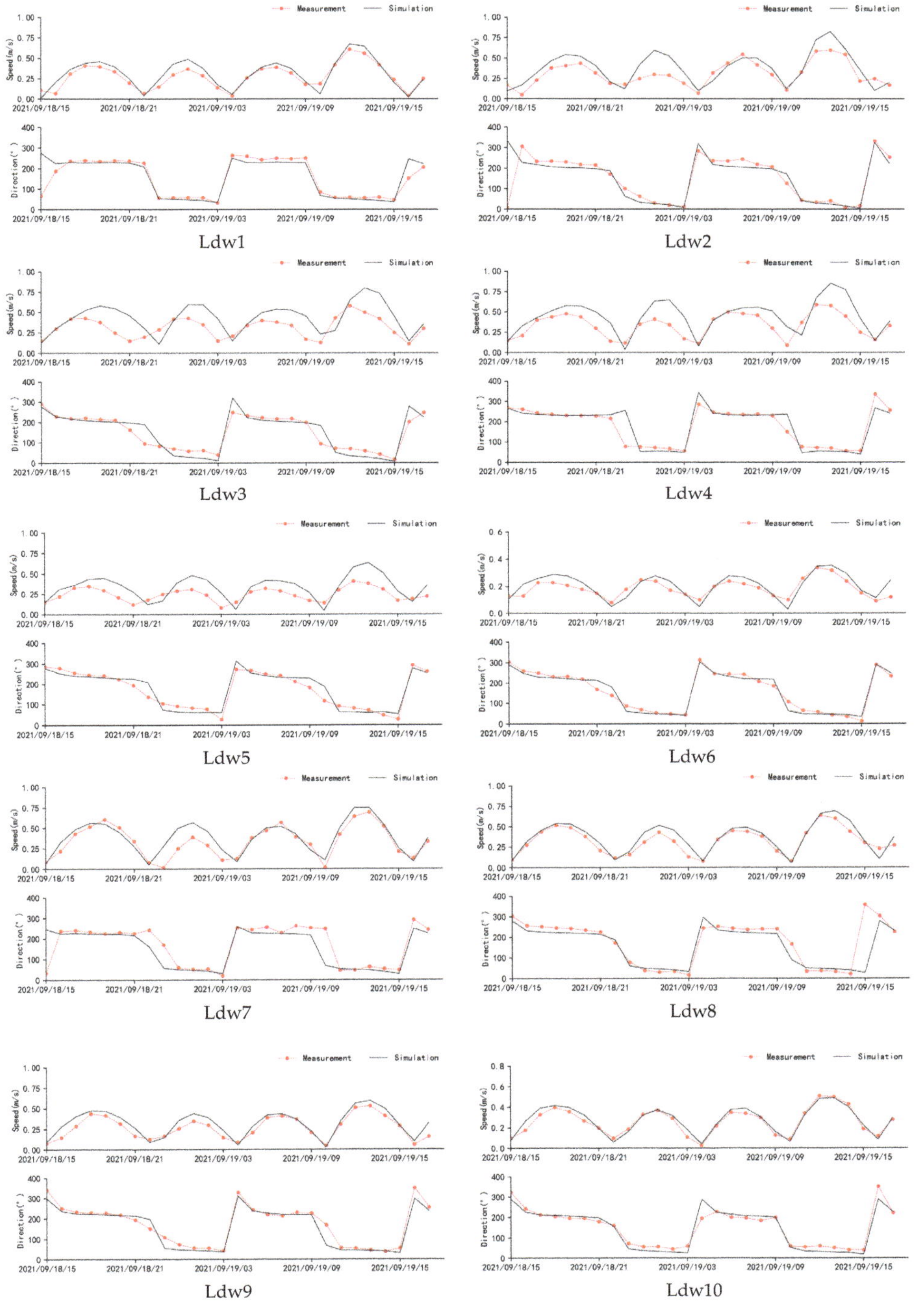

Figure 4. Comparison between calculated and measured data.

3.3. Liaodong Bay Tidal Current

The TC field of Liaodong Bay at the time of the flood tide and ebb tide can be calculated (Figure 5). At the time of the flood tide, the water flows from the Bohai Sea to the bay and converges at the estuary of the Liao River; the TC velocity gradually increases from the mouth of the bay to the estuary, and the maximum TC velocity is about 0.2 m/s at the mouth of the bay, 0.3 m/s in the central area of Liaodong Bay, and about 0.9 m/s at the estuary of the Liao River. During the ebb tide, the TC flows from Liaodong Bay to the Bohai Sea, contrary to the time of the flood tide, the maximum TC velocity is about 0.3 m/s at the estuary of the Liao River, 0.5 m/s at the center of Liaodong Bay, and 0.6 m/s at the mouth of the bay.

3.4. Euler Residual Current

Based on the calculated TC field and the definition of ERC in Section 2.2, taking 750 h (15 October to 15 November 2021) as the averaging time, this paper calculated the ERC field of Liaodong Bay for the same period as the COD observations, as shown in Figure 6. The ERC in Liaodong Bay is larger in the nearshore and estuary regions, and smaller in the central region of Liaodong Bay, which is also consistent with the mechanism of ERC caused by the non-linear effects of the shoreline. In addition, the distribution of the circulation is also a distinctive feature of the ERC field in Liaodong Bay, especially in the estuarine area at the north and the east coast of Liaodong Bay. There is a circulation current in the Liao River estuary near the west side of Panjin, the south side of Panjin, and the estuary of Daliao River, in the counterclockwise, clockwise, and clockwise direction, respectively, and with ERC velocities ranging from 0.02 to 0.1 m/s. Between the city of Dalian and Yingkou on the east coast of Liaodong Bay, three circulation currents also exist from south to north, in clockwise, counterclockwise, and clockwise directions, and with ERC velocities ranging from 0.01 to 0.3 m/s.

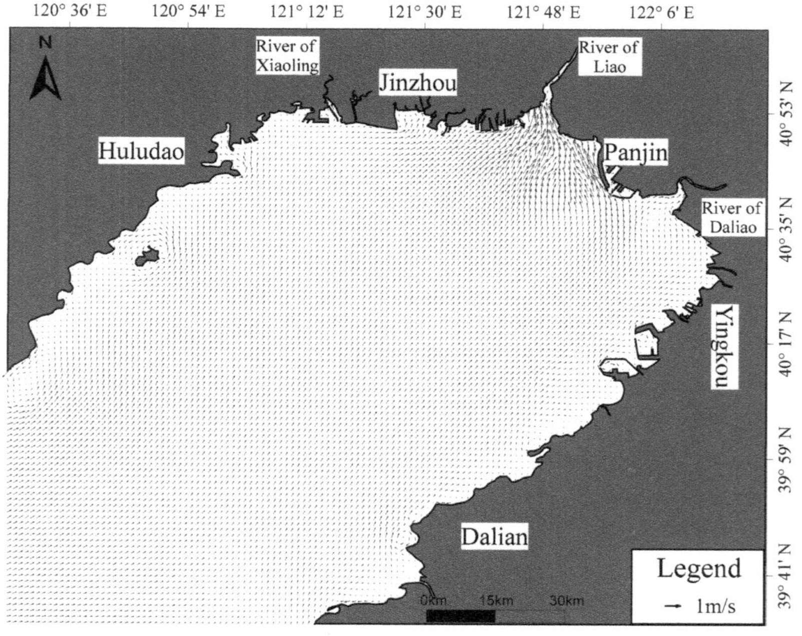

(**a**) Flood tide

Figure 5. *Cont.*

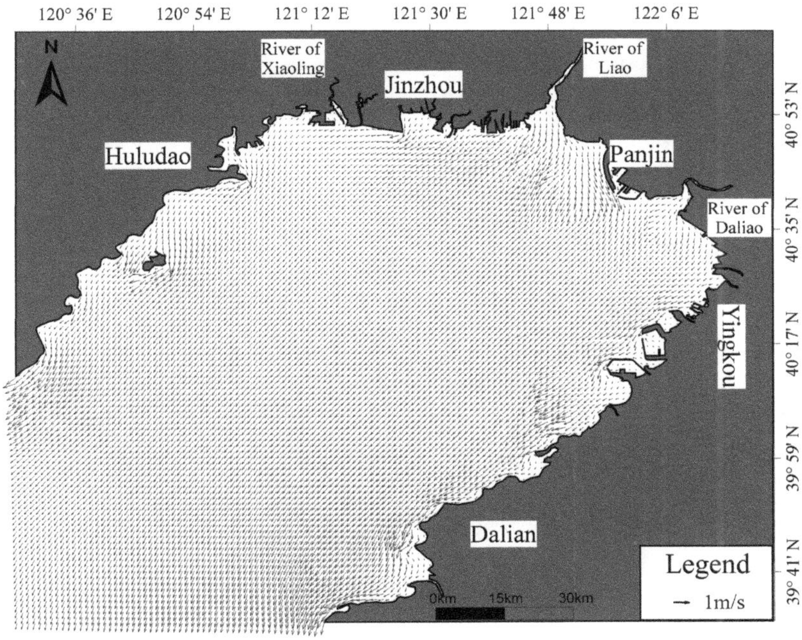

(**b**) Ebb tide

Figure 5. TC field in Liaodong Bay.

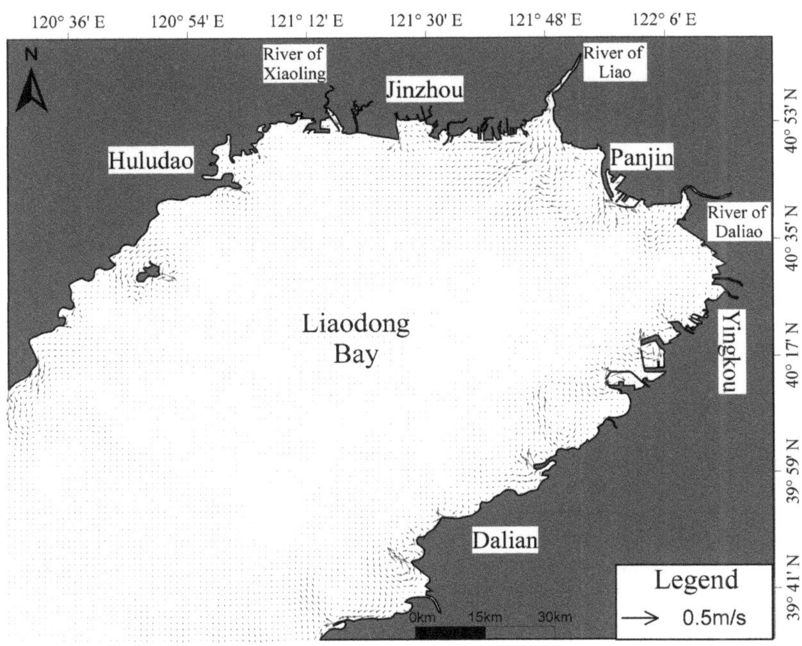

Figure 6. ERC field in Liaodong Bay.

3.5. *Lagrangian Residual Current*

According to the method introduced in Section 2.2, the LRC field in Liaodong Bay is calculated, and the results are shown in Figure 7. The calculated results of the LRC and ERC are similar, and the LRC velocity in the nearshore and estuarine regions was greater than that in the central region. The LRC velocity in the estuarine region of the Liao River and the Daliao River is between 0.05~0.1 m/s. The maximum velocity on the east coast of Liaodong Bay is about 0.2 m/s. There is also an area of relatively high current velocity, around 0.12 m/s, between the islands south of Huludao and the coastline. LRC flows from the two estuaries to Jinzhou and the Xiaoling River at the top of the bay, and the current velocity gradually decreases. The LRC on the west coast of Liaodong Bay flows along the shoreline to the Bohai Sea, while on the east coast the opposite happens, flowing from Dalian to Yingkou and finally flowing to the two estuary areas of the Liao River and the Daliao River. In addition, one of the distinctive features of the LRC, along the east coast of Liaodong Bay, is that there are several circulations with different radii and flow velocities in the nearshore area.

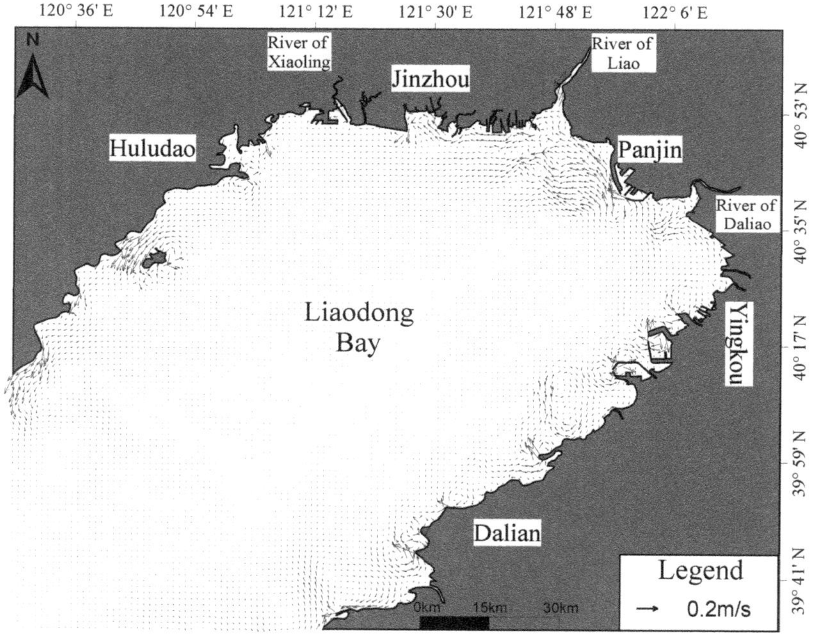

Figure 7. LRC field in Liaodong Bay.

3.6. *Particle Tracking*

The positions and displacements of uniformly distributed particles in Liaodong Bay, after 750 h, for the calculated TC, are given in Figure 8. After 750 h under the influence of the TC, uniformly distributed free particles in the two river estuaries and adjacent sea areas mostly migrate toward the center of the bay. Meanwhile, the particle displacement after 750 h indicates that the particle displacement in the two estuaries and near the shore is larger than that in the center of Liaodong Bay. In addition, the displacement of particles on the east coast of Liaodong Bay points to the two estuaries, while that in the two estuaries points to the Xiaoling estuary and the sea near Huludao on the west coast, and the displacement of particles is inversely proportional to the distance from the two estuaries, namely the farther away from the river mouth, the smaller the displacement.

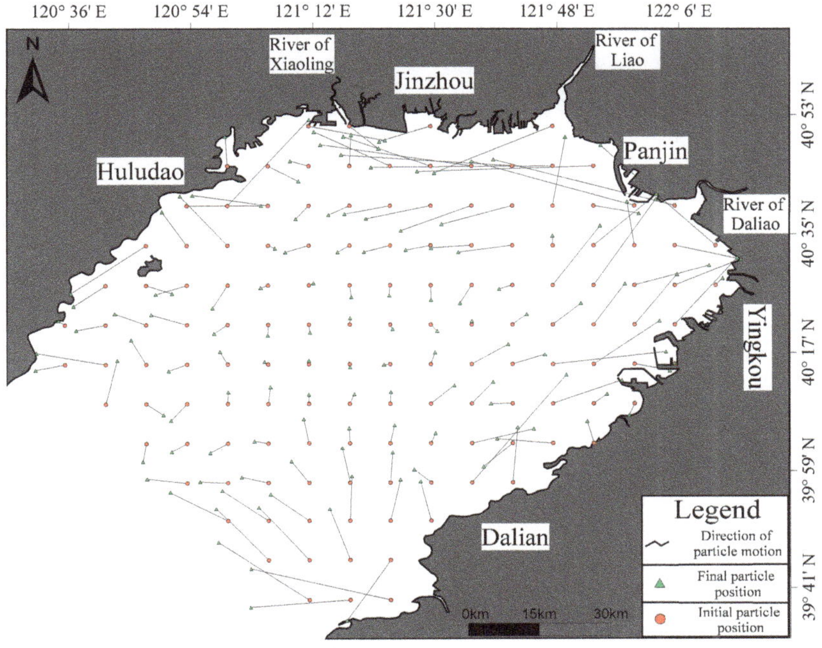

Figure 8. Particle tracking results in Liaodong Bay.

4. Discussion

The COD concentration distribution in Liaodong Bay show that the Liao River and the Daliao River, as the main sea-entering rivers in Liaodong Bay, are the main sources of COD pollution in Liaodong Bay. The pollutants originating from the rivers spread to all parts of the bay under the flow of tidal currents. Therefore, it is crucial to fully understand the hydrodynamic process of Liaodong Bay and analyze the dynamic mechanisms of pollutant transportation to solve the transportation problem of riverine COD into Liaodong Bay. This article compares the COD observation results with the calculated Euler residual current, Lagrangian residual current, and particle tracking results, to explore the hydrodynamic mechanisms behind pollutant transportation represented by COD.

4.1. COD and Euler Residual Current

Figures 2 and 6 show that there is good agreement between the COD concentration distribution and ERC circulation. The COD concentration at the monitoring stations between the Liao River and Daliao River estuaries is the highest in Liaodong Bay, and there are also three circulations of different sizes in the ERC field here. This overlap is partly due to the fact that the Liao River and the Daliao River are the main sources of COD in Liaodong Bay; the ERC plays a noteworthy role in causing a significant amount of COD to be retained in this area. At the same time, there are three ERCs with a slightly smaller radius in the east coast of Liaodong Bay away from the estuaries, and the COD concentration in the center of the currents and the neighboring areas is obviously higher than that of the surrounding stations. This indicates that the ERC circulation has a certain impact on the distribution of COD. In the central area of the circulation, the recirculation of COD within this area results in a lower diffusion rate compared to the surrounding sea areas, making it easy to form COD hotspots, with concentrations higher than the background level. The conclusion by Kong is consistent with the findings of this study [29].

4.2. Lagrangian Residual Current and COD

The ERC circulation has a certain impact on local COD hotspots, while the influence of LRC on COD distribution is manifested in long-distance transportation. The area with >1.5 mg/L COD, in Figure 2, shows that the COD in this concentration range is mainly distributed in the west side of the two estuaries, which indicates that the COD is mainly transported to the west bank of Liaodong Bay after leaving the estuarine region. Because the COD concentration in this area is higher than the background value in the middle of the bay and the bay mouth away from the estuary, the contour pattern reflects the gradual reduction in COD from the high concentration area at the source of the two estuaries to the low concentration area in the middle of the bay and the bay mouth, which is a good indication of the direction of pollutant transportation. The flow field of the LRC in the two estuarine regions and the diffusion direction of COD outward from the estuarine region are pretty consistent in Figures 2 and 7, indicating that the LRC plays an important role in the long-distance transportation of COD with a higher concentration than the background concentration, and it can effectively indicate the transportation direction of substances.

4.3. Particle Tracking and COD Transport

Comparing the COD concentration distribution and the PT results, there are two findings. Firstly, the ending position and distribution density of the particles after drifting for 750 h under the action of TC largely explains the COD concentration distribution; Figure 8 shows that the particle density in the western part of the two estuaries is generally higher than that in the eastern part, which is consistent with the fact that the COD concentration shows a higher value in the west than in the east of the estuaries, according to the analysis in Section 4.2. Meanwhile, the final position of most of the particles from the river estuaries and adjacent coastal areas are predominantly located in the central part of Liaodong Bay, which is consistent with the fundamental understanding of COD transportation from river estuaries into the inner Liaodong Bay. Secondly, the displacement of the particles, as shown in Figure 8, is a key factor that provides the transportation process of the substances in the region with a vector and has a good level of representation in terms of the direction and capacity of COD transportation in various places in the bay. This result is verified by the COD transportation process from the two estuaries to the western part of the sea. Moreover, due to the influence of stochastic terms in the PT results, the magnitude of particle displacement is more significant than its direction. The magnitude of particle displacement can be considered as an indicator of the bay's material transportation capacity. However, in the area near the mouths of the two rivers, where TCs are strong, the direction of particle displacement is relatively consistent, which can also indicate the direction of material transportation.

5. Conclusions

In this paper, a numerical model verified by measured current data was established in the Bohai Sea, China. The TC, ERC, LRC, and PT in Liaodong Bay and the Bohai Sea were calculated through the numerical model, and the concentration distribution and transportation pattern of COD, monitored during the same period, were analyzed from the perspective of hydrodynamic mechanisms. The following findings were made:

(1) COD in Liaodong Bay has highest concentration in the estuary of the Liao River and the Daliao River and the surrounding areas, and the concentration gradually decreases from the two estuaries to the sea. The distribution of the area with >1.5 mg/L COD in the east coast and west coast differs. Overall, the west coast has more area with >1.5 mg/L COD than the east coast, but there is also a local high concentration area between Dalian and Yingkou in this area;

(2) The COD enrichment area between the Liao River and the Daliao River and on the east coast is consistent with the distribution of the ERC circulation, the recirculation of COD within this area results in a lower diffusion rate compared to the surrounding

sea areas, making it easy to form COD hotspots, with concentrations higher than the background level;

(3) The LRC mainly influences the overall long-distance transportation of COD and plays a dominant role in the transportation of COD with a concentration that is higher than the background value in the center of the bay, but less than the peak concentration in the estuarine regions. The direction of COD transportation, indicated by the 1.5 mg/L~2 mg/L COD contour pattern, is consistent with the flow velocity and direction of the LRC field in the same region;

(4) The displacement of the particles from the initial and final position can be used as a scalar to indicate the transportation capacity of different regions of the bay. The end position and distribution density of the particles can indicate the distribution characteristics of COD after leaving the estuarine regions. The final position of the particles is mainly concentrated in the central and west side of the bay, which is consistent with the common knowledge on COD transportation from the estuary to the sea and the main transportation direction analyzed in this paper.

Author Contributions: Methodology, Software, Writing—Original draft preparation, X.L.; Writing—Review & Editing, J.D., L.Z., G.Z., X.D. and J.S.; Conceptualization, Data curation, Investigation, P.W.; Funding acquisition, G.Z. All authors have read and agreed to the published version of the manuscript.

Funding: This work was supported by the National Key Research and Development Program of China (2023YFC3208400)] and the National Natural Science Foundation of China [Grant No. 42201070].

Data Availability Statement: The original contributions presented in the study are included in the article, further inquiries can be directed to the corresponding author.

Conflicts of Interest: The authors have no relevant financial or non-financial interests to disclose.

References

1. Wei, H.; Sun, J.; Moll, A.; Zhao, L. Phytoplankton dynamics in the Bohai Sea—Observations and modelling. *J. Mar. Syst.* **2004**, *44*, 233–251. [CrossRef]
2. Li, K.; Ma, Y.; Dai, A.; Wang, X. Degradation dynamics and bioavailability of land-based dissolved organic nitrogen in the Bohai Sea: Linking experiment with modeling. *Mar. Pollut. Bull.* **2017**, *124*, 856–870. [CrossRef] [PubMed]
3. Chen, C.; Liang, J.; Yang, G.; Sun, W. Spatio-temporal distribution of harmful algal blooms and their correlations with marine hydrological elements in offshore areas, China. *Ocean Coast. Manag.* **2023**, *238*, 106554. [CrossRef]
4. Yang, H.F.; Zhu, Q.Y.; Liu, J.A.; Zhang, Z.L.; Yang, S.L.; Shi, B.W.; Zhang, W.X.; Wang, Y.P. Historic changes in nutrient fluxes from the Yangtze River to the sea: Recent response to catchment regulation and potential linkage to maritime red tides. *J. Hydrol.* **2023**, *617*, 129024. [CrossRef]
5. Gajardo, G.; Morón-López, J.; Vergara, K.; Ueki, S.; Guzmán, L.; Espinoza-González, O.; Sandoval, A.; Fuenzalida, G.; Murillo, A.A.; Riquelme, C.; et al. The holobiome of marine harmful algal blooms (HABs): A novel ecosystem-based approach for implementing predictive capabilities and managing decisions. *Environ. Sci. Policy* **2023**, *143*, 44–54. [CrossRef]
6. Wang, K.; Wei, Q.; Jian, H.; Gao, Z.; Yao, Q. Variations in the nutrient concentration and composition in Liaodong Bay under long-term human activities. *Mar. Pollut. Bull.* **2022**, *182*, 114016. [CrossRef] [PubMed]
7. Kong, X.; Ye, S. The impact of water temperature on water quality indexes in north of Liaodong Bay. *Mar. Pollut. Bull.* **2014**, *80*, 245–249. [CrossRef]
8. Dai, A.; Li, K.; Ding, D.; Li, Y.; Liang, S.; Li, Y.; Su, Y.; Wang, X. Total maximum allocated load calculation of nitrogen pollutants by linking a 3D biogeochemical–hydrodynamic model with a programming model in Bohai Sea. *Cont. Shelf Res.* **2015**, *111*, 197–210. [CrossRef]
9. Yang, F.; Su, R.; Jian, H.; Yao, Q. Using algal biomass to evaluate numeric nutrient criteria in an estuary: A case study of Daliaohe Estuary in China. *Ecol. Indic.* **2016**, *70*, 373–381. [CrossRef]
10. Ghaemi, M.; Mohammadpour, G.; Hamzei, S.; Gholamipour, S. Spatial and temporal characterizations of seawater quality on marine waters area of the Persian Gulf. *Reg. Stud. Mar. Sci.* **2022**, *53*, 102407. [CrossRef]
11. Lega, M.; Kosmatka, J.; Casazza, M.; De Maio, L.; Severino, D.; Teta, R. Environmental monitoring and assessment: A multi-scale and multiparameter case study in Campania Region (S Italy) on the recreational seawater quality evaluation. *Reg. Stud. Mar. Sci.* **2020**, *39*, 101465. [CrossRef]
12. Li, X.; Lei, G.; Li, Y.; Wang, Y.; Tan, Z. Assessing hydrodynamic effects of ecological restoration scenarios for a tidal-dominated wetland in Liaodong Bay (China). *Sci. Total Environ.* **2021**, *752*, 142339. [CrossRef]

13. Wei, H.; Hainbucher, D.; Pohlmann, T.; Feng, S.; Suendermann, J. Tidal-induced Lagrangian and Eulerian mean circulation in the Bohai Sea. *J. Mar. Syst.* **2004**, *44*, 141–151. [CrossRef]
14. Yang, S.; Li, J.; Ye, S.; Mackenzie, L.; Yuan, H.; He, L.; Zhao, G.; Pei, S.; Ding, X. Pollen distribution and transportation patterns in surface sediments of Liaodong Bay, China. *Sci. Total Environ.* **2021**, *771*, 144883. [CrossRef]
15. Zhang, C.; Lei, M.; Liu, X.; Zhou, Z.; Liu, M.; Chen, H.; Yang, W.; Wang, X. Presence and inputs of legacy and novel per- and polyfluoroalkyl substances from rivers and drainage outlets to Liaodong Bay, China. *Reg. Stud. Mar. Sci.* **2022**, *56*, 102684. [CrossRef]
16. Wu, J.; Wang, Z.; Tian, J.; Wang, K.; Li, A.; Li, N.; Song, L.; Song, G. Comparison of nutrients status in Liaodong Bay and Northern Yellow Sea, China: Controlling factors and nutrient budgets. *Mar. Environ. Res.* **2024**, *194*, 106338. [CrossRef]
17. Xu, Y.; Gao, H.; Li, R.; Lou, Y.; Li, B.; Cheng, G.; Na, G. Occurrence and distribution of antibiotics and antibiotic resistance genes from the land to ocean in Daliao River-Liaodong Bay, China. *Mar. Environ. Res.* **2024**, *197*, 106470. [CrossRef]
18. Ye, Y.; Zhang, A.; Teng, J.; Yang, X.; Yuan, X.; Wang, Q.; Zhao, J.; Zhang, B.; Zhang, T.; Chen, X.; et al. Pollution characteristics and ecological risk of microplastic in sediments of Liaodong Bay from the northern Bohai Sea in China. *Mar. Pollut. Bull.* **2023**, *187*, 114505. [CrossRef]
19. Luo, M.; Zhang, Y.; Xiao, K.; Wang, X.; Zhang, X.; Li, G.; Li, H. Effect of submarine groundwater discharge on nutrient distribution and eutrophication in Liaodong Bay, China. *Water Res.* **2023**, *247*, 120732. [CrossRef] [PubMed]
20. Pei, S.; Laws, E.A.; Zhu, Y.; Zhang, H.; Ye, S.; Yuan, H.; Ding, X. Nutrient dynamics and their interaction with phytoplankton growth during autumn in Liaodong Bay, China. *Cont. Shelf Res.* **2019**, *186*, 34–47. [CrossRef]
21. Ma, Q.; Liang, S.; Sun, Z.; Zhang, R.; Wang, P. Development and evaluation of a GPU-based coupled three-dimensional hydrodynamic and water quality model. *Mar. Pollut. Bull.* **2023**, *187*, 114494. [CrossRef] [PubMed]
22. Liu, H.; Yin, B. Numerical investigation of nutrient limitations in the Bohai Sea. *Mar. Environ. Res.* **2010**, *70*, 308–317. [CrossRef] [PubMed]
23. Wang, P.; Zhang, L.; Lin, X.; Yan, J.; Zhang, P.; Zhao, B.; Zhang, C.; Yu, Y. Spatial distribution, control factors and sources of heavy metal in the surface sediments of Fudu Estuary waters, East Liaodong Bay, China. *Mar. Pollut. Bull.* **2020**, *156*, 111279. [CrossRef] [PubMed]
24. Zhang, A.; Wang, L.; Zhao, S.; Yang, X.; Zhao, Q.; Zhang, X.; Yuan, X. 2017. Heavy metals in seawater and sediments from the northern Liaodong Bay of China: Levels, distribution and potential risks. *Reg. Stud. Mar. Sci.* **2020**, *11*, 32–42.
25. Wu, J.; Wang, Z.; Zhang, Y.; Tian, J.; Song, L.; Han, J.; Yu, J.; Zhang, Y. Spatial distribution and ecological risks of polycyclic aromatic hydrocarbons in sea ice and seawater from northern Liaodong Bay, China. *Mar. Pollut. Bull.* **2022**, *174*, 113319. [CrossRef] [PubMed]
26. Hou, W.; Chen, X.; Wu, J.; Zhang, C.; Yu, J.; Bai, J.; Chen, T. Sources and spatiotemporal variations of nitrogen and phosphorus in Liaodong Bay, China. *Mar. Pollut. Bull.* **2022**, *185*, 114191. [CrossRef] [PubMed]
27. Rodríguez, P.A.; Carbajal, N.; Rodríguez, J.H.G. Lagrangian trajectories, residual currents and rectification process in the Northern Gulf of California. *Estuar. Coast. Shelf Sci.* **2017**, *194*, 263–275. [CrossRef]
28. Liubartseva, S.; Coppini, G.; Lecci, R.; Creti, S. Regional approach to modeling the transport of floating plastic debris in the Adriatic Sea. *Mar. Pollut. Bull.* **2016**, *103*, 115–127. [CrossRef]
29. Kong, X. Influence of inter-river water diversion on estuary pollution control: A case study of Liaodong Bay. *Reg. Stud. Mar. Sci.* **2024**, *70*, 103362. [CrossRef]

sustainability

MDPI

Article

Effective Placement Strategies for Portable Fishways in Agricultural Drainage Canals: A Numerical Investigation

Shigeya Maeda [1,*], Yuri Yoshida [2], Kanuka Yoshinari [3] and Naoki Takahashi [4]

[1] College of Agriculture, Ibaraki University, 3-21-1 Chuuo, Ami, Inashiki 300-0393, Ibaraki, Japan
[2] Fukushima Prefectural Northern Agriculture and Forestry Office, 2-16 Sugitsuma, Fukushima 960-8670, Fukushima, Japan
[3] Independent Researcher, Tsukuba 305-0035, Ibaraki, Japan
[4] Department of Civil Engineering, National Institute of Technology, Kagawa College, 355 Chokushi, Takamatsu 761-8058, Kagawa, Japan; ntakahashi@t.kagawa-nct.ac.jp
* Correspondence: shigeya.maeda.15@vc.ibaraki.ac.jp; Tel.: +81-29-888-8601

Abstract: Agricultural drainage canals that connect upstream fish spawning areas to downstream rivers and lakes serve as crucial habitats for migrating fish. However, disconnections, such as drops and chutes, have been constructed in these canals due to agricultural modernization and flood control measures, hindering the movement of fish that find it difficult to ascend in fast-flowing currents. Portable fishways offer a promising solution to reconnect waterbodies in agricultural canals, as they can be easily removed during high water discharges to avoid impeding the canals' drainage function. In addition to experimental assessments of fishway functionality, employing a hydrodynamic model to explore effective placement strategies for portable fishways is essential to maximize their effectiveness. This study presents a method for determining the best horizontal location of a portable fishway in an agricultural drainage canal using two-dimensional hydrodynamic simulations within the specified cases. The applicability of this method is demonstrated by addressing the positioning challenge of a portable fishway on a chute in an agricultural drainage canal in Japan. The results indicate that the proposed method allows for the selection of a suitable location, considering preferable hydraulic conditions both within the portable fishway and around its entrance.

Keywords: portable fishway; hydraulic environment; agricultural drainage canal; chute; effective positioning; two-dimensional hydrodynamic simulation

1. Introduction

Conservation of fish habitat and spawning sites in rural areas is crucial for sustainable agricultural development. Agricultural drainage canals, which link spawning sites in upper reaches to downstream rivers and lakes, play a pivotal role in providing vital habitats for migrating fish. However, due to advancements in agriculture and flood control efforts, these canals have been subjected to disruptions such as drops and chutes. Reconnecting the water network has become an essential issue for the conservation of fish habitats in rural areas. These disruptions often result in fast-flowing currents, making it challenging for fish to swim against them. To address these challenges, portable fishways offer a promising solution. They can be easily installed in these canals and conveniently removed during periods of high water flow, ensuring that the canals retain their drainage function. As a result, both fish conservation and agricultural practices can be harmoniously integrated.

Many studies on installing fishways have been conducted around the world. However, most of the previous studies employed a comparatively large-scale fishway, intended for long-term use, made of concrete, which can be called a "permanent fishway". A permanent fishway requires considerable labor, capital costs, and time for installation. Therefore, it is difficult to conserve migratory aquatic organisms in waterbodies using only permanent fishways. In contrast, a portable fishway can be manufactured at a low cost and does not require specialized knowledge for installation. Takahashi et al. (2021) developed a portable fishway system that can be used in agricultural canals and demonstrated its effectiveness through laboratory and field experiments [1]. A portable fishway can be attached and detached by farmers and local residents to help fish pass through a disconnected area caused by a chute or a drop in an agricultural canal during a particular season or day.

Fishways can be classified into the following four types: vertical-slot, pool-weir, Denil, and culvert fishways [2]. Experimental and numerical investigations of the flow in permanent fishways were conducted. For example, An et al. (2016) simulated the three-dimensional flow to facilitate the design of a vertical-slot fishway [3]. Gao et al. (2016) developed a numerical model to simulate the trajectories of a virtual fish through a vertical-slot fishway by computing a turbulent flow field [4]. Zhong et al. (2021) investigated the hydraulic performance of a semi-frustum weir in a pool-weir fishway using both experimental and numerical methods [2]. Amaral et al. (2019) assessed several types of retrofitted ramped weirs to improve the passage of potamodromous fish using experiments and numerical simulations of flow [5]. Chorda et al. (2019) conducted numerical modeling of the free surface flow across a steeply sloped ramp covered with staggered surface emergent cylinders [6].

Very few studies have been conducted on the numerical simulation of flow in a portable fishway. Sudo (2020) applied a portable fishway on a chute (steep slope) of an agricultural drainage canal in Ibaraki, Japan and conducted a demonstration test to determine whether it promoted the ascent of small aquatic animals [7]. The results showed that the flow velocity in the fishway decreased to less than half of the flow velocity in the chute of the drainage canal and that medaka (*Oryzias latipes*) and Yoshinobori goby (orange type; *Rhinogobius* sp. OR) [8] went up the fishway. The authors of this paper have simulated the two-dimensional flow in the portable fishway of Sudo (2020) [7] and identified regions of high current velocity and low water depth under several longitudinal gradients of fishway and discharge conditions. However, the hydraulic influence of the portable fishway downstream in the canal and its effective location for fish migration have not been investigated.

Experimental assessments of the functionality of a portable fishway are insufficient for its effective use because several key issues remain unexamined. One of these problems is the positioning of the portable fishway within a target agricultural drainage canal. In this study, a method for numerically examining the transverse location for installing a portable fishway on an agricultural drainage canal was developed using the two-dimensional hydrodynamic model iRIC Nays2DH Ver. 3.0 [9,10]. This method was applied to determine the installation area of a portable fishway on a chute in a drainage canal in the Ibaraki Prefecture, Japan.

2. Materials and Methods

2.1. Portable Fishway

Takahashi et al. (2021) developed a portable fishway as an improvement to existing fishways [1]. The plan and cross-sectional views of the fishway are shown in Figure 1. Because the fishway was intended to be used temporarily, it was made of wood, and, thus, it was not expensive or heavy to carry. The fishway consists of two steps and three pools. The height of each step is 45 mm. Each step contains two "blocks", which are spherical and permeable objects made of net sponge (approximately 130 mm in height), that meander the flow and reduce the current velocity within the fishway. The pool can be used to allow the fish to ascend to rest. Because the trapezoidal cross section of the fishway in shallow water produces varying water depths at the water's edge, the created low-water region is expected to function as a passage for smaller fish.

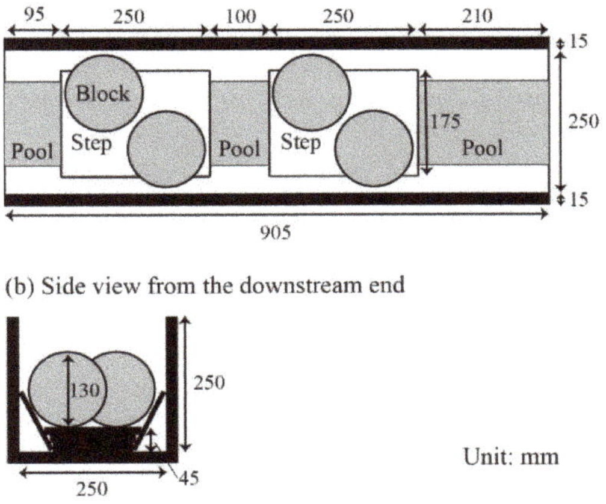

Figure 1. Plan and side view of a portable fishway.

2.2. Study Area

Figure 2 illustrates the target section in the agricultural drainage canal in Japan, where an artificial chute is located in the middle of the drainage canal. In this figure, the shapes of the longitudinal cross section of the canal and the two transverse cross sections, that is, upstream and downstream of the artificial chute, are also given. Note that the slope in the chute was so severe, 41.6%, that smaller fish with low swimming ability could not go upstream easily. It was observed that fish having suckers could ascend upstream by clinging to the canal's sidewalls using their suckers.

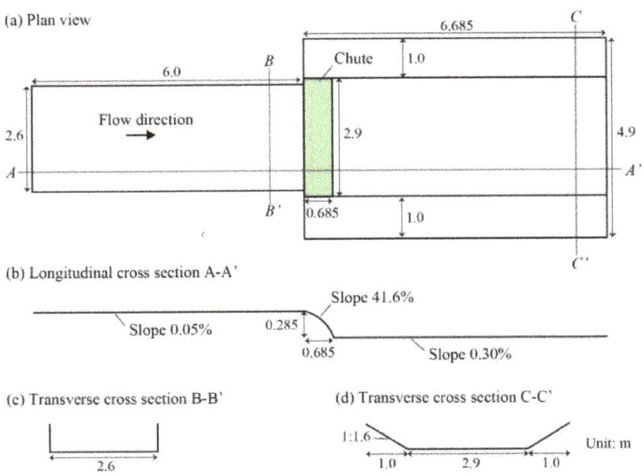

Figure 2. Computational domain of the drainage canal without a fishway. (**a**) Plan view of the whole domain. (**b**) Longitudinal cross section A-A′. (**c**) Transverse cross section B-B′. (**d**) Transverse cross section C-C′.

2.3. Field Observations

A portable fishway was applied to the chute in the study area on 10 June 2021 (Figures 3 and 4). Fish moving upstream through the fishway were captured, and the hydraulic conditions inside and around them were observed. The fishway was placed at a distance of 0.45 m from the left (Figure 5) and right banks. The bed slope of the fishway in the longitudinal direction was set to 12.3%. A catching wire net (0.3 m long and 0.35 m wide) was attached to the upstream end of the fishway to capture the upstream-going fish. The installation of the fishway with the net continued from 10:20 to 12:20 on the left-hand side of the bank and from 13:01 to 14:19 on the right-hand side of the bank on 10 June 2021. The fish captured were photographed and measured for their total length, then promptly released back into the canal.

Figure 3. A portable fishway with a catching wire net located on the chute on the right side of the bank. The flow is from back to front. The hanging plants along the left-bank sidewall in the upstream canal section reach the water surface.

Figure 4. Zoom-in photo of the portable fishway positioned on the left side of the bank of the drainage canal.

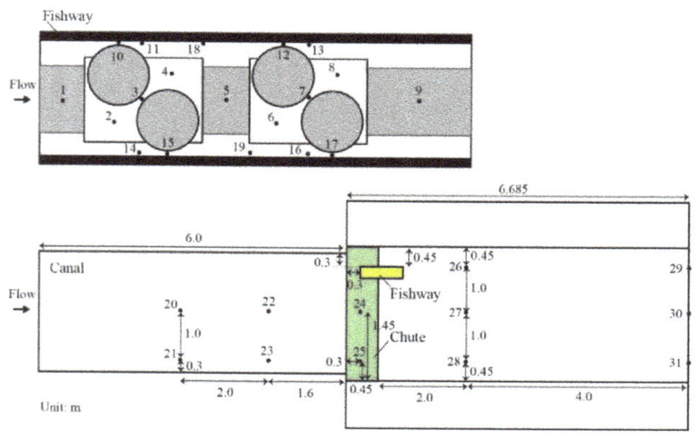

Figure 5. Monitoring points in the portable fishway (1–19) and the canal (20–31).

The water depth and current velocity were recorded at 31 points, as illustrated in Figure 5. The water depth was measured using a stainless steel ruler with 1 mm increments. The values of current velocity averaged in 10 s were observed twice at a 60% water depth with a portable electromagnetic current meter, VP3500 (Kenek). The specifications of the meter are listed in Table 1. It should be noted that the observation points along the left-bank side in the upstream canal section were omitted from Figure 5 because the cernuous plants in the upper-left corner of the section (seen in Figure 3) affected the hydraulic conditions. The hydraulic observations were conducted from 10:31 to 11:16 in the canal and from 11:44 to 12:10 in the fishway on 10 June 2021.

Table 1. Specifications of the portable electromagnetic current meter used in this study.

Item	Explanation
Measuring direction	Three axes and six directions
Flow range	−2 to +2 m/s
Accuracy	±0.010 m/s (velocity range: 0 to ±0.499 m/s), ±0.020 m/s (velocity range: ±0.500 to ±0.999 m/s), ±0.040 m/s (velocity range: ±1.000 to ±2.000 m/s)
Noise	Within ±0.005 m/s under hydrostatic condition
Response time	0.5 s

2.4. Numerical Computation of the Flow in the Canal and Fishway

Nays2DH, the 2D hydrodynamic model, was applied to the canal sections with and without the fishway, as shown in Figures 2 and 6, respectively. Although Nays2DH can simulate sediment transport and morphological changes of riverbeds and banks, only the function of simulating current flow was used in this study because of the short time period considered. The unsteady 2D Reynolds-averaged Navier–Stokes equations and the continuity equation were solved to estimate the spatial distribution of water depth and current velocity. The zero-equation model was employed for turbulence, where kinetic eddy viscosity was expressed as a linear function of the von Kármán constant and water depth. The bottom friction was formulated using Manning's roughness coefficient. The roughness coefficient is the calibration parameter of the flow model.

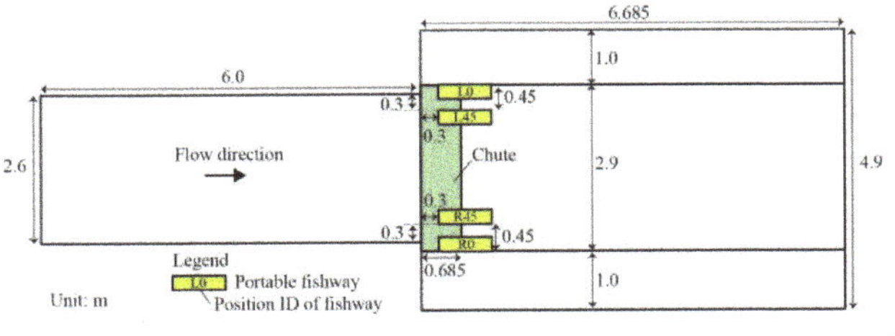

Figure 6. Computational domain for the various cases specified in Table 2, featuring the portable fishway positioned on the chute. This figure delineates each case's fishway placement, with position IDs marked on the fishway corresponding to those detailed in Table 2. Note that, in each case, only one portable fishway is installed on the chute.

Nays2DH can be used with the orthogonal curvilinear system. It employs time stepping with a differencing scheme. For the advection of momentum, the cubic interpolated pseudo-particle scheme was employed. The pressure (water surface elevation for this hydrostatic approach) was calculated using a successive relaxation technique [10]. Please refer to the Nays2DH solver manual (https://i-irc.org/en, accessed on 22 November 2023) for details.

The computational domain consists of a portable fishway installed on an artificial chute midstream of the domain. The domain in Figure 6 consists of three parts, as follows: (i) portable fishway, (ii) chute excluding the fishway, and (iii) other parts. Using the cross-sectional topography data that were prepared using results of the topographical survey on 21 October 2020, a rectangle embracing the entire domain in Figure 6 was discretized into quadrilateral cells with the nodes, as listed in Table 2. An example of the generated mesh around the fishway located on the left-bank side, Case L0_0.265 in Table 2, is given in Figure 7. Note that the finer grid cells are placed along the sidewalls and blocks within the fishway. The area around the inlet of the fishway is also represented in the fine-grid resolution, where the spatial variation of the hydraulic variables should be identified by computation.

Table 2. Computational conditions for the model calibration, validation, and configuration of the portable fishway.

Case	Upstream Discharge (m^3s^{-1})	Downstream Water Depth (m)	Position of Fishway	Number of Nodes	Note
N_0.140	0.140	0.143	No fishway	11,685 (=123 × 95)	
N_0.210	0.210	0.149	No fishway	11,685 (=123 × 95)	Validation
N_0.265	0.265	0.187	No fishway	11,685 (=123 × 95)	
L0_0.140	0.140	0.143	On left bank	50,024 (=338 × 148)	
L0_0.265	0.265	Uniform depth	On left bank	50,024 (=338 × 148)	
L45_0.140	0.140	0.143	0.45 m away from left bank	59,312 (=337 × 176)	Calibration
L45_0.265	0.265	Uniform depth	0.45 m away from left bank	59,488 (=338 × 176)	
R0_0.140	0.140	0.143	On right bank	50,024 (=338 × 148)	
R0_0.265	0.265	Uniform depth	On right bank	50,024 (=338 × 148)	
R45_0.140	0.140	0.143	0.45 m away from right bank	71,136 (=342 × 208)	
R45_0.265	0.265	Uniform depth	0.45 m away from right bank	71,136 (=342 × 208)	

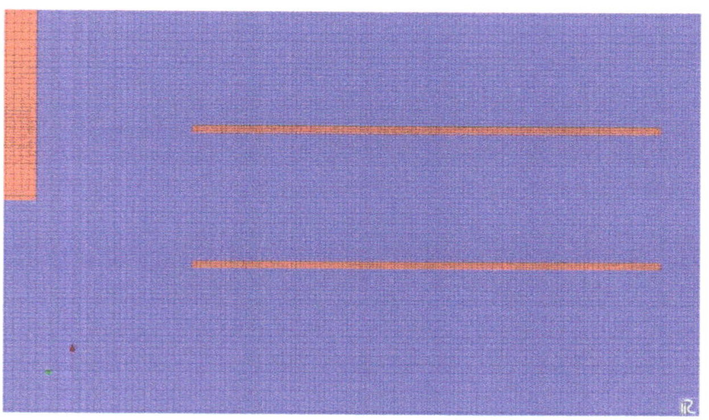

Figure 7. Computational mesh around the portable fishway in Case L0_0.265. The flow direction is from left to right. The two red parallel lines indicate the sidewalls of the fishway.

The numerical computation cases are summarized in Table 2. The entire computational domain is refined by subtracting the unnecessary regions, e.g., the red regions appearing in

Figure 7 as "obstacles" in the Nays2DH, from the rectangular area that encompasses the target computational region. It is noteworthy that the "number of nodes" shown in Table 2 is calculated with respect to the rectangular area.

Case L45_0.140 provides a numerical computation for the model's calibration. The 2D hydrodynamic model was calibrated at 0.140 m^3s^{-1} using water depth and current velocity data measured on 10 June 2021. The fishway in Case L45_0.140 was installed 0.45 m away from the left bank (Figure 5) at a slope of 12.3% on the chute in the canal. The downstream boundary condition featured a water depth of 0.143 m during calibration. The computational time step was 10^{-4} s. The total computational time was 300 s, which is a reasonable value assuming that the flow reached a steady state. The values of the Manning's roughness coefficients were determined for each of the three areas through model calibration.

Case N_0.210 in Table 1 was prepared to validate the hydrodynamic model. Regarding the model parameters of the fishway, Manning's roughness coefficient, the drag coefficient, and the density of the permeable object were determined based on our numerical computation of the flow in the fishway. The hydrodynamic model developed in this study was validated for flow simulation in a canal section without a fishway, as shown in Figure 2, using hydraulic data collected on 21 October 2020. The typical mesh size used for the validation was approximately 0.06 × 0.13 m. In the validation, the upstream discharge and downstream water depth in the canal were 0.210 m^3s^{-1} and 0.149 m, respectively, with the same time step and computational time as those in the calibration.

Using the validated hydrodynamic model, a total of eleven cases, including three cases without a fishway, as shown in Table 2, were investigated. The fishway was arranged on the chute in the canal in the following two ways: (i) on or near the left bank (L0 or L45 in Figure 6) and (ii) on or near the right bank (R0 or R45 in Figure 6). Two types of canal discharges, 0.140 and 0.265 m^3s^{-1}, were assumed in the numerical computation.

The computational time step was 10^{-4} s in all cases except for N_0.140 and N_0.265, where the time step was 10^{-3} s, because a smaller time step was required owing to the necessity of a finer mesh to represent the complex fishway structure.

3. Results

3.1. Fish Capture

Twenty and forty-eight juvenile Yoshinobori goby (orange type, *Rhinogobius* sp. OR, Figure 8), with total lengths of 0.020 ± 0.001 m and 0.021 ± 0.003 m, respectively, were caught using both the upstream net of the fishway and the fishway itself, placed in the L45 and R45 positions on 10 June 2021. This demonstrated the effectiveness of the portable fishway.

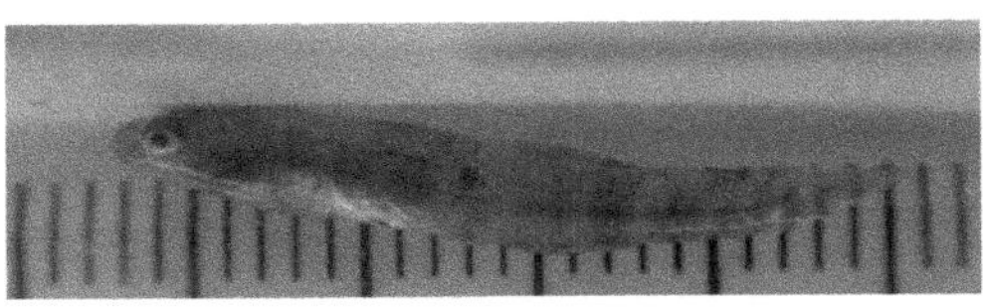

Figure 8. Juvenile *Rhinogobius* sp. OR caught in the portable fishway system on 10 June 2021.

3.2. Calibration of the Flow Model

The observed water depth and current velocity at the monitoring points (Figure 5) on 10 June 2021 are shown in Figure 9. The computed water depths and current velocities for the four cases at a discharge rate of 0.140 m^3s^{-1} within the portable fishway are shown in Figure 10. A comparison of the computational results for Case L45_0.140 with the observed results showed that the average relative errors for the water depth and velocity at the monitoring points 20–31 were 21.0 and 25.2%, respectively. Their root-mean-square errors were 0.03 m and 0.21 ms^{-1}, respectively. The difference between the observed and computed values seemed to be due to two main reasons. First, the upstream discharge in the canal varied because of agricultural activities around the canal during the hydraulic observations conducted from 10:31 to 11:16 on 10 June 2021. Since the observation time differed at each monitoring point, some measured values were based on discharges different from the given value of 0.140 m^3s^{-1} used in the numerical simulation. The second reason is that the supercritical flow on the chute led to temporal variations in the water surface, making accurate water depth measurements challenging. Given these difficulties in hydraulic measurements in the agricultural canal, it is not meaningful to aim for perfect error minimization between the observed and computed values in this context. Overall, the errors in the depth and velocity on the chute were within acceptable limits, and the spatial distributions of the hydraulic variables appeared reasonable. This was also the case for the validation, Case N_0.210. Hence, the numerical model was validated for analyzing an optimal configuration of the portable fishway. The identified Manning's roughness coefficients were 0.015, 0.030, and 0.086 $m^{-1/3}$ s for the fishway, chute, and other canal regions, respectively.

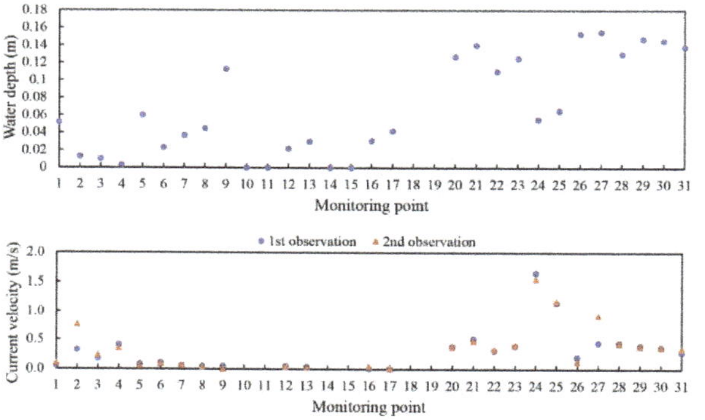

Figure 9. Observed water depth and current velocity under a discharge of 0.140 m^3s^{-1} in the portable fishway (monitoring points 1–19) and canal (monitoring points 20–31).

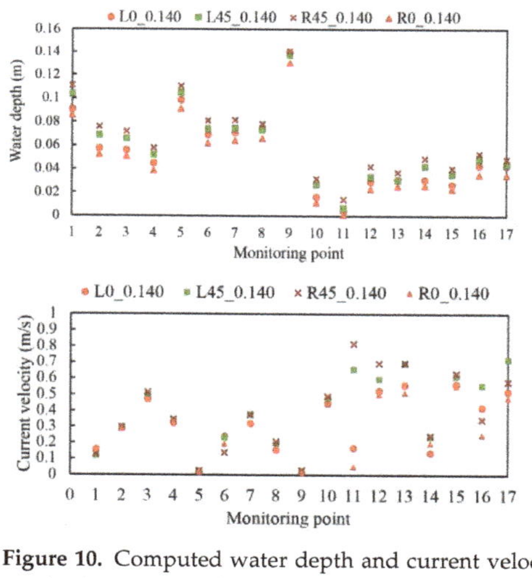

Figure 10. Computed water depth and current velocity under a discharge of 0.140 m^3s^{-1} at the monitoring points in the portable fishway in Cases L0_0.140, L45_0.140, R0_0.140, and R45_0.140.

3.3. *Effect of Introducing the Portable Fishway to the Chute*

The computed water depth and current velocity in the canal with and without the portable fishway at 0.140 and 0.265 m^3s^{-1} are summarized in Figures 11–14. Because the velocity in the area around the chute is critical for considering the reconnection of waterbodies and assisting fish migration, such areas were focused on and shown in Figures 12 and 14. The depicted streamlines, based on the computational results around the fishway at 0.140 and 0.265 m^3s^{-1}, are shown in Figures 15 and 16, respectively. The vectors of the current velocity and its magnitude under a discharge of 0.265 m^3s^{-1} in and around the fishway are shown in Figure 17.

As shown in Figure 12a (Case N_0.140), the computed flow on the chute was supercritical, and the maximum current velocity was 1.48 ms^{-1}. After crossing the chute, the flow changed from supercritical to subcritical. A hydraulic jump, which is depicted as a spatially irregular variation in velocity along the downstream end of the chute, can be observed during the transition of flows.

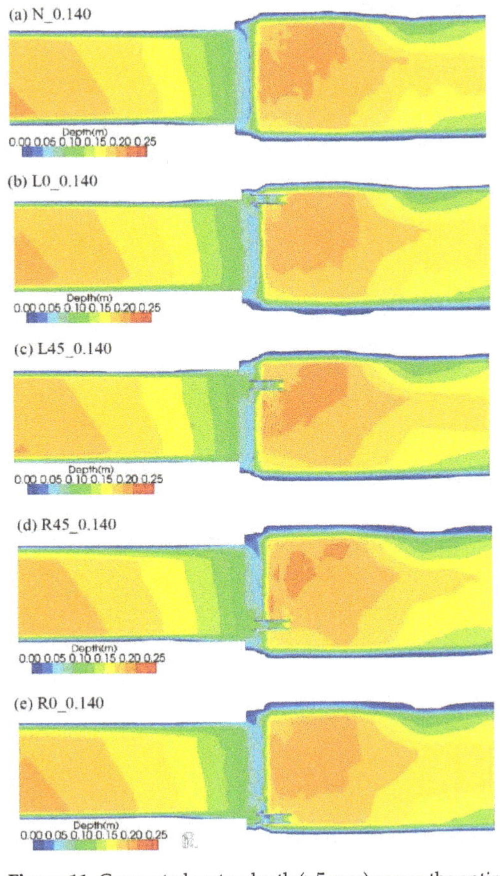

Figure 11. Computed water depth (>5 mm) across the entire computational domain under a discharge of 0.140 m^3s^{-1}. In each panel, the central part represents the chute in the canal. In panels (**b**) through (**e**), the sidewalls of the portable fishway are depicted using two horizontal, parallel white lines on the chute.

When the fishway was located on the chute on the left bank side in Case L0_0.140, a low-velocity region appeared in the fishway (Figure 12b). Within the fishway, the overall current velocity was suppressed to less than half of that on the chute. A relatively rapid flow was generated between the blocks located on the step (i.e., monitoring points 2–4 and 6–8 in Figure 5) and along the inner-bank side of the fishway (Figure 10), which may induce fish migration. In addition, the streamlines from the downstream fishway entrance were less complicated than those in Case L45_0.140 (see Figure 15a,b). Therefore, the fish could more easily detect the entrance of the fishway in Case L0_0.140. Upstream of the fishway in Case L0_0.140, a low-current velocity region existed along the left bank of the canal, which could be the preferential path for fish. The passage of *Rhinogobius* sp. OR along both banks of the canal was often observed.

When the fishway was located 0.45 m apart from the left bank of the canal in Case L45_0.140, the velocity within the fishway was larger than that at some points in Case L0_0.140 (Figures 10 and 12). Additionally, a fast flow was produced between the fishway and the left bank of the canal (Figure 12c). In this case, the streamlines downstream of the fishway became more complicated (Figure 15b). Thus, it may be more difficult for fish to find the entrance to the fishway than in Case L0_0.140. In particular, because juvenile or small adult fish with low swimming capabilities tend to move upstream along the bank of the canal, a wide region of low current velocity could be important.

Similar discussions apply for the case where the fishway was configured on the right-bank side of the canal in Case R45_0.140. In Case R0_0.140, the fishway settled on the right bank of the canal and a wide region of low current velocity was generated (Figure 12e), which was appropriate for assisting smaller fish to migrate. However, the estimated streamlines around the fishway entrance were more complicated than those at L0_0.140 (Figure 15).

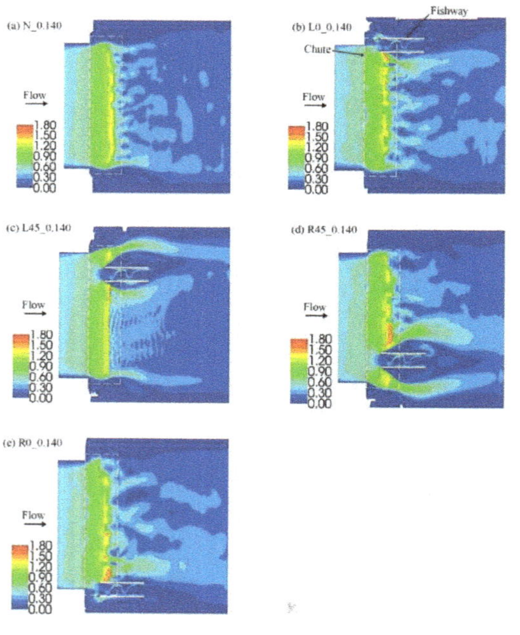

Figure 12. Computed current velocity in the canal in and around the chute under a discharge of 0.140 m^3s^{-1}.

Under a discharge of 0.265 m^3s^{-1}, the spatial variation in the computed current velocity (Figure 14a) was smaller than that at 0.140 m^3s^{-1}. Although the current velocity on the chute increased in proportion to the canal discharge, the influence of the fishway configuration on the flow was comparable to that at 0.140 m^3s^{-1}.

The streamlines around the portable fishway in the canal in Case_L0_0.265 and Case_L45_0.265 are compared in Figure 16. Note that the streamlines entering the downstream end of the fishway were significantly different in these cases. In Case_L0_0.265, where the fishway was located along the left bank of the canal, the streamlines on the left side of the fishway were almost attached to its sidewall. The streamlines originating from the fishway entrance were comparatively more straightforward than those in Case_L45_0.265. Because smaller fish tend to migrate to shallow regions near the bank of the canal, especially under a high discharge, such fish could find the inlet of the fishway along the streamlines more easily in Case_L0_0.265. Therefore, in view of the ease of finding the fishway entrance, Case_L0_0.265 seems to be better.

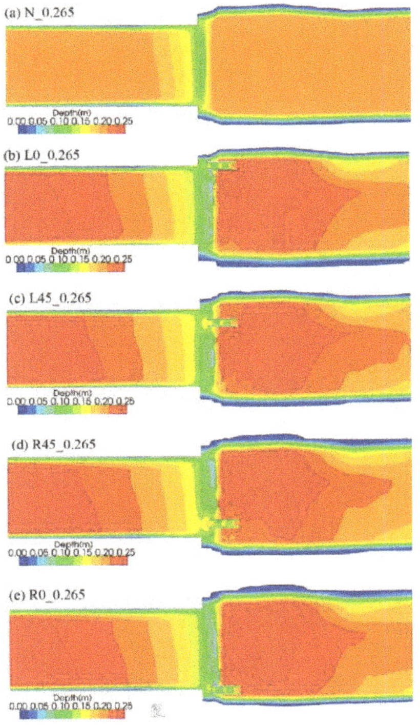

Figure 13. Computed water depth (>5 mm) across the entire computational domain under a discharge of 0.265 m^3s^{-1}. In each panel, the central part represents the chute in the canal. In panels (**b**) through (**e**), the sidewalls of the portable fishway are depicted using two horizontal, parallel white lines on the chute.

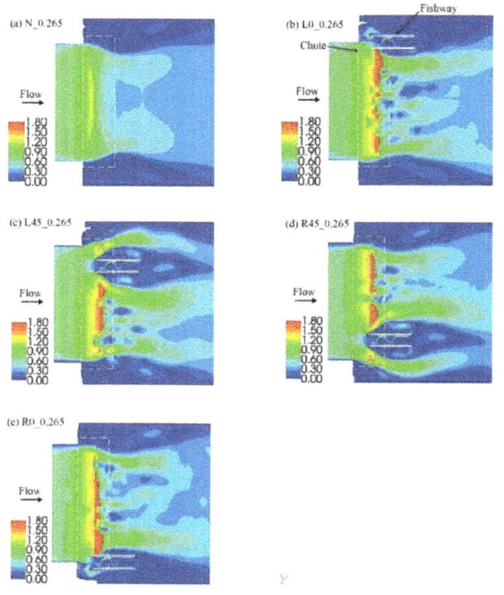

Figure 14. Computed current velocity in the canal in and around the chute under a discharge of 0.265 m^3s^{-1}.

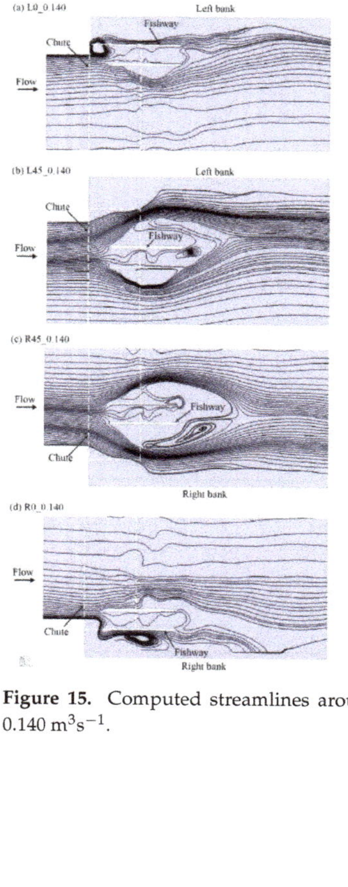

Figure 15. Computed streamlines around the portable fishway in the canal at a discharge of 0.140 m^3s^{-1}.

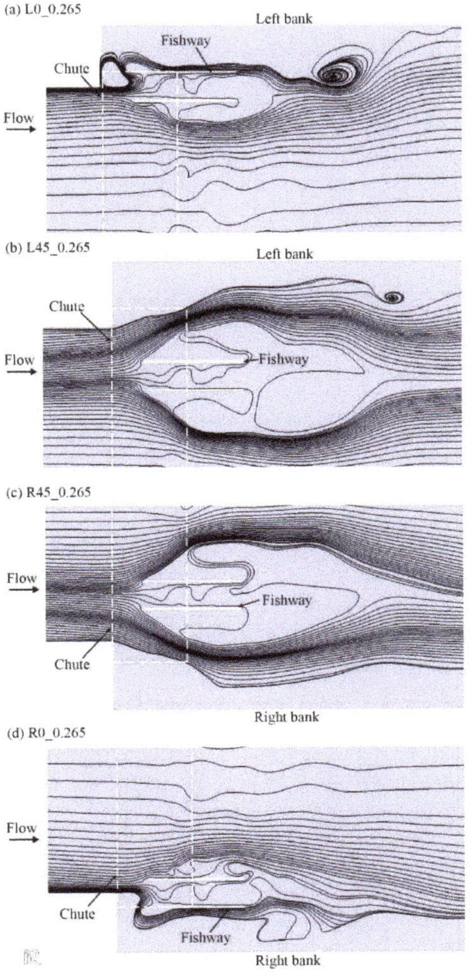

Figure 16. Computed streamlines around the portable fishway in the canal at a discharge of $0.265\ m^3s^{-1}$.

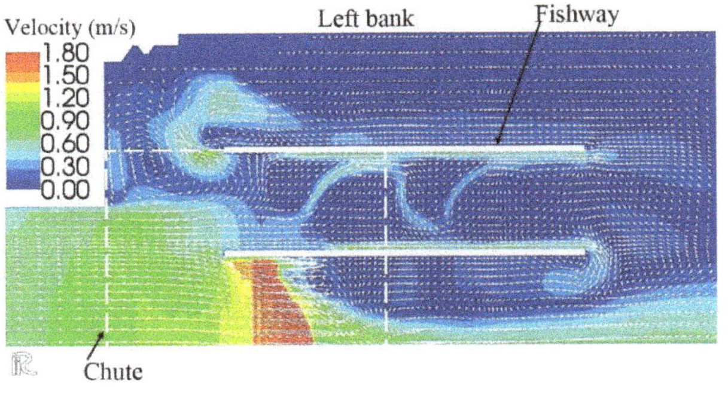

Figure 17. Computed velocity vectors and velocity magnitude focused in and around the fishway in Case L0_0.265.

4. Discussion

The steep slope of the chute, at 41.6%, presents a significant barrier for smaller fish with a limited swimming ability, hindering their upstream migration. Despite observations of fish with suckers ascending by clinging to the canal's sidewalls, the introduction of the portable fishway in this study has facilitated an easier upstream migration for these species.

The observed use of the portable fishway predominantly by Yoshinobori goby, with no evidence of use by other species, can be attributed to several factors specific to the study area. Unpublished eDNA surveys and netting have indicated a higher population density of Yoshinobori goby compared to other species. Additionally, larger fish, such as those in the Cyprinid family, may be able to ascend without the aid of the fishway, as their physical capabilities allow them to navigate the steep slopes. The limited width of the fishway likely renders it unsuitable for larger fish, thus favoring smaller species like the Yoshinobori goby.

For successful fish passage in canals, the current velocity must be maintained well-below the burst speed of the fish, as noted by [11]. Burst speed, which is powered by an anaerobic white muscle, is typically sustainable for only a short duration, such as 15 s [11]. Zhong et al. (2021) [2] summarized the burst speeds for common fish species; for instance, the burst speed for ayu and eel are 1.6 ms^{-1} and 0.61–1.22 ms^{-1}, respectively. Given that the observed velocity on the chute ranged from 1.15 to 1.66 ms^{-1}, as shown in Figure 9, the upstream migration of these species without a fishway is likely challenging. However, the installation of the portable fishway, as demonstrated in Cases L0_0.140 and L0_0.265, is expected to reduce the maximum velocity in the chute to less than 0.67 ms^{-1} (refer to Figures 10 and 12), which is below the burst speed threshold for these fish.

5. Conclusions

This study focuses on the numerical method for assisting the better placement of portable fishways in canals to facilitate fish migration. Portable fishways are easily attachable and detachable structures designed to aid fish in moving upstream. One of the notable features of these portable fishways is their flexibility, allowing for the adjustment of fishway placement based on fish life-stage and hydraulic conditions, resulting in sustainable agricultural development. This functionality proves effective when hydrodynamic simulations are conducted before the actual application of portable fishways in canals and rivers, ensuring the appropriate determination of their final locations.

In this case study, the fishways located 0.45 m from the left or right bank were found to be less attractive to fish and exhibited faster currents compared to those placed along the bank. This study demonstrates that the utility of portable fishways, which have already been validated in laboratory hydraulic experiments, can be further enhanced through the integration of 2D hydrodynamic simulations to determine the best fishway placement within the specified cases. Another critical factor in creating a favorable flow field for fish is the bed slope of the portable fishway. While this factor was not examined in this study, the methodology for doing so is straightforward.

In the future, it is imperative to investigate the impact of turbulence features such as Reynolds shear stress, turbulent intensity, and kinetic energy in the water flow in canals on fish passage, both in the field and through numerical computations. The application of a three-dimensional hydrodynamic model to this problem may also be necessary to analyze the detailed characteristics of flow in and around portable fishways.

Author Contributions: Conceptualization, S.M. and N.T.; methodology, S.M., Y.Y. and N.T.; validation, S.M. and Y.Y., formal analysis, S.M, Y.Y. and K.Y.; investigation, S.M., Y.Y. and K.Y.; resources, S.M. and N.T.; data curation, S.M.; writing—original draft preparation, Y.Y.; writing—review and editing, S.M. and N.T.; supervision, S.M.; project administration, S.M.; and funding acquisition, S.M. and N.T. All authors have read and agreed to the published version of the manuscript.

Funding: This study was partially funded by the River Fund of the River Foundation, Japan (Grant Number 2020-5211-031), and the Japan Society for the Promotion of Science KAKENHI (Grant Number 20H03095).

Data Availability Statement: The data presented in this study are available on request from the corresponding author.

Acknowledgments: The authors appreciate the support provided by the Miho Town Office, the Okitsu District Land Improvement Association, and the Suigo Tukuba Agricultural Cooperative during the field observations. The authors would like to thank the members of the Laboratory of Agricultural Water Use of Ibaraki University for their assistance with the fieldwork and data organization. Our sincere thanks go to the anonymous reviewers whose valuable comments significantly contributed to the improvement of this manuscript.

Conflicts of Interest: The authors declare no conflict of interest. The funders had no role in the following: in the design of the study; in the collection, analyses, or interpretation of data; in the writing of the manuscript; or in the decision to publish the results.

References

1. Takahashi, N.; Misawa, Y.; Honzu, M.; Yanagawa, R.; Tagawa, T.; Nakata, K. Proposal of portable fishway system applicable to agricultural waterways. *Trans. Jpn. Soc. Irrig. Drain. Rur. Eng.* **2021**, *89*, I_29–I_35. [CrossRef]
2. Zhong, Z.; Ruan, T.; Hu, Y.; Liu, J.; Liu, B.; Xu, W. Experimental and numerical assessment of hydraulic characteristic of a new semi-frustum weir in the pool-weir fishway. *Ecol. Eng.* **2021**, *170*, 106362. [CrossRef]
3. An, R.; Li, J.; Liang, R.; Tuo, Y. Three-dimensional simulation and experimental study for optimizing a vertical slot fishway. *J. Hydro-Environ. Res.* **2016**, *12*, 119–129. [CrossRef]
4. Gao, Z.; Andersson, H.I.; Dai, H.; Jiang, F.; Zhao, L. A new Eulerian–Lagrangian agent method to model fish paths in a vertical slot fishway. *Ecol. Eng.* **2016**, *88*, 217–225. [CrossRef]
5. Amaral, S.D.; Quaresma, A.L.; Branco, P.; Romão, F.; Katopodis, C.; Ferreira, M.T.; Pinheiro, A.N.; Santos, J.M. Assessment of retrofitted ramped weirs to improve passage of potamodromous fish. *Water* **2019**, *11*, 2441. [CrossRef]
6. Chorda, J.; Cassan, L.; Laurens, P. Modeling steep-slope flow across staggered emergent cylinders: Application to fish passes. *J. Hydraul. Eng.* **2019**, *145*, 04019038. [CrossRef]
7. Sudo, S. Observation of Fish Passage and Flow Analysis in Portable Fishway Located on Chute in Canal. Bachelor's Thesis, Ibaraki University, Mito, Japan, 2020; pp. 20–24. (In Japanese).
8. Hosoya, K.; Uchiyama, R.; Fujita, T.; Takeuchi, H.; Kawase, S. *Japanese Freshwater Fish*; Yama-Kei Publishers: Minato, Japan, 2019; pp. 474–476. (In Japanese)
9. Nays2DH Development Team. *Nays2DH. Solver Manual*. IRIC Project. 2011. Available online: https://i-irc.org/en (accessed on 20 October 2023).
10. Nelson, J.M.; Shimizu, Y.; Abe, T.; Asahi, K.; Gamou, M.; Inoue, T.; Iwasaki, T.; Kakinuma, T.; Kawamura, S.; Kimura, I.; et al. The international river interface cooperative: Public domain flow and morphodynamics software for education and applications. *Adv. Water Resour.* **2016**, *93*, 62–74. [CrossRef]
11. Yagci, O. Hydraulic aspects of pool-weir fishways as ecologically friendly water structure. *Ecol. Eng.* **2010**, *36*, 36–46. [CrossRef]

water

MDPI

Article

Large Eddy Simulations of Flow Past Circular Cylinders to Determine Head Loss Coefficients of Circular Bar Trash Racks with Perpendicular Inflow Conditions

Hannes Zöschg

Unit of Hydraulic Engineering, Department of Infrastructure, University of Innsbruck, 6020 Innsbruck, Austria; hannes.zoeschg@uibk.ac.at

Abstract: Trash racks installed at hydropower plants cause head losses that reduce energy output. Previous research has thoroughly investigated head losses through both experimental and field studies. However, only a limited number of numerical studies have been performed, which have shown significant simplifications in terms of model complexity. In this study, the head loss coefficients ξ of circular bar trash racks (CBTRs) were analyzed using 3D Large Eddy Simulation (LES). Specifically, a single submerged bar oriented perpendicular to the flow direction was studied under homogeneous inflow conditions while (i) the blocking ratio P was varied between 0.043 and 0.444, and (ii) the flow velocity U was varied between 0.3 and 1.0 m/s. The model parameters were selected primarily based on the extensive literature on flow past circular cylinders, particularly at a Reynolds bar number Re_b of 3900. To ensure the validity of the parameters, systematic independence tests were performed, including simulations with three and five bars in the computational domain. The results confirmed the suitability of 3D LES as an appropriate tool to determine ξ of CBTRs. In general, ξ decreased continuously with decreasing P and increased with increasing U when $Re_b \geq 3981$, which is consistent with comparable flow parameters observed in previous studies of flow past circular cylinders. Notably, the study found that the empirical formulas used for comparison tended to underestimate ξ when P was relatively low. Finally, the potential of the presented approach for future applications was discussed in detail.

Keywords: 3D numerical modeling; Computational Fluid Dynamics (CFD); circular bar trash racks; flow past a cylinder; head loss; hydraulics; hydropower plants; Large Eddy Simulation; trash racks

1. Introduction

As essential components of hydropower plants (HPPs), trash racks located upstream of turbine intakes primarily serve to protect turbines and other electromechanical equipment from solids transported by the river, particularly floating debris, driftwood, ice drift, and bedload transport [1–3]. These materials can accumulate on trash racks and cause partial or complete blockage or clogging of the intakes, with potential adverse effects including increased static and dynamic loads [3,4], increased upstream flooding potential [5,6], and reduced energy production [7,8]. Generally, such conventional trash racks consist of an array of horizontally or vertically arranged bars held together by support structures. Rack cleaning machines are commonly used to remove accumulated debris. Values for the clear bar spacing b of conventional trash racks typically range from about 20 to 300 mm [2]. By reducing b, trash racks can also serve to protect fish during their downstream migration, as the majority of fish cannot physically pass the rack due to their body proportions [9,10]. Consequently, they cannot enter the turbines, which is associated with a high risk of injury or mortality due to direct contact with blades or other turbine structures, sudden pressure changes, or increased shear or turbulence [11,12]. Typical b values of physical barriers for fish protection range from 10 to 30 mm [10,13], with the recommendation that b should be determined based on the size of the target fish species in the particular river section [14,15]. Another type of barrier that can be used for fish protection is the behavioral barrier, which uses external stimuli such as electricity, light, sound, or the creation of turbulent currents to elicit an avoidance or escape response in fish [16–18]. More recently, hybrid barriers have been increasingly studied, which combine conventional racks or racks for fish protection with an electric field to add an additional behavioral effect to an existing or newly constructed rack [19–23]. Hybrid barriers offer a distinct advantage over racks which provide only a physical barrier effect, allowing the use of larger b values, while still effectively protecting fish [19,20].

When a trash rack is installed upstream of an HPP, some of the kinetic energy of the water is converted into heat as it flows through the rack, which can be considered a loss to the mechanical system and is therefore referred to as an energy or hydraulic loss [2]. These losses are mainly influenced by the geometry of the rack and its arrangement relative to the flow direction, as well as the flow conditions [2]. In practical applications, such as the design phase of a trash rack at an HPP, hydraulic losses are typically calculated using empirical formulas derived from experimental models or empirical data sets. Kirschmer [24] was one of the first to conduct experiments with metal and wooden racks with vertical bars in a flume, varying key parameters such as b, the bar thickness s, and the rack angle relative to the ground plane β. In addition, different bar cross-sectional shapes were investigated, including circular bar trash racks (CBTRs). To determine the head loss coefficient ξ of the studied racks, Kirschmer [24] proposed the formula

$$\xi = k_F \left(\frac{s}{b}\right)^{\frac{4}{3}} \sin\beta, \tag{1}$$

where k_F is the bar shape coefficient. Since then, many researchers have conducted experiments to assess the hydraulic losses of trash racks by analyzing the effect of unit discharge q, flow velocity U, rack angle relative to the side wall α, bar orientation (horizontal or vertical), and other parameters beyond those studied by Kirschmer [24] that affect ξ [2,12,15,25–35]. For example, Meusburger [2] extended Kirschmer's [24] formula by including the area of spacers and support structures along with the area of the bars through the blocking ratio P. Furthermore, the influence of sectional clogging of the rack, expressed by the loss factor k_V, and the horizontal inflow angle to the rack δ, defined as $\delta = 90^\circ - \alpha$ for undisturbed homogeneous flow conditions upstream of the rack, were also considered. The formula proposed by Meusburger [2] can be expressed as

$$\xi = k_F \left(\frac{P}{1-P}\right)^{\frac{3}{2}} \left(1 - \frac{\delta}{90^\circ}\right) P^{-1.4\tan\delta} k_V \sin\beta. \tag{2}$$

Although the experiments were limited to rectangular bars, other bar shapes such as CBTRs can be considered using the k_F values given by Kirschmer [24] in Equation (2) [2]. Böttcher et al. [29] conducted experiments in a flume using horizontally oriented CBTRs and flexible steel cables. Based on the results, an adapted version of Equation (2) was proposed for both rack options, which for CBTRs is given by

$$\xi = 1.80 \left(\frac{P}{1-P}\right)^{1.3} (\sin\alpha)^{1.7}. \tag{3}$$

Further empirical formulas for determining ξ of CBTRs have been proposed by Josiah et al. [27], Zayed et al. [28,36], Spangler [37], Berezinski [38], and the United States Army Corps of Engineers (USACE) [39]. In addition, other researchers, including Tsikata et al. [12], Fellenius and Lindquist [40], Escande [41], Orsborn [42], Zimmermann [43], the Bureau of Reclamation [44], and Clark et al. [45], have suggested formulas for conventional racks with non-circular bar shapes based on physical model experiments, field studies, and empirical data sets. Likewise, formulas for fish protection racks have been proposed by Albayrak et al. [15,46], Raynal et al. [25,26], Beck et al. [30], and Meister et al. [32], among others.

Advances in computational power and the growth of high-performance computing (HPC) clusters have made numerical simulations an important tool for studying hydraulic problems over the last decades. However, studies investigating ξ of trash racks have been limited to a small number of cases [12,31,34], using significant simplifications in terms of model complexity. For example, Raynal et al. [47] performed 2D numerical simulations of angled bar racks by solving the Reynolds-averaged Navier–Stokes (RANS) equations and compared the results with data from previous physical model tests [26,48]. The study found that the ξ values obtained from the numerical simulations were at least 7% lower than those obtained from the physical model. The discrepancies were attributed to the lack of spacers and the inability to replicate 3D effects in 2D simulations. Similarly, Åkerstedt et al. [49] and Leuch et al. [50] used comparable approaches and showed mostly divergent results for ξ when compared to physical model tests. In the former study, the discrepancies were related to the absence of 3D effects, as already noted by Raynal et al. [47], and the application of the RANS approach. The authors recommended that future research should implement more sophisticated models, such as Large Eddy Simulation (LES). Lučin et al. [31] performed 2D and 3D simulations of vertically arranged angled bar racks using the RANS approach based on the physical models of Albayrak et al. [46]. The study found that ξ was underestimated by about 15% compared to the physical model tests, while the 2D and 3D simulations produced comparable ξ values. Again, the use of LES was suggested for future research rather than the RANS approach, which relies on mean flow characteristics [51]. LES, on the other hand, applies a spatial filtering technique to resolve large-scale eddies, while smaller eddies are accounted for using a subgrid-scale (SGS) turbulence transport model [51–53]. By incorporating large turbulent parameters into the numerical solver, LES generally achieves higher accuracy. However, a finer mesh discretization is required [51], resulting in a higher computational cost. To reduce the computational cost, the number of mesh elements can be reduced by focusing on a limited number of bars within the numerical domain. For example, Nascimento et al. [54] performed simulations using the LES model with only two rectangular bars in a 2D domain to investigate the vortex shedding frequencies induced by the rack, but they excluded ξ in the analysis. Other researchers, including Ghamry and Katopodis [55], Paul and Adaramola [56], Jeethulakshmi et al. [57], and Čarija et al. [58], also performed simulations with the number of bars n ranging from 1 to 14, mainly using the RANS approach and 3D domains, and calculated ξ. Typically, the results overestimated ξ compared to the values derived from empirical formulas and previous model tests. Recently, Latif et al. [34] performed 3D RANS simulations of three rectangular bars and proposed a prediction formula for ξ as a function of b, P, and β based on the results. The obtained values were in good agreement with ξ measured at a real HPP, but tended to overestimate the calculated ξ values of empirical formulas, including Kirschmer [24].

Although, to the best of the author's knowledge, no simulations of resolved bars in a 3D domain have been performed using the LES model to determine ξ of trash racks, it has been widely used in other engineering applications, such as flows around bridge piers [59], ground vehicles [60], and high-rise buildings [61]. Many of these studies are mainly based on a classical fluid mechanics problem, the flow past a bluff body, in particular a circular cylinder [62]. Flows in the subcritical regime with bar Reynolds numbers

$$Re_b = \frac{U_{ref}\, s}{\nu}, \tag{4}$$

where U_{ref} is the free-stream velocity, and ν is the kinematic viscosity, ranging from 300 to 140,000 [63], are of significant interest due to the complexity around the cylinder wake [64] with the transition from laminar to turbulent within the separated shear layer [65–67]. Specifically, the flow around a circular cylinder at Re_b = 3900 has been extensively studied through a variety of experimental and numerical methods (e.g., [66,68–72]). Besides LES, researchers have also used Direct Numerical Simulation (DNS, e.g., [73–75]) and Detached Eddy Simulation (DES, e.g., [76,77]). The former method is known to provide the most accurate and reliable results, but requires relatively high computational effort even at low

Re_b [51,64]. The latter method combines the RANS and LES methods by applying the RANS approach in the near-wall region and the LES model in the distant flow regions [52,53,77]. In general, the results of the aforementioned studies using LES, DNS, and DES were found to be in good agreement with experimental studies. Conversely, the use of the RANS method or the unsteady RANS (URANS) method showed larger discrepancies with experimental data in previous studies (e.g., [76,78]). These results suggest that the URANS approach is not capable of accurately predicting the transient flow characteristics of flow past a bluff body [79,80]. Furthermore, previous studies have shown that 2D numerical simulations provide inaccurate values for most flow parameters when $Re_b \geq 250$ [66,67,81]. Therefore, to account for the 3D wake instabilities in the subcritical regime and to obtain accurate results, 3D numerical simulations are recommended [66,82]. In addition to the flow past a single cylinder, the literature contains numerous studies of the flow past two or more circular cylinders arranged in various tandem, side-by-side, or staggered patterns [83,84]. The relevant flow parameters in these cases are mainly influenced by b and Re_b [83–85]. Finally, several studies have investigated the flow past a circular cylinder placed between two parallel walls [86]. Small wall distances can significantly affect the periodic vortex shedding behind the cylinder, as well as other flow structures, indicating that the flow parameters of a confined compared to an unconfined cylinder can vary strongly as a function of P [86–88].

In the present study, 3D numerical simulations using the LES model were performed for a single bar and three or five bars of CBTRs arranged in a side-by-side pattern to determine ξ, mainly based on recommendations from previous research on flow past one or more circular cylinders in confined or unconfined channels and preliminary simulations. The resulting data were then compared with previous studies. The objective of the study was to evaluate the applicability of the presented approach in determining ξ of CBTRs with perpendicular inflow conditions. Furthermore, the effects of b and U on ξ of CBTRs were investigated.

2. Materials and Methods

2.1. Overview and Test Cases

The commercial Computational Fluid Dynamics (CFD) code ANSYS Fluent 19.2 was used for the present study. In order to reduce the high computational cost of the numerical simulations, this study focused on investigating relatively short sections of single bars or arrays of three or five bars of CBTRs in a flume, as schematically shown in Figure 1. The inflow conditions to these sections were idealized (i.e., ideally homogeneous and perpendicular with $\alpha = 90°$ and $\beta = 90°$, no clogging, and no rack damage). The rack sections were fully submerged in water to avoid free surface effects and positioned away from walls to avoid friction effects. For these reasons, as well as for the selected boundary conditions described in Section 2.2, the investigated sections are independent of the orientation of the bars. Horizontally oriented bars are considered in the following, although the results are also applicable to vertically oriented bars.

The rack configurations investigated were inspired by physical models from previous studies, in particular, Böttcher et al. [29] and Meister et al. [32]. The dimensions were approximately in the prototype range (scale 1:1), with s = 8 mm and b = 10, 20, and 30 mm. These b values were investigated in computational domains with n = 1, 3, and 5 bars. Larger values of b, ranging from 40 to 120 mm in increments of 10 mm, as well as 150 mm and 180 mm, were also investigated, but only for the case of a single bar, since the total number of elements in the computational domain N_t increases linearly with n. Instead, additional simulations were performed with periodic boundary conditions at the top and bottom of the computational domains to investigate a similar effect (see Section 2.2). Since the present study did not consider spacers or other support structures, the ratio of the blocked area to the total rack area was calculated as $P = s/(s + b)$. Consequently, P ranged from 0.043 (for b = 180 mm) to 0.444 (for b = 10 mm). Furthermore, the effect of U on ξ was investigated by varying the flow velocity at the inlet of the computational domain U_1 from

0.3 to 1.0 m/s in steps of 0.1 m/s for a single bar with b values of 10, 20, and 30 mm. The lower limit of U_1 was set at 0.3 m/s (corresponding to Re_b = 2389) to avoid perturbations caused by undersized computational domains, as reported by Jiang and Cheng [89] for flow past a circular cylinder at $Re_b \leq 2000$. To keep critical computational parameters that could affect the results within an acceptable range, the upper limit of U_1 was set at 1.0 m/s (corresponding to Re_b = 7962). In addition, to investigate the effect of U_1 in a setting where the influence of the top and bottom walls of the computational domain is negligible, U_1 was also varied for a case with a larger b value. For this purpose, b = 90 mm was chosen, resulting in P = 0.082, which is less than 0.100 as recommended by Norberg [90] to avoid top and bottom wall effects.

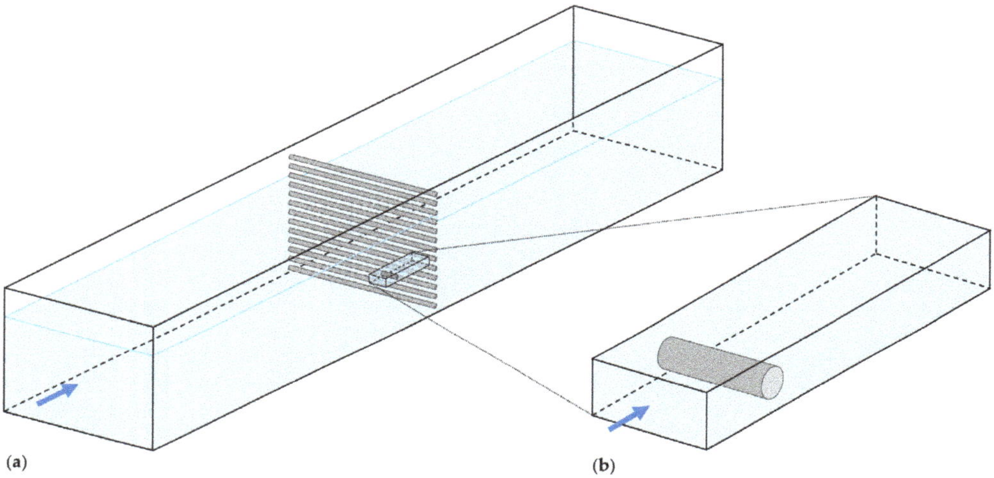

Figure 1. Schematics of (**a**) a complete model of a horizontally arranged circular bar trash rack (CBTR) in a flume, and (**b**) a waterbody segment consisting of a single bar section taken from (**a**): Such segments served as the basis for defining the computational domains in this study. The arrows indicate the direction of flow.

In this study, water density ρ was defined as 998.2 kg/m^3 at a constant temperature of 20 °C, and ν was defined as 0.0000010048 m^2/s. When U_{ref} or U_1 (interchangeable terms in this study) is equal to 0.5 m/s, Equation (4) calculates Re_b to be 3981. This value is close to the frequently used Re_b value of 3900 in previous research on flow past a circular cylinder, as noted in Section 1. Due to the similarities between the present problem and this classic topic, along with the large amount of data available for Re_b = 3900, the basic settings and recommendations for the numerical simulations in this study were adopted. To assess the appropriateness of these settings and recommendations for the determination of ξ, systematic independence tests were performed for the single-bar case with the lowest b value of 10 mm or the highest P of 0.444 at Re_b = 3981, as well as in selected cases with b = 90 mm (P = 0.082).

2.2. Computational Domains and Boundary Conditions

Hexahedral domains in a Cartesian coordinate system were used for the simulations in this study. The x, y, and z axes represented the streamwise, spanwise, and vertical directions, respectively. In the case of a single bar, the bar was placed symmetrically between two parallel walls in the vertical direction, each equidistant from the boundaries by $b/2$. Figure 2 shows a schematic representation of such a computational domain. In the cases where n = 3 and 5 were used, the equivalent number of single-bar domains were superimposed, with b as the distance between two bars. The dimensions of the domains in the streamwise, spanwise, and vertical directions are denoted as L_x, L_y, and L_z, respectively. L_x can be further divided into upstream and downstream parts separated by the examined rack section (i.e., $L_x = L_{x,us} + s + L_{x,ds}$). $L_{x,ds}$, $L_{x,us}$, and L_y were initially defined based on the literature data and preliminary simulations. These lengths were then validated in the independence test using the lengths given in Section 3.1.2. In contrast, the value of L_z is a function of the chosen values of n and P, described as $L_z = n\ (s + b)$.

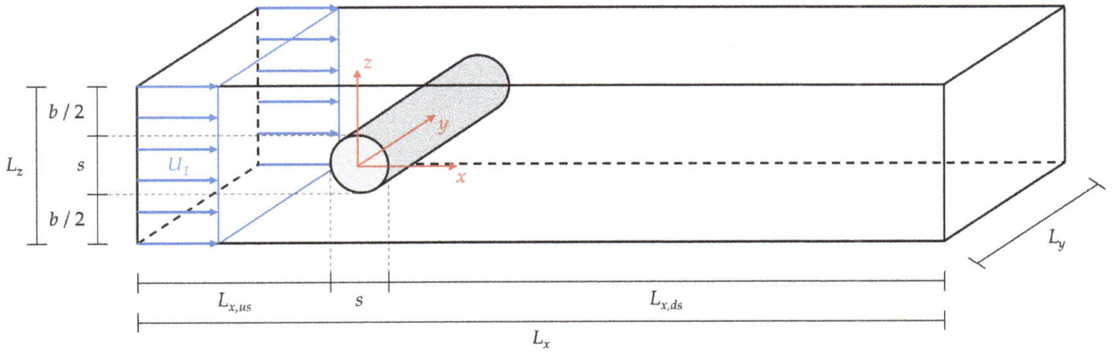

Figure 2. Schematic of a computational domain with a single bar, where b is the clear bar spacing, L_x, L_y, and L_z are the lengths of the domain in the streamwise, spanwise, and vertical directions, respectively, $L_{x,ds}$ and $L_{x,us}$ are the lengths of the domain downstream and upstream of the bar, respectively, in the streamwise direction, s is the bar thickness, and U_1 is the flow velocity at the inlet.

At the inlet of the domains, the velocity inlet boundary condition was used with a constant value of U_1. At the outlet, the pressure outlet boundary condition was applied with zero static gauge pressure. Periodic boundary conditions were used in the spanwise direction to assume an infinite length of the bars, as often used in previous studies of flow past circular cylinders (e.g., [69,85,88]), thereby reducing the end wall effects on the three-dimensional flow conditions. Unlike most studies of flow past a confined cylinder (e.g., [88,91]), which used no-slip wall boundary conditions for the top and bottom of the computational domains, this study mainly used free-slip wall boundary conditions. The application of free-slip wall and periodic boundary conditions in the vertical and spanwise directions, respectively, resulted in no frictional losses along the walls, thus producing a uniform velocity profile with U_1 across the cross section far upstream of the rack section, as shown in Figure 2. The surfaces of the bars were subjected to a no-slip wall condition to represent the velocity gradient, which is zero at the surfaces and reaches free-flow velocity at a sufficient distance from the bars. It should also be noted that periodic top and bottom boundary conditions were used for a limited number of simulations, as described in Section 2.1.

2.3. Spatial and Temporal Discretization

Following the methodology used in several previous studies of flow past a cylinder [64], a hybrid mesh was used as the mesh pattern, consisting of an O-grid in the near-field grid around each bar and an H-grid in the outer regions. An example of the mesh structure in the x–z plane is shown in Figure 3, which shows the front part of a computational domain with $n = 3$ and the surface of one of the bars in detail. The near-field grid of each bar consisted of a concentric ring shape with an outer diameter of $2s$ starting from the center of each bar. This size was chosen to ensure homogeneity in the near-field grid construction across all simulations, regardless of L_z. In the near-field grid, the resolution in the x–z plane was defined by the number of elements on the bar circumference and in the radial direction, denoted as N_c and N_r, respectively. To resolve the viscous sublayer of the near-wall region, the dimensionless distance from the wall y^+ should be less than one. To achieve this, the height of the first cell adjacent to the bars was set to about 0.00001 m, or about 0.00125*s* in all simulations. The grid resolution was then increased radially based on N_r. In the outer regions, a maximum grid stretch ratio based on the ratio in the near-field grid was used. By applying periodic and free-slip conditions in the spanwise and vertical directions, respectively, no refined grid resolution was required near the outer boundaries of the domains. In contrast to the *x*–*z* plane, the mesh was uniformly structured in the spanwise direction, i.e., the mesh generated in the streamwise and vertical directions was extruded in the spanwise direction with uniform resolution L_y/N_y, where N_y is the number of elements in the spanwise direction. The effects of varying N_c, N_r, and N_y on the results were evaluated in the independence test described in Section 3.1.3. In addition, N_t is reported for all simulations in Section 3.1.

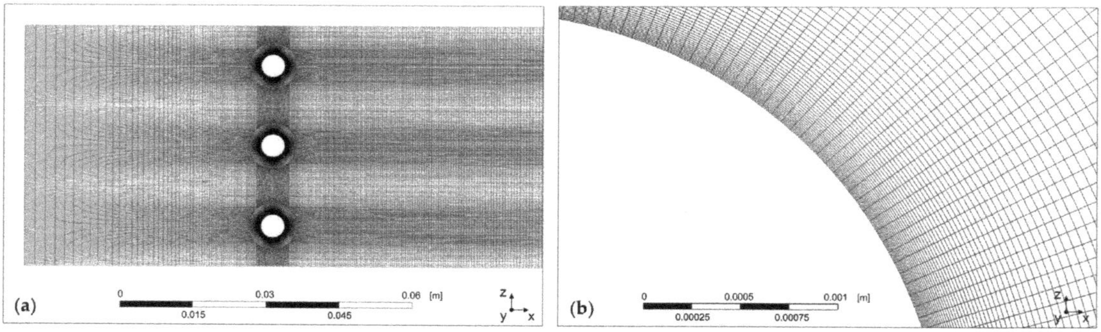

Figure 3. Computational mesh of a simulation consisting of three bars with a clear bar spacing of b = 20 mm and a domain length in the streamwise direction of L_x = 111*s* in the *x*–*z* plane: (**a**) overview of the front part of the domain, and (**b**) close-up view of the surface of one of the bars.

As a second criterion to evaluate the quality of the meshes, the LES index of quality LES_{IQ} was used, which is defined as the ratio of the resolved turbulent kinetic energy k_r to the total turbulent kinetic energy k_t, and is calculated as

$$LES_{IQ} = \frac{k_r}{k_t} = \frac{1}{1 + 0.05\left(\frac{\nu+\nu_{sgs}}{\nu}\right)^{0.53}}, \tag{5}$$

where ν_{sgs} is the SGS eddy viscosity (see Section 2.4). According to Celik et al. [92], LES_{IQ} values in the range of 75% to 85% are appropriate for most engineering applications. Pope [51] recommends a percentage of k_r above 80% for well-resolved LES.

All simulations in this study used a fixed time step Δt. The effect of Δt on the results was evaluated in the independence test described in Section 3.1.4. Furthermore, the total simulation time was defined based on the evaluation time t_e for relevant flow parameters and is therefore explained in Section 2.5.

2.4. Numerical Methods

The 3D numerical simulations in this study use the incompressible form of the Navier–Stokes equations (i.e., ρ is constant). The LES model relies on the filtered form of the continuity and momentum equations, obtained by a filter operation and expressed as

$$\frac{\partial \overline{u}_i}{\partial x_i} = 0, \tag{6}$$

$$\frac{\partial \overline{u}_i}{\partial t} + \frac{\partial \overline{u_i u_j}}{\partial x_j} = -\frac{1}{\rho}\frac{\partial \overline{p}}{\partial x_i} + \nu\frac{\partial^2 \overline{u}_i}{\partial x_i \partial x_j} - \frac{\partial \tau_{ij}}{\partial x_i}, \tag{7}$$

where u_i are the flow velocity components, x_i are the Cartesian coordinates, t is the time, and p is the pressure. In addition, the macron denotes the grid-scale filter operation, and τ_{ij} represents the SGS stress tensor, defined as

$$\tau_{ij} = \overline{u_i u_j} - \overline{u}_i \overline{u}_j. \tag{8}$$

Modeling is required to account for the unknown SGS stresses [51]. The Smagorinsky SGS model [93] was used in this study for this purpose, which calculates the SGS turbulent stresses as

$$\tau_{ij} - \frac{1}{3}\tau_{kk}\delta_{ij} = -2\nu_{sgs}\overline{S}_{ij}, \tag{9}$$

where δ_{ij} is the Kronecker delta symbol, and S_{ij} is the resolved rate of the strain tensor. ν_{sgs} is defined as

$$\nu_{sgs} = \left(C_s\overline{\Delta}\right)^2|\overline{S}|, \tag{10}$$

where C_s is the Smagorinsky coefficient, Δ is the grid size defining the subgrid length scale, and $|S|$ is the characteristic filtered rate of strain. In ANSYS Fluent 19.2, C_s can be defined either by a constant value (hereafter referred to as the Smagorinsky–Lilly model) or dynamically based on data obtained from the resolved motion scales (hereafter referred to as the dynamic Smagorinsky–Lilly model) [94]. Previous studies have shown that the latter is more accurate in predicting flow parameters for flow past a cylinder (e.g., [66,69]). Therefore, the dynamic model was primarily used in this study. Nevertheless, in Section 3.1.6, results using the Smagorinsky–Lilly model with a fixed C_s value of 0.1, which is the default value in ANSYS Fluent 19.2 [94], were compared with those using dynamic C_s values. In addition, simulations using the URANS approach and the hybrid model DES were performed to evaluate their ability to predict ξ and other flow parameters in a manner comparable to LES. The Shear Stress Transport (SST) k-ω model proposed by Menter [95] was used to model turbulence for both the DES model and the URANS approach. Since these two methods are only used for comparison in the present work, the governing equations are not included here, but can be found in the literature [51,94], along with additional information on the LES model.

In all numerical simulations, the Semi-Implicit Method for Pressure-Linked Equations (SIMPLE) algorithm was used for the pressure–velocity coupling, which has previously been used in studies to determine ξ (e.g., [31,57]) and to study flow past a cylinder (e.g., [64,80]). Specifically, in the LES model, Equations (6) and (7) are solved using the SIMPLE algorithm. Furthermore, the LES and DES models used bounded central differencing and bounded second-order implicit schemes for spatial discretization and transient formulation, respectively. In contrast, the URANS simulations used second-order upwind and second-order implicit schemes for spatial discretization and transient formulation, respectively.

The numerical simulations were performed on the LEO HPC infrastructure of the University of Innsbruck with 64 cores each. To avoid rounding errors, all simulations were performed in double precision.

2.5. Post-Processing

After performing the numerical simulations, the results were subjected to a thorough error check, including visual inspection of the velocity and pressure profiles and distributions along with other numerical results. MATLAB R2018b software was used for further analysis of the results, except for the determination of the Strouhal number St, which was performed using ANSYS Fluent 19.2.

The calculation of ξ was determined using the Bernoulli equation with the cross-sectional averaged parameters at the inlet (subscript 1) and outlet (subscript 2) of the domain. Due to the chosen boundary conditions, there were no frictional losses at the domain boundaries and thus no continuous head losses between the cross sections. Consequently, the local head loss h_v is equal to the total head loss, and can be calculated as

$$h_v = z_1 - z_2 + \frac{p_1 - p_2}{\rho\, g} + \frac{\alpha_{kin,1}\, U_1^2 - \alpha_{kin,2}\, U_2^2}{2g} = \xi \frac{U_{ref}^2}{2g}, \tag{11}$$

where z_1 and z_2 are the elevations, g is the gravitational acceleration, and $\alpha_{kin,1}$ and $\alpha_{kin,2}$ are the kinetic energy correction factors. In previous studies (e.g., [2,29,32]), U_{ref} was defined as the undisturbed cross-sectional average flow velocity upstream of the rack, assuming that the velocities upstream and downstream of the rack do not differ significantly. In this study, this approach is adopted due to the defined computational domains and boundary conditions described in Section 2.2, which imply that U_1 is equal to U_{ref} and, except for non-uniformities in the velocity profile, is also equal to U_2. To account for non-uniformities in the respective cross sections, α_{kin} was used. In general, α_{kin} is defined as the ratio of the kinetic energies between the actual velocity profile and the theoretical profile with constant mean flow velocity $\overline{U}$ over the cross-sectional area A [96,97], and is calculated as

$$\alpha_{kin} = \frac{1}{\overline{U}^3\, A} \int_A U^3(A) dA. \tag{12}$$

Typical values of α_{kin} in a pipe range from 1.00 for turbulent velocity distributions with a uniform constant velocity profile to 2.00 for laminar conditions with a parabolic profile [97–99]. Previous studies have generally ignored the effect of α_{kin} on ξ due to its relatively low values close to unity, such as between 1.01 and 1.02 in Tsikata et al. [12]. In the simulations of the present study, $\alpha_{kin,1}$ was exactly unity (excluding rounding errors) due to the ideal homogeneous flow field at the inlet. Conversely, $\alpha_{kin,2}$ ranged between 1.001 and 1.011, as shown in the domain independence test in Section 3.1.2. Therefore, α_{kin} was not included in the calculation of ξ. Furthermore, the inlet and outlet of the water body segments were assumed to be at the same height (i.e., $z_1 = z_2$). For p_1 and p_2, the cross-sectional averaged total pressure at the inlet and outlet were used, respectively. Considering all these factors, Equation (11) can be expressed as

$$\xi = \frac{2(p_1 - p_2)}{\rho\, U_1^2}. \tag{13}$$

In addition to Equation (13), ξ can also be determined using the drag coefficient C_D and P [2,99] by

$$\xi = C_D\, P. \tag{14}$$

C_D is a parameter often studied in the analysis of flow past a cylinder [63,86] and is defined as

$$C_D = \frac{F_D}{0.5 U_1^2\, s\, L_y}, \tag{15}$$

where F_D is the drag force. Another commonly studied flow parameter is St, which is a dimensionless number that characterizes the shedding frequency independent of time. St can be calculated as

$$St = \frac{f_v s}{U_1}, \tag{16}$$

where f_v is the vortex shedding frequency determined by calculating the frequency corresponding to the maximum energy of the spectra obtained by a Fast Fourier Transform (FFT) of the time series of the lift coefficient C_L, which is defined as

$$C_L = \frac{F_L}{0.5 U_1^2\, s\, L_y}, \tag{17}$$

where F_L is the lift force on the cylinder. Both C_D and St were calculated for all simulations in this study to allow comparison with previous research on flow past a cylinder.

In post-processing, the collected data from each time step were averaged using the mean value over a specified period after an initial period, which was necessary to ensure that the flow was statistically quasi-steady and to remove initial transients. In previous studies, these periods were usually defined based on a certain number of vortex shedding cycles $T = 1/f_v$ (e.g., [70,71]). Due to significant variations in St ranging from 0.208 to 0.340, as observed in the simulations of this study (see Section 3), standardized values were used to define the critical periods, with $St = 0.200$ and $U_1 = 0.5$ m/s. The former is a typical value for flow past an unconfined cylinder in the subcritical regime [63] and is lower than the minimum St value obtained in this study. Consequently, the data averaging period was set to a minimum of 100T, resulting in $t_e = 8$ s according to Equation (16), after an initial period of 50T, which is equivalent to 4 s. The total simulation time was 12 s, except for simulations where $U_1 < 0.5$ m/s. In these cases, the total simulation time, including t_e, was extended to maintain a consistent minimum number of T across all simulations. Notably, the use of the same time period for different P values, rather than adjusting the data averaging period to f_v, is consistent with previous experimental studies aimed at determining ξ (e.g., [29,32]). Furthermore, the effect of extending t_e was examined in the independence test described in Section 3.1.4.

Finally, it is important to clarify that the parameters obtained from the numerical calculations presented in the following sections, such as ξ and C_D, represent time–mean values. For values that fluctuate over time, additional subscripts are added to denote them, such as ξ_f and C_{Df}. In addition, Equation (13) was used to calculate ξ unless otherwise noted.

3. Results

3.1. Independence Tests

3.1.1. General

Independence tests are critical when performing time-consuming and computationally expensive numerical simulations to evaluate the effects of selected model settings on the results. For the independence tests in this study, the reference case parameters listed in Table 1 were used, with U_1 held constant at 0.5 m/s (corresponding to $Re_b = 3981$). Unless otherwise specified, $b = 10$ mm ($P = 0.444$) was used. The parameters in Table 1 were then individually adjusted to evaluate their effect on ξ, C_D, St, and other key parameters through a sensitivity analysis. To quantify the differences, the mean relative error (MRE), defined as the ratio of the mean absolute error to the mean reference value, was calculated for ξ and C_D. It should be noted that key parameters such as ξ and C_D are susceptible to time-dependent fluctuations (see Section 3.1.4). In order to assess the influence of these fluctuations on the results, selected simulations with different values of b were repeated. As a result, the time–mean values for both ξ and C_D showed minimal differences, except for a maximum MRE of 1.4% in one simulation. Therefore, 1.4% was used as a threshold for an acceptable range of MRE in the following.

Table 1. Selected values for the parameters of the reference case and the selected reference LES model in the independence tests, independent of the studied clear bar spacings b, where $L_{x,ds}$ and $L_{x,ds}$ are the lengths of the domain downstream and upstream of the studied rack section, respectively, in the streamwise direction, L_y is the length of the domain in the spanwise direction, n is the number of bars in the domain, N_c, N_r, and N_y are the number of elements in the bar circumference, the radial direction of the near-field grid, and the spanwise direction, respectively, t_e is the evaluation time, and Δt is the time step.

$L_{x,us}$ [-]	$L_{x,ds}$ [-]	L_y [-]	n [-]	N_c [-]	N_r [-]	N_y [-]	t_e [s]	Δt [s]	LES Model
10s	100s	4s	1	240	64	80	8	0.00004	Dynamic Smagorinsky–Lilly

3.1.2. Domain Independence

Defining sufficiently large domains is essential to reduce potential influences from both domain size and boundary conditions, which is particularly important for L_x and L_y in this study. Table 2 shows the results of the domain independence test for the determination of ξ, C_D, St, and $\alpha_{kin,2}$. Doubling the reference case value of $L_{x,us}$ resulted in insignificant changes in the results (−0.3% for ξ and −0.4% for C_D), indicating that $L_{x,us}$ = 10s is sufficient to establish homogeneous inflow conditions to the studied rack section. Similarly, $L_{x,ds}$ must be sufficient to provide uniform flow conditions at the outlet without significant recirculation zones. Halving and doubling $L_{x,ds}$ resulted in negligible effects (−0.9% and +0.3%, respectively) on both ξ and C_D. However, when $L_{x,ds}$ was set to 50s, small recirculation zones were observed at the outlet, resulting in a higher value of $\alpha_{kin,2}$ compared to the reference case. Nevertheless, $\alpha_{kin,2}$ = 1.011 is still within an acceptable range as described in Section 2.5. Furthermore, doubling L_y had no significant effect on the results, but halving L_y increased ξ by 1.9% and C_D by 1.7%. These results suggest that the reference case values have a satisfactory degree of domain independence.

Table 2. Results of the domain independence test to determine the drag coefficient C_D, the Strouhal number St, the kinetic energy correction factor at the outlet $\alpha_{kin,2}$, and the head loss coefficient ξ, calculated using Equation (13), where L_v is used as a variable for the length of the domain upstream and downstream of the studied rack section in the streamwise direction $L_{x,us}$ and $L_{x,ds}$, respectively, and in the spanwise direction L_y. In addition, N_t is the total number of elements in the domain.

Parameter	L_v [-]	N_t [-]	ξ (MRE) [-]	C_D (MRE) [-]	St [-]	$\alpha_{kin,2}$ [-]
$L_{x,us}$	10s	4,312,000	1.262 (Ref.)	2.840 (Ref.)	0.296	1.002
	20s	4,408,320	1.258 (−0.3%)	2.830 (−0.4%)	0.296	1.002
$L_{x,ds}$	50s	3,706,560	1.251 (−0.9%)	2.815 (−0.9%)	0.298	1.011
	100s	4,312,000	1.262 (Ref.)	2.840 (Ref.)	0.296	1.002
	200s	4,697,280	1.266 (+0.3%)	2.848 (+0.3%)	0.296	1.001
L_y	2s	2,156,000	1.285 (+1.9%)	2.889 (+1.7%)	0.294	1.007
	4s	4,312,000	1.262 (Ref.)	2.840 (Ref.)	0.296	1.002
	8s	8,624,000	1.257 (−0.3%)	2.833 (−0.3%)	0.296	1.003

3.1.3. Grid Independence

Table 3 shows the results of the grid independence test for the determination of ξ, C_D, and St. In addition, the volume-averaged mean value of LES_{IQ} and the minimum value of LES_{IQ} over all time steps of t_e, denoted as $LES_{IQ,avg}$ and $LES_{IQ,min}$, respectively, are listed in Table 3. Increasing N_c or N_r had minimal effects on the results, while decreasing them caused changes in ξ and C_D equal to or greater than 1.5%. Thus, the values chosen for N_c and N_r in the reference case appear to be appropriate. It should be noted that N_r = 32, 48, 64, and 80 correspond to mesh growth rates of 1.133, 1.074, 1.048, and 1.034, respectively. Furthermore, Table 3 shows that a large number of simulations were performed with different values of N_y because of its importance for the results, as reported in previous studies (e.g., [64,69]). Halving N_y from 80 to 40 resulted in a 1.3% decrease in both ξ and C_D. Conversely, increasing N_y also resulted in decreased values of ξ and C_D, but to different extents. For N_y = 120 and 160, both ξ and C_D decreased by 0.5% and 2.1%, respectively, compared to the reference case. In addition, for N_y = 240, ξ decreased by 0.9% and C_D decreased by 1.0%. These results indicate that ξ and C_D tend to vary depending on the chosen N_y value, but generally remain close to an average value. Thus, the choice of N_y = 80 as the reference value seems to produce satisfactory results while keeping the computational effort at an acceptable level. Moreover, a simulation with N_y = 1 was performed, which basically corresponds to a simplified 2D case, since 3D vortices cannot occur due to the use of only one element in the spanwise direction. The results showed a 22.4% increase in ξ and a 13.7% increase in C_D compared to the reference case, well above the threshold of acceptable values for the MRE. When LES_{IQ} was used as a complementary metric to evaluate mesh quality, a correlation between LES_{IQ} and mesh resolution was observed, with a significant increase in $LES_{IQ,min}$ as N_y increased. However, only a small number of elements exhibited LES_{IQ} values below the 80% threshold recommended by

Pope [51], as detailed in Section 2.3, over the entire t_e. Except for $N_y = 1$, $LES_{IQ,avg}$ remained approximately constant around 0.948 for all simulations. The results suggest that k_r is in the appropriate range for most simulations. Finally, it is noteworthy that y^+ was consistently less than one for all simulations in the grid independence test.

Table 3. Results of the grid independence test to determine the drag coefficient C_D, the volume-averaged mean value of the LES index of quality $LES_{IQ,avg}$, the minimum value of the LES index of quality over all time steps $LES_{IQ,min}$, the Strouhal number St, and the head loss coefficient ξ, calculated using Equation (13), where N_v is used as a variable for the number of elements on the bar circumference N_c, in the radial direction of the near-field grid N_r, and in the spanwise direction N_y. In addition, N_t is the total number of elements in the domain.

Parameter	N_v [-]	N_t [-]	ξ (MRE) [-]	C_D (MRE) [-]	St [-]	$LES_{IQ,avg}$ [-]	$LES_{IQ,min}$ [-]
N_c	160	3,022,400	1.243 (−1.5%)	2.798 (−1.5%)	0.298	0.947	0.780
	240	4,312,000	1.262 (Ref.)	2.840 (Ref.)	0.296	0.948	0.777
	320	5,614,400	1.262 (±0.0%)	2.839 (±0.0%)	0.296	0.948	0.786
N_r	32	1,824,000	1.304 (+3.3%)	2.935 (+3.3%)	0.292	0.946	0.763
	48	3,254,400	1.280 (+1.5%)	2.882 (+1.5%)	0.294	0.947	0.776
	64	4,312,000	1.262 (Ref.)	2.840 (Ref.)	0.296	0.948	0.777
	80	5,369,600	1.256 (−0.5%)	2.825 (−0.5%)	0.292	0.948	0.794
N_y	1	53,900	1.545 (+22.4%)	3.231 (+13.7%)	0.296	0.919	0.687
	40	2,156,000	1.246 (−1.3%)	2.804 (−1.3%)	0.298	0.946	0.753
	80	4,312,000	1.262 (Ref.)	2.840 (Ref.)	0.296	0.948	0.777
	120	6,468,000	1.256 (−0.5%)	2.827 (−0.5%)	0.292	0.949	0.792
	160	8,624,000	1.236 (−2.1%)	2.782 (−2.1%)	0.300	0.949	0.798
	240	12,936,000	1.250 (−0.9%)	2.813 (−1.0%)	0.298	0.950	0.808

3.1.4. Time Independence

In this study, time independence was tested in two different ways. First, to investigate the influence of the time-dependent fluctuations shown in Figure 4a for ξ_f and C_{Df} on the time–mean results, the total simulation time was extended from 12 to 20 s, thus doubling t_e from 8 to 16 s. Figure 4b shows the time–mean normalized head loss coefficient ξ_n as a function of time for the reference case with b = 10 and 90 mm, while indicating t_e = 8 and 16 s. The calculation of time–mean values started at the beginning of t_e after 4 s of simulation, as defined in Section 2.5. In addition, Table 4 shows the results for the determination of ξ, C_D, and St. When t_e was doubled, both ξ and C_D increased slightly by 0.2% for b = 10 mm (P = 0.444). For b = 90 mm (P = 0.082), the MRE was even smaller for both parameters. Overall, the extension of t_e did not significantly affect the results, indicating that the defined time periods are sufficient to obtain time-independent results at an acceptable level. Second, the reference value of Δt was varied. The results in Table 4 show that halving or doubling Δt had a minimal effect on both ξ and C_D. It is worth noting, however, that instability problems occurred during the simulation with Δt = 0.00008 s. Thus, Δt had to be reduced to 0.00004 s for the first 2 s of the simulation before it was increased to 0.00008 s. The transient instability problems may have been caused by the violation of the Courant–Friedrichs–Lewy condition in some elements, which was not observed in the reference case with Δt = 0.00004 s.

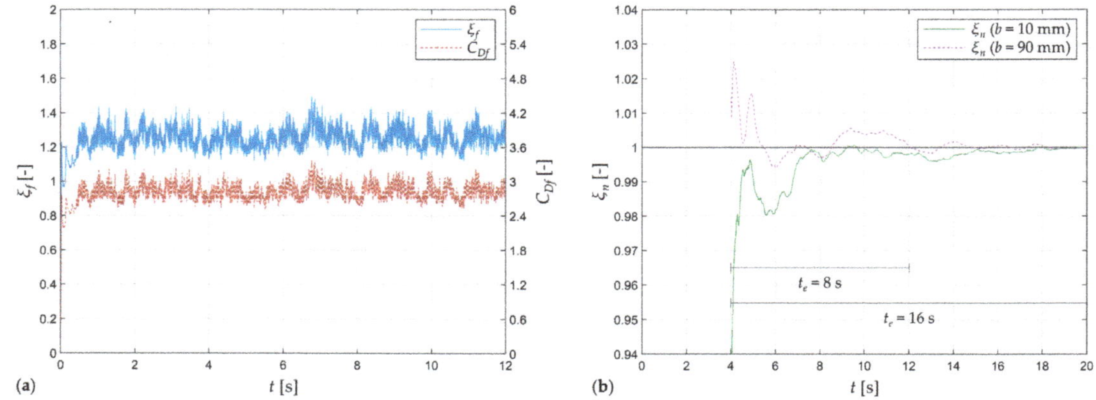

Figure 4. Time histories of (**a**) the fluctuating drag coefficient C_{Df} and the fluctuating head loss coefficient ξ_f for the reference case with clear bar spacing b = 10 mm, and (**b**) the time–mean normalized head loss coefficient ξ_n, starting to compute time–mean values at 4 s, for the reference case values with b = 10 and 90 mm, where t_e are the evaluation times examined as part of the time independence test. ξ_f and ξ_n were calculated using Equation (13).

Table 4. Results of the time independence test to determine the drag coefficient C_D, the Strouhal number St, and the head loss coefficient ξ, calculated using Equation (13), where t_v is used as a variable for the evaluation time t_e and the time step Δt. In addition, b is the clear bar spacing, N_t is the total number of elements in the domain, and P is the blocking ratio.

Parameter	t_v [s]	b [mm]	P [-]	N_t [-]	ξ (MRE) [-]	C_D (MRE) [-]	St [-]
t_e	8 16	10	0.444	4,312,000	1.262 (Ref.) 1.264 (+0.2%)	2.840 (Ref.) 2.845 (+0.2%)	0.296 0.296
	8 16	90	0.082	7,369,600	0.085 (Ref.) 0.085 (±0.0%)	1.026 (Ref.) 1.025 (±0.0%)	0.212 0.215
Δt	0.00002 0.00004 0.00008	10	0.444	4,312,000	1.270 (+0.6%) 1.262 (Ref.) 1.267 (+0.4%)	2.858 (+0.6%) 2.840 (Ref.) 2.850 (+0.4%)	0.300 0.296 0.296

3.1.5. Independence of the Number of Bars in the Computational Domain

Table 5 shows the results of the numerical simulations with b = 10, 20, and 30 mm and n = 3 and 5 for the determination of ξ, C_D, and St compared to those with the corresponding b value and the reference value n = 1. The findings indicate significant single digit percentage differences when n was greater than one. The maximum deviation for both ξ and C_D was 6.0% for b = 20 mm (P = 0.286) and n = 5. As noted in Section 2.1, an increase in n resulted in a corresponding increase in N_t. For example, N_t reached 29,008,000 for b = 30 mm (P = 0.211) and n = 5, as shown in Table 5. Due to the increased N_t values, substantial computational resources were required despite the use of an HPC cluster. Therefore, to facilitate a simplified assessment of the effect of n on the results, additional simulations were performed using periodic boundary conditions at the top and bottom of the domains with n = 1, while b ranged from 10 to 90 mm in 10 mm increments (i.e., P values between 0.082 and 0.444 were examined). The results presented in Table 5 show that the values of ξ and C_D tended to approach the values obtained under free-slip wall conditions as b increased. For b = 10 mm (P = 0.444), the MRE for both ξ and C_D was 15.9%, but decreased to 1.2% for b = 60 mm (P = 0.118), and continued to decrease with increasing b, except for the outlier at b = 90 mm (P = 0.082), which was still within the defined acceptable range of the MRE. Thus, the independence of n was only observed for larger b values. Nevertheless, the feasibility of using the results with $n = 1$ to determine ξ of CBTRs is further discussed in Sections 4.1 and 4.2.

Table 5. Results of the independence test of the number of bars in the computational domain by varying the number of bars n and using different top and bottom boundary conditions to determine the drag coefficient C_D, the Strouhal number St, and the head loss coefficient ξ, calculated using Equation (13), where b is the clear bar spacing, N_t is the total number of elements in the domain, and P is the blocking ratio.

n [-]	Top and Bottom Boundary Condition	b [mm]	P [-]	N_t [-]	ξ (MRE) [-]	C_D (MRE) [-]	St [-]
1	Free-slip wall	10	0.444	4,312,000	1.262 (Ref.)	2.840 (Ref.)	0.296
3	Free-slip wall			12,936,000	1.242 (−1.6%)	2.794 (−1.6%)	0.300
5	Free-slip wall			21,560,000	1.209 (−4.2%)	2.719 (−4.3%)	0.302
1	Periodic			4,312,000	1.061 (−15.9%)	2.390 (−15.9%)	0.340
1	Free-slip wall	20	0.286	5,174,400	0.474 (Ref.)	1.659 (Ref.)	0.244
3	Free-slip wall			15,523,200	0.459 (−3.0%)	1.609 (−3.0%)	0.252
5	Free-slip wall			25,872,000	0.445 (−6.0%)	1.559 (−6.0%)	0.256
1	Periodic			5,174,400	0.409 (−13.6%)	1.434 (−13.6%)	0.276
1	Free-slip wall	30	0.211	5,801,600	0.266 (Ref.)	1.262 (Ref.)	0.240
3	Free-slip wall			17,404,800	0.261 (−1.6%)	1.243 (−1.5%)	0.242
5	Free-slip wall			29,008,000	0.260 (−2.1%)	1.237 (−2.0%)	0.244
1	Periodic			5,801,600	0.256 (−3.8%)	1.216 (−3.7%)	0.246
1	Free-slip wall	40	0.167	6,193,600	0.191 (Ref.)	1.149 (Ref.)	0.230
	Periodic				0.188 (−1.6%)	1.131 (−1.6%)	0.232
1	Free-slip wall	50	0.138	6,507,200	0.150 (Ref.)	1.084 (Ref.)	0.226
	Periodic				0.148 (−1.2%)	1.071 (−1.2%)	0.224
1	Free-slip wall	60	0.118	6,820,800	0.127 (Ref.)	1.068 (Ref.)	0.224
	Periodic				0.125 (−1.2%)	1.057 (−1.1%)	0.222
1	Free-slip wall	70	0.103	7,056,000	0.109 (Ref.)	1.051 (Ref.)	0.216
	Periodic				0.108 (−0.9%)	1.041 (−0.9%)	0.218
1	Free-slip wall	80	0.091	7,212,800	0.095 (Ref.)	1.027 (Ref.)	0.216
	Periodic				0.094 (−0.3%)	1.026 (−0.3%)	0.218
1	Free-slip wall	90	0.082	7,369,600	0.085 (Ref.)	1.026 (Ref.)	0.212
	Periodic				0.084 (−0.8%)	1.018 (−0.8%)	0.216

3.1.6. Method Independence

The results of the comparison with the DES model, the LES Smagorinsky–Lilly model, and the URANS approach are shown in Table 6 for b values of 10 and 90 mm. For $b = 10$ mm ($P = 0.444$), the LES Smagorinsky–Lilly model and the URANS approach showed slight differences in ξ and C_D compared to the reference case, with a maximum MRE of −0.8%. Conversely, the DES model produced larger MREs of −4.3%. For $b = 90$ mm ($P = 0.082$), the LES (both Smagorinsky–Lilly and dynamic Smagorinsky–Lilly) and DES models yielded comparable results for ξ and C_D with minor differences. However, the URANS simulation produced values that were 23.6% higher for ξ and 23.3% higher for C_D compared to the reference case. The results for $b = 90$ mm ($P = 0.082$) were then used to validate the method, which gave results similar to the literature values from previous experimental and numerical studies of flow past a circular cylinder at $Re_b = 3900$ [64,65,68–72,74,85,89]. The comparison showed that all C_D values were within the literature range of 0.97 to 1.04, except for the value of 1.265 obtained with the URANS approach. In addition, the St values for all simulations were within the literature range of 0.203 to 0.218. Consequently, the results indicate that the reference model with $b = 90$ mm ($P = 0.082$) produced comparable values to the literature data for both C_D and St, thereby validating the use of the selected reference case values for simulating flow past a cylinder.

Table 6. Results of the comparison of the DES model, the LES Smagorinsky–Lilly and dynamic Smagorinsky–Lilly models, and the URANS approach to determine the drag coefficient C_D, the Strouhal number St, and the head loss coefficient ξ, calculated using Equation (13), where b is the clear bar spacing, N_t is the total number of elements in the domain, and P is the blocking ratio.

Method	b [mm]	P [-]	N_t [-]	ξ (MRE) [-]	C_D (MRE) [-]	St [-]
DES	10	0.444	4,312,000	1.208 (−4.3%)	2.717 (−4.3%)	0.298
LES (Dynamic Smagorinsky–Lilly)				1.262 (Ref.)	2.840 (Ref.)	0.296
LES (Smagorinsky–Lilly)				1.253 (−0.7%)	2.819 (−0.7%)	0.298
URANS				1.252 (−0.8%)	2.817 (−0.8%)	0.296
DES	90	0.082	7,369,600	0.086 (+0.8%)	1.033 (+0.6%)	0.216
LES (Dynamic Smagorinsky–Lilly)				0.085 (Ref.)	1.026 (Ref.)	0.212
LES (Smagorinsky–Lilly)				0.085 (+0.2%)	1.028 (+0.2%)	0.214
URANS				0.105 (+23.6%)	1.265 (+23.3%)	0.212

3.2. Effect of Clear Bar Spacing or Blocking Ratio

Figure 5 and Table 7 show the effects of varying b or P on ξ and C_D. Increasing b led to a decrease in ξ (Figure 5a), while increasing P led to an increase in ξ (Figure 5b). Similarly, C_D decreased with increasing b or decreasing P, but for b values greater than or equal to 80 mm ($P \leq 0.091$), C_D remained in the range of 0.998 to 1.027. Notably, C_D was close to 1.000 for b = 150 mm (P = 0.051) and b = 180 mm (P = 0.043). The St values in Table 7 show a similar pattern to C_D, reaching a nearly constant value at high b values or low P values. The results suggest that C_D and St are largely unaffected beyond certain b or P values, while ξ shows a consistent tendency to either increase or decrease. For comparison, Figure 5 and Table 7 show additional ξ values calculated using Equations (1), (2), and (3) proposed by Kirschmer [24], Meusburger [2], and Böttcher et al. [29], respectively. In Equations (1) and (2), k_F was defined as 1.79, as proposed by Kirschmer [24] for circular bar shapes. Furthermore, due to the perpendicular inflow conditions, α and β were both set to 90°, and δ was set to 0°. Since the study did not examine rack clogging, k_V was set to 1. The comparison showed minimal variation in ξ. Specifically, the ξ values calculated by LES for low b or high P were generally lower than those obtained by the empirical formulas. Conversely, the empirical formulas resulted in lower ξ values for high b or low P values. To quantify the differences, the root mean square relative errors (RMSREs) for ξ were calculated and were 18.5%, 39.1%, and 15.3% using Equation (13) compared to Equations (1), (2), and (3), respectively. In addition to Equation (13), ξ was also calculated based on C_D using Equation (14), with the results presented in Table 7. A high degree of agreement was obtained with an RMSRE of 1.3% using Equations (13) and (14), with the most notable deviations found at relatively high b or low P values.

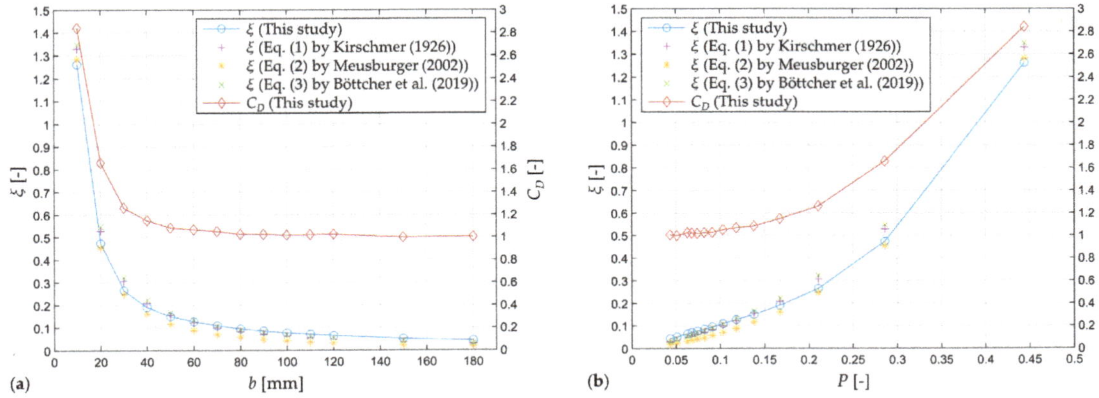

Figure 5. The drag coefficient C_D and the head loss coefficient ξ, calculated using Equation (13), as functions of (**a**) the clear bar spacing b and (**b**) the blocking ratio P, compared to ξ, calculated using Equations (1), (2), and (3) proposed by Kirschmer [24], Meusburger [2], and Böttcher et al. [29], respectively.

Table 7. Results of varying the clear bar spacing b or the blocking ratio P to determine the drag coefficient C_D, the Strouhal number St, and the head loss coefficient ξ, calculated using Equations (1), (2), and (3) proposed by Kirschmer [24], Meusburger [2], and Böttcher et al. [29], respectively, and Equations (13) and (14) based on the results of the numerical simulations.

b [mm]	P [-]	ξ [-] Equation (1)	Equation (2)	Equation (3)	Equation (13)	Equation (14)	C_D [-]	St [-]
10	0.444	1.329	1.281	1.347	1.262	1.262	2.840	0.296
20	0.286	0.528	0.453	0.547	0.474	0.474	1.659	0.244
30	0.211	0.307	0.246	0.323	0.266	0.266	1.262	0.240
40	0.167	0.209	0.160	0.222	0.191	0.191	1.149	0.230
50	0.138	0.155	0.115	0.166	0.150	0.150	1.084	0.226
60	0.118	0.122	0.087	0.131	0.127	0.126	1.068	0.224
70	0.103	0.099	0.069	0.107	0.109	0.108	1.051	0.216
80	0.091	0.083	0.057	0.090	0.095	0.093	1.027	0.216
90	0.082	0.071	0.047	0.077	0.085	0.084	1.026	0.212
100	0.074	0.062	0.041	0.067	0.077	0.075	1.018	0.214
110	0.068	0.054	0.035	0.060	0.071	0.069	1.021	0.212
120	0.063	0.048	0.031	0.053	0.065	0.064	1.025	0.212
150	0.051	0.036	0.022	0.040	0.052	0.051	0.998	0.212
180	0.043	0.028	0.017	0.031	0.044	0.043	1.005	0.212

3.3. *Effect of Flow Velocity*

In the previous sections, U_1 was fixed at 0.5 m/s, resulting in $Re_b = 3981$. Here, U_1 was varied to examine its effect on the results. Figures 6a and 6b show the results for ξ and C_D, respectively. In addition, the results for the case of $b = 90$ mm are listed in Table 8, while the results for $b = 10$, 20, and 30 mm are given in Appendix A. The results show a velocity-dependent behavior within the velocity ranges tested. Reducing U_1 to 0.3 and 0.4 m/s for $b = 10$, 20, and 90 mm ($P = 0.444$, 0.286, and 0.082, respectively) led to a decrease in both ξ and C_D compared to $U_1 = 0.5$ m/s, which was not observed for $b = 30$ mm ($P = 0.211$). For all b values tested when $U_1 \geq 0.5$ m/s, ξ and C_D generally increased. For example, doubling U_1 from 0.5 to 1.0 m/s for $b = 90$ mm ($P = 0.082$) resulted in a 15.8% increase in ξ and a 15.7% increase in C_D. For comparison, Figure 6a also shows the ξ values calculated using Equation (1) proposed by Kirschmer [24], which does not take flow velocity into account, similar to Equation (2) by Meusburger [2] and (3) by Böttcher et al. [29]. Thus, ξ

remained constant as U_1 was varied. Consistent with Section 3.2, the use of Equation (14) to determine ξ showed excellent agreement with an RMSRE of 0.7% compared to Equation (13) for all b and U_1 values tested. In contrast to ξ and C_D, St mainly decreased or remained nearly constant with increasing U_1.

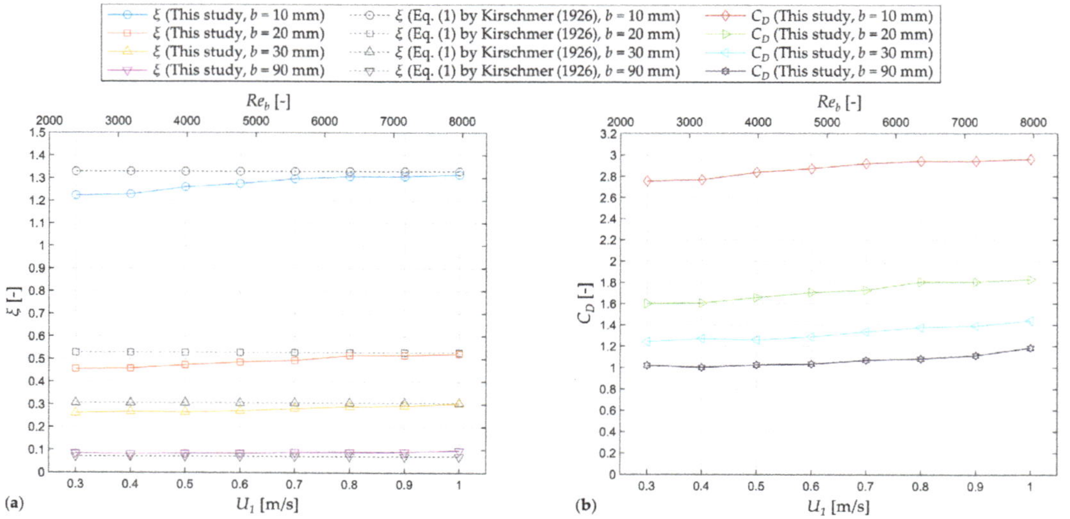

Figure 6. (**a**) The drag coefficient C_D and (**b**) the head loss coefficient ξ, calculated using Equation (13), as functions of the flow velocity at the inlet U_1 and the bar Reynolds number Re_b, compared to ξ calculated using Equation (1) proposed by Kirschmer [24].

Table 8. Results of varying the flow velocity at the inlet U_1 or the bar Reynolds number Re_b for the clear bar spacing b = 90 mm or the blocking ratio P = 0.082 to determine the drag coefficient C_D, the Strouhal number St, and the head loss coefficient ξ, calculated using Equation (1) proposed by Kirschmer [24] and Equations (13) and (14) based on the results of the numerical simulations.

U_1 [m/s]	Re_b [-]	b [mm]	P [-]	ξ (MRE) [-] Equation (1)	Equation (13)	Equation (14)	C_D (MRE) [-]	St [-]
0.3	2,389	90	0.082	0.071 (±0.0%)	0.085 (−0.5%)	0.083 (−0.6%)	1.020 (−0.6%)	0.216
0.4	3,185			0.071 (±0.0%)	0.083 (−2.1%)	0.082 (−2.2%)	1.003 (−2.2%)	0.216
0.5	3,981			0.071 (Ref.)	0.085 (Ref.)	0.084 (Ref.)	1.026 (Ref.)	0.212
0.6	4,777			0.071 (±0.0%)	0.086 (+0.8%)	0.084 (+0.8%)	1.034 (+0.8%)	0.212
0.7	5,573			0.071 (±0.0%)	0.089 (+4.4%)	0.087 (+4.3%)	1.070 (+4.3%)	0.210
0.8	6,369			0.071 (±0.0%)	0.090 (+5.8%)	0.089 (+5.7%)	1.084 (+5.7%)	0.211
0.9	7,166			0.071 (±0.0%)	0.092 (+8.6%)	0.091 (+8.6%)	1.115 (+8.6%)	0.208
1.0	7,962			0.071 (±0.0%)	0.098 (+15.8%)	0.097 (+15.7%)	1.187 (+15.7%)	0.208

4. Discussion

4.1. Interpretation of the Results in Relation to Previous Studies of Flow Past Circular Cylinders

The availability of a considerable amount of previous research on flow past circular cylinders with comparable Re_b was a major advantage of the present study. Thus, several settings and model parameters were adopted in this study, including the definition of $L_{x,us}$ as 10s (similar to, e.g., [85,91]), L_y as 4s (similar to, e.g., [75,100]), N_c as 240, N_r as 64 (both similar to, e.g., [64]), and N_y as 80 (similar resolution to, e.g., [89]). In contrast, a value of 100s was used for $L_{x,ds}$, which is significantly higher than in previous studies, because in this study, ξ was calculated based on p_1 and p_2 at the inlet and outlet, respectively, using

Equation (13), and to keep $\alpha_{kin,2}$ as close to unity as possible. In Section 3.1, independence tests confirmed the suitability of the selected reference case values. Furthermore, a simplified 2D scenario was investigated by defining $N_y = 1$. The results were comparable to those of Khan et al. [64], as the C_D value was overestimated. This is in accordance with the findings of Beaudan and Moin [66] that 3D effects occur at $Re_b = 3900$, such as pairs of counter-rotating streamwise vortices, which cannot be represented in 2D simulations. In addition, Lu et al. [100] observed both streamwise and spanwise vortices in the wake of the cylinder. It should also be noted that the ANSYS Fluent User's Guide [101] recommends the use of 3D numerical simulations for LES. As shown in Figure 4a and comparable to previous research (e.g., [72,100]), unsteady flow conditions in the wake of the cylinder resulted in time-dependent fluctuations of relevant flow parameters such as ξ_f and C_{Df}. Therefore, the calculation of time–mean (or time-averaged) values over an appropriate time period is crucial. Interestingly, increased turbulence in the same region has led to the measurement of fluctuating flow parameters downstream of CBTRs in model experiments, as reported by Böttcher et al. [29].

The use of multiple bars or periodic top and bottom boundary conditions produced divergent results compared to the single-bar reference case at low b or high P values, but this effect diminished with increasing b or decreasing P. According to Afgan et al. [85], two side-by-side cylinders at $Re_b = 3000$ and $P \leq 0.500$ behave independently, exhibiting symmetric wake patterns and antiphase vortex shedding. However, the interaction of the Kármán vortex streets in the wake of the cylinders can still affect C_D and St [83]. As P continues to decrease, the results of two side-by-side cylinders tend to approach those of a single cylinder, with no perturbation observed at P values less than 0.200 to 0.250 [84]. The results listed in Table 7 support this trend, although an MRE below the predefined threshold was only achieved for $b \geq 50$ mm or $P \leq 0.138$. Nevertheless, this is strongly related to the threshold definition and other factors such as the definition of $L_{x,ds}$. Significant differences were found when comparing the results of the DES model, the Smagorinsky–Lilly and dynamic Smagorinsky–Lilly LES models, and the URANS approach in Section 3.1.6, especially for $b = 90$ mm ($P = 0.082$). Thus, it can be concluded that the URANS approach is not suitable to accurately predict C_D, while the DES and LES models give comparable results, as stated in Section 1. However, for $b = 10$ mm ($P = 0.444$), the C_D and St values calculated using the URANS approach were close to those of the reference case, suggesting that this conclusion may not be valid for low b or high P values and highlighting the need for further research to address this issue.

As shown in Section 3.2, a decrease in b or an increase in P resulted in an increase in C_D and St values when $P \geq 0.103$, which is consistent with previous research on flow past confined cylinders [86–88]. Conversely, C_D remained relatively constant when $P \leq 0.091$, except for the cases when b was 150 and 180 mm ($P = 0.051$ and 0.043, respectively), where slightly lower C_D values were obtained. This may align with the recommendation of Nguyen et al. [86] that P should be limited to less than 0.05 for confined cylinders with no-slip walls to minimize wall effects on the results. In Section 3.3, a velocity-dependent behavior of C_D and St was shown. Specifically, at lower velocities between 0.3 and 0.5 m/s, C_D remained almost constant, but increased from 0.5 m/s (corresponding to $Re_b = 3981$) to 1.0 m/s (corresponding to $Re_b = 7962$) for all b values tested. Similar results have previously been found in this Re_b range for flow past a circular cylinder [102,103]. Typically, as Re_b continues to increase, C_D maintains a relatively constant value of about 1.2 [102] until it reaches the so-called drag crisis at $Re_b \approx 200{,}000$ to 300,000, where C_D drops abruptly [104–106]. In addition, increasing Re_b in this study generally led to a decrease in St, which has also been observed in previous studies [107]. This behavior can be attributed to the transition from laminar to turbulent flow, which shifts from the wake region to the point of flow separation at the cylinder surface with increasing Re_b [67].

4.2. Interpretation of the Results in Relation to Previous Studies for Determining Head Loss Coefficients of Trash Racks

Consistent with previous experimental studies (e.g., [2,24,29]), the results in Section 3.2 showed a progressive decrease in ξ with increasing b or decreasing P. The ξ values obtained from the numerical simulations were generally in good agreement with those obtained using the empirical formulas, with RMSREs ranging from 15.3% (with respect to Equation (3)) to 39.1% (with respect to Equation (2)). For comparison, Albayrak et al. [46] reported a mean prediction error of 16% when comparing their proposed prediction equation for ξ with the corresponding measurement results. The differences observed in this study can be attributed, on the one hand, to the P values between 0.134 and 0.535 [24], 0.190 and 0.550 [2], and 0.250 and 0.500 (without transversal elements) [29] tested in the studies from which the Equations (1), (2), and (3), respectively, were derived. Accordingly, these empirical formulas were not intended for low P values, such as those between 0.043 and 0.118 (b between 60 and 180 mm) examined in this study. Based on the good agreement of the C_D and St results with previous research on flow past a cylinder, as discussed in Section 4.1, this may imply that the numerically obtained ξ values are more accurate for relatively low P ranges. Nevertheless, further research is needed to confirm this. On the other hand, in the experiments with CBTRs by Böttcher et al. [29], flow-induced vibrations of the bars were observed, especially at high P values. Although additional spacers were used, it could not be excluded that ξ was affected [29]. Typically, circular bars are highly susceptible to flow-induced vibrations [108]. However, the rigid and stationary bars used in this study did not allow for any effects of flow-induced vibrations on ξ. This may explain why the numerically obtained ξ values at low b or high P values are lower than those obtained using the empirical formulas. Overall, the relative differences in ξ were higher for high b or low P values.

The use of three or five bars in Section 3.1.5 resulted in a decrease in ξ compared to the use of a single bar in the computational domain. While the results appear to be significantly different, with MREs up to 6.0% for ξ, the deviations are in a comparable range to those obtained numerically and using the empirical formulas. For example, for $b = 10$ mm ($P = 0.444$) and $n = 1$, a deviation of 6.7% was observed compared to the values obtained using Equation (3). Thus, the use of $n = 1$ seems to be appropriate for an initial estimation of ξ. However, larger differences were observed when periodic top and bottom boundary conditions were applied, especially for $b = 10$ mm, which resulted in an MRE of 15.9% for ξ. Increasing n led to further reductions in ξ, but only for relatively low b or high P values. The results in Section 3.1.5 indicate a correlation between n and ξ that appears to decrease as b increases or P decreases, which is in line with the behavior of n and C_D discussed in Section 4.1. Implementing periodic boundary conditions can generally reduce the influence of the walls on the wake behind the cylinders, but it may not fully account for potential interactions between individual bars compared to using multiple bars. A similar approach was used by Åkerstedt et al. [49], where a single-bar domain with periodic boundary conditions in the symmetry plane between the bars was able to produce comparable ξ values to domains consisting of more than 20 bars. Whether this finding holds true for the approach presented in this study requires further investigation.

Many commonly used empirical formulas for determining ξ do not account for the effect of U_1 or Re_b on the results [33], including Equation (3) by Böttcher et al. [29]. Through experimental studies with CBTRs, Böttcher et al. [29] found that ξ is largely unaffected by U_1 when $Re_b \geq 750$. However, their investigation only covered Re_b values between 750 and 3000. In this study, the Re_b values tested ranged from 2389 to 7962. No clear trend for ξ was found for $Re_b \leq 3185$ (corresponding to $U_1 \leq 0.4$ m/s), which is in agreement with the results of Böttcher et al. [29]. Conversely, a velocity-dependent behavior was observed for $Re_b \geq 3981$ (corresponding to $U_1 \geq 0.5$ m/s), as described in Section 3.3. Specifically, as U_1 increased from 0.5 to 1.0 m/s (corresponding to Re_b from 3981 to 7962), ξ increased by 15.8% for $b = 90$ mm ($P = 0.444$). As discussed in Section 4.1, these increases are consistent with the velocity-dependent behavior of C_D in previous studies of flow past a circular cylinder. It is important to note that this behavior is not universally applicable to non-circular bar shapes. Bar shapes with sharp edges, such as rectangular, square, or triangular shapes, typically experience flow separation at the fixed sharp edges due to the abrupt change in geometry, making the flow characteristics relatively insensitive to Re_b [107,109,110]. Circular shapes, on the other hand, exhibit a back-and-forth oscillation of the flow separation point on the cylinder surface [109] and have therefore received more attention in the past [63,107]. In terms of head losses, rectangular bar shapes generally give higher ξ values, while rounded edges give considerably lower ξ values [15,24,32,45]. Taking into account previous findings on flow past non-circular cylinders (e.g., [110–112]), future research should include numerical simulations using non-circular bar shapes. This may lead to the development of customized formulas for different bar shapes, similar to Meister et al. [32], who provided two ξ prediction formulas, one for rectangular bar shapes and another for streamlined bar shapes.

The numerical simulations in this study focused on submerged bars, thus ignoring the effect of the free surface on the results. In contrast, most previous experimental studies used open-channel flow conditions (e.g., [24,29]). Meusburger [2] analyzed ξ in both open-channel flow and closed-pipe flow under laboratory conditions and found that the ξ values for both types of flow are comparable, with insignificant deviations within the range of measurement accuracy. In addition, Clark et al. [45] performed experiments in pressurized rectangular conduits and compared the results with ξ values of established empirical formulas developed based on open-channel flow conditions. Their results indicate that the formulas are also applicable to submerged trash racks. Therefore, it can be assumed that the results of this study are also valid for CBTRs in open-channel flow conditions. However, since neither Meusburger [2] nor Clark et al. [45] studied the circular bar shape, future research should investigate CBTRs in open-channel flow conditions, taking into account existing studies on flow past a circular cylinder near a free surface (e.g., [113]).

Although this study was only superficially concerned with flow behavior, the results showed pressure curves in the streamwise direction analogous to those of Clark et al. [45], indicating a constant pressure level upstream and mostly uniform flow conditions far downstream of the rack section. For a more comprehensive understanding of the flow behavior near CBTRs, the extensive literature on flow past cylinders can be a helpful reference, where the flow behavior has been studied in detail (e.g., [67,69,100]).

4.3. Further Limitations and Potential Future Research Topics

Cost-effective numerical analysis of complex problems requires the setting of assumptions and constraints. At the same time, simulations and their results can be significantly affected by various factors such as input data, model parameters, SGS, computational domain, and grid resolution [114]. For instance, the findings of this study suggest that the use of a more sophisticated model than the URANS approach in 3D numerical simulations would be beneficial to the problem at hand. In this context, it is important to note that the numerical simulations for the method comparison in Section 3.1.6 used the same model settings as the reference model in terms of domain, grid, and time issues, without performing additional independence studies. Consequently, there were also no significant differences in computational time between the simulations. Overall, all simulations required a considerable amount of time to complete. The reference case specified in Section 3.1 with b = 10 mm (P = 0.444) took about twelve days, while the simulation with the longest computation time, which also included the largest N_t (i.e., the case described in Section 3.1.5 with b = 30 mm or P = 0.211 and n = 5), took about 100 days. The computational cost generally increased with increasing b or decreasing P, as well as with increasing U_1. Regarding the variations in U_1 described in Section 3.3, it is noteworthy that U_1 has a direct correlation with several numerical parameters, including y^+, which increased almost linearly with U_1. As a result, y^+ values of one or even higher were observed for a small number of elements in the computational domains, specifically for b = 10 mm (P = 0.444) at $U_1 \geq 0.7$ m/s, b = 20 mm (P = 0.286) at $U_1 \geq 0.9$ m/s, and b = 30 and 90 mm (P = 0.211 and

0.082, respectively) at U_1 = 1.0 m/s. However, since the number of elements with $y^+ \geq 1$ was very small, it can be assumed that the effect on the results is negligible. In addition to U_1, the friction coefficients also affect the y^+ values. Therefore, the friction coefficients for the bars surfaces were kept constant in all simulations by using the default friction value in ANSYS Fluent 19.2 [101].

Previous experimental studies have analyzed the effect of spacers and other support structures on ξ (e.g., [2,115]), which were not considered in the present study. Similarly, this study did not examine the effects of clogging (e.g., [2,9]) or bottom and top overlays (e.g., [15,32]), nor did it consider bars with non-perpendicular inflow conditions. Such bars are important for fish guidance racks, which typically consist of bars angled to either the side wall or the ground plane to guide downstream migrating fish to an appropriate bypass [9,10,13,14,18]. Furthermore, many empirical equations are based on undisturbed, uniform inflow conditions [116]. However, inflow conditions to trash racks at HPPs are much more complex [117,118] and also depend on the design of the HPP [2,32]. This highlights the importance of future studies investigating cylinders angled with respect to the approach flow direction. Previous research, especially on flow past inclined or yawed cylinders (e.g., [119–121]), can serve as a basis. Beyond that, the implementation of rack configurations that deflect the flow, including angled bar racks, louvers [25,46,122], and curved bar racks (CBRs) [30,123], may present challenges in the current approach and will require additional research in the future.

Few experimental studies have investigated scale effects in the context of trash racks, resulting in only anecdotal findings (e.g., [15,46]). In general, achieving satisfactory similarity between physical scale models and their real-world counterparts can be challenging [124]. Therefore, scale effects represent another potential area for further research. This includes the effect of s on the results, which in this study was defined as a constant of 8 mm to reduce computational cost and achieve a Re_b value close to 3900. According to Kirschmer [24], the effect of s on the results is negligible for bars under frontal inflow conditions if the ratio of s to b remains constant. Thus, the use of 8 mm seems appropriate for this study, but its applicability should be re-evaluated in possible future studies.

4.4. Transferability to Applications in Research and Engineering Practice

Besides the limitations described in the previous sections, the presented approach also offers advantages over experimental studies. In physical model tests under open-channel flow conditions, trash racks result in small head losses relative to water depth, in some cases, comparable to the measurement accuracy of the instruments used. Water level fluctuations during high discharges further complicate measurements [29]. Thus, the determination of head losses can be difficult. Conversely, under pressurized pipe flow conditions, piezometers are generally used to measure the pressure differences upstream and downstream of the rack, as performed by Clark et al. [45], thereby determining head losses or ξ in a manner comparable to this study. This also allows the neglect of U_2 in the calculation of ξ, which can have a significant effect on the results in open-channel experiments [33]. However, for the determination of ξ in experimental studies, it is critical to account for continuous losses due to friction and local losses due to spacers or support structures. Lučin et al. [31] stated that experimental studies using open-channel flow conditions tend to overestimate ξ due to shallow water depths compared to the actual geometries at HPPs, resulting in a more pronounced influence of bottom flow resistance on ξ. Furthermore, especially in scenarios with considerable lateral flow deflections, the effect of restricted width in physical model tests on ξ is worth investigating. The critical question is whether the use of periodic boundary conditions in numerical simulations provides a more accurate representation of reality than the restricted width in experimental studies.

The combination of physical model tests and numerical simulations can offer notable advantages, including increased availability of hydraulic parameter data and reduced cost and time, compared to relying solely on physical model tests. For instance, Leuch et al. [50] validated their numerical simulations using results from previous experimental model tests [30,122] and numerically investigated a more streamlined bar shape for a CBR. Similarly, future research can determine ξ values for new or modified bar shapes, such as bars with electrodes attached to the side or front for upgrading existing racks to hybrid barriers [125]. In terms of numerical analysis, this study showed that the implementation of trash racks in 3D numerical models of HPPs is only feasible in a simplified manner due to the requirement of very fine grid discretization and hence significant computational effort when using the LES model. Alternatively, practical solutions involve the use of baffles or porous media with ξ values as input data [118,126–128]. In general, the most appropriate approach depends on the level of detail required for the specific problem.

While authorities in some countries require low b values for physical barriers to improve fish protection, such as several German federal states with $b \leq 20$ mm [129], higher b values may become decisive when hybrid or mechanical barriers are used. However, commonly used empirical formulas for the calculation of ξ are mainly based on a limited number of investigated parameters, as discussed in Section 4.2 for b or P. Consequently, the formulas are not universally applicable to all rack configurations. The findings of this study provide a better understanding of ξ for relatively high b values. Nevertheless, further research is needed to propose new prediction formulas for ξ or to adjust existing empirical formulas to account for high b values. Equation (14) provides a simplified approach for determining ξ in cases with high b or low P values, as indicated by the low RMSREs for ξ using Equations (13) and (14) in this study. Considering that C_D remains relatively constant in certain P ranges, such as $P \leq 0.091$ for circular bar shapes at $U_1 = 0.5$ m/s, as described in Section 3.2, ξ can be estimated by referring to established C_D values from the existing literature and using the P value of the corresponding trash rack.

Finally, this study focused on trash racks located upstream of HPPs. However, racks or similar structures are used in various engineering settings, such as wastewater treatment plants and pumping stations [33]. The results of this study may be also applicable to other types of racks. In addition, the results of the independence tests presented in Section 3.1 provide a valuable resource for future studies of flow past (confined) cylinders, particularly when free-slip wall conditions are applied at the top and bottom of the computational domain.

5. Conclusions

The present study used 3D numerical simulations to analyze the head loss coefficients ξ of circular bar trash racks (CBTRs), mainly including a single bar with a bar thickness s of 8 mm in the computational domain and applying the Large Eddy Simulation (LES) model. In detail, bars oriented perpendicular to the flow direction were studied under homogeneous inflow conditions. Furthermore, the effects of the clear bar spacing b and the flow velocity at the inlet of the computational domain U_1 on the results were investigated. The model settings for the simulations were established through an extensive analysis of the available literature on flow past circular cylinders, specifically at a Reynolds bar number Re_b of 3900, in addition to preliminary studies. These settings were then validated by systematic independence tests. Based on the results and their interpretation in relation to previous studies of flow past cylinders and experimental studies to determine ξ of trash racks, the following key findings were obtained:

- Well-resolved simulations using the LES model were able to calculate ξ of CBTRs with a degree of accuracy similar to that of the empirical formulas used for comparison. The Detached Eddy Simulation (DES) model, although used only briefly in this study, showed similar potential. In contrast, the unsteady Reynolds-averaged Navier–Stokes (URANS) approach produced divergent results, especially when compared to previous research on flow past unconfined cylinders. Furthermore, the flow parameters analyzed indicate that the problem is primarily 3D in nature. Therefore, future research in this area should be based on 3D numerical simulations rather than 2D models.
- For relatively low blocking ratios P, the presented approach accurately calculated ξ of CBTRs, as confirmed by comparison with previous studies on flow past circular

cylinders. Conversely, the empirical formulas of Kirschmer [24], Meusburger [2], and Böttcher et al. [29] tended to underestimate ξ in such cases. The fact that these formulas were not intended to be used when $P < 0.134$ may account for these differences. In addition, Equation (14) can be used for a simple calculation of ξ when the drag coefficient C_D is known, for example, from previous studies of flow past cylinders, especially for low P values. In this study, C_D remained relatively constant for $U_1 = 0.5$ m/s when $P \leq 0.091$ ($b \geq 80$ mm).

- For relatively high P values, especially for $P = 0.444$ ($b = 10$ mm), ξ was underestimated compared to the empirical formulas used for comparison. The differences can be attributed to the neglect of important effects in the numerical simulations, such as flow-induced vibrations. Furthermore, comparisons with simulations using three and five bars in the domain or periodic boundary conditions at the top and bottom of the domain showed that an acceptable level of independence for high P values could not be achieved with the presented approach.
- While ξ remained relatively constant for U_1 between 0.3 and 0.5 m/s, it increased continuously for $U_1 \geq 0.5$ m/s (corresponding to $Re_b \geq 3981$) and all b values tested. This finding is consistent with previous research on flow past circular cylinders in terms of C_D, but cannot be directly applied to non-circular bar shapes due to the different flow patterns that occur depending on the cross-sectional shape [63,107].

In addition to the limitations of the study, the potential of the presented approach for future research was discussed in detail in Section 4, which may include detailed studies of angled bars or novel bar shapes. Advancements in this approach may help to determine ξ in various applications, thereby enhancing the overall understanding of head losses at trash racks.

Funding: This research received no external funding.

Data Availability Statement: The data presented in this study are available on request from the author. The data are not publicly available due to ongoing research.

Acknowledgments: The computational results presented here have been achieved using the LEO HPC infrastructure of the University of Innsbruck.

Conflicts of Interest: The author declares no conflicts of interest.

Abbreviations

CBR	Curved bar rack
CBTR	Circular bar trash rack
CFD	Computational Fluid Dynamics
DES	Detached Eddy Simulation
DNS	Direct Numerical Simulation
FFT	Fast Fourier Transform
HPC	High-performance computing
HPP	Hydropower plant
LES	Large Eddy Simulation
MRE	Mean relative error
RANS	Reynolds-averaged Navier–Stokes
RMSRE	Root mean square relative error
SGS	Subgrid-scale
SIMPLE	Semi-Implicit Method for Pressure-Linked Equations
SST	Shear Stress Transport
URANS	Unsteady Reynolds-averaged Navier–Stokes

Notations

A	Area [m^2]
b	Clear bar spacing [m]
C_D	Drag coefficient [-]
C_L	Lift coefficient [-]
C_s	Smagorinsky coefficient [-]
F_D	Drag force on the cylinder [N]
F_L	Lift force on the cylinder [N]
f_v	Vortex shedding frequency [1/s]
g	Gravitational acceleration [m/s^2]
h_v	Local head loss [m]
k_F	Bar shape coefficient by Kirschmer [24] [-]
k_r	Resolved turbulent kinetic energy [m^2/s^2]
k_t	Total turbulent kinetic energy [m^2/s^2]
k_V	Loss factor to account for sectional clogging of the rack [-]
LES_{IQ}	LES index of quality [-]
$LES_{IQ,avg}$	Volume-averaged mean value of LES index of quality [-]
$LES_{IQ,min}$	Minimum value of LES index of quality for all evaluated time steps [-]
L_x	Length of the domain in the streamwise direction [m]
$L_{x,ds}$	Length of the domain downstream of the rack section in the streamwise direction [m]
$L_{x,us}$	Length of the domain upstream of the rack section in the streamwise direction [m]
L_y	Length of the domain in the spanwise direction [m]
L_z	Length of the domain in the vertical direction [m]
n	Number of bars in the computational domain [-]
N_c	Number of elements on the bar circumference [-]
N_r	Number of elements in the radial direction of the near-field grid [-]
N_t	Total number of elements in the computational domain [-]
N_y	Number of elements in the spanwise direction [-]
P	Blocking ratio [-]
p	Pressure [N/m^2]
p_1, p_2	Pressure at the inlet and outlet of the computational domain, respectively [N/m^2]
q	Unit discharge [m^2/s]
Re_b	Bar Reynolds number [-]
$\mid S \mid$	Characteristic filtered rate of strain [1/s]
s	Bar thickness [m]
S_{ij}	Resolved rate of the strain tensor [1/s]
St	Strouhal number [-]
T	Vortex shedding cycle [s]
t	Time [s]
t_e	Evaluation time [s]
U	Flow velocity [m/s]
U_{ref}	Free-stream velocity or the undisturbed flow velocity upstream of the rack [m/s]
U_1, U_2	Flow velocity at the inlet and outlet of the computational domain, respectively [m/s]
u, v, w	Flow velocity components in x-, y- and z-direction, respectively [m/s]
x, y, z	Cartesian coordinates [-]
y^+	Dimensionless distance from the wall [-]
z_1, z_2	Elevation at the inlet and outlet of the computational domain, respectively [m]
α	Rack angle relative to the side wall [°]
α_{kin}	Kinetic energy correction factor [-]
$\alpha_{kin,1}, \alpha_{kin,2}$	Kinetic energy correction factor at the inlet and outlet of the computational domain, respectively [-]
β	Rack angle relative to the ground plane [°]
Δ	Grid size defining the subgrid length scale [m]
Δt	Time step [s]
δ	Horizontal inflow angle to the rack [°]
δ_{ij}	Kronecker delta symbol [-]

ν	Kinematic viscosity [m^2/s]
ν_{sgs}	Subgrid-scale eddy viscosity [m^2/s]
ξ	Head loss coefficient [-]
ξ_n	Time-averaged normalized head loss coefficient [-]
ρ	Fluid density [kg/m^3]
τ_{ij}	Subgrid-scale (SGS) stress tensor [N/m^2]

Appendix A

Table A1. Results of varying the flow velocity at the inlet U_1 or the bar Reynolds number Re_b for the clear bar spacings b = 10, 20, and 30 mm or the blocking ratios P = 0.444, 0.286, and 0.211 to determine the drag coefficient C_D, the Strouhal number St, and the head loss coefficient ξ, calculated using Equation (1) proposed by Kirschmer [24] and Equations (13) and (14) based on the results of the numerical simulations.

U_1 [m/s]	Re_b [-]	b [mm]	P [-]	ξ (MRE) [-] Equation (1)	Equation (13)	Equation (14)	C_D (MRE) [-]	St [-]
0.3	2389	10	0.444	1.329 (±0.0%)	1.225 (−2.9%)	1.225 (−3.0%)	2.756 (−3.0%)	0.306
0.4	3185			1.329 (±0.0%)	1.230 (−2.5%)	1.231 (−2.5%)	2.769 (−2.5%)	0.302
0.5	3981			1.329 (Ref.)	1.262 (Ref.)	1.262 (Ref.)	2.840 (Ref.)	0.296
0.6	4777			1.329 (±0.0%)	1.277 (+1.2%)	1.277 (+1.2%)	2.874 (+1.2%)	0.290
0.7	5573			1.329 (±0.0%)	1.298 (+2.9%)	1.299 (+2.9%)	2.922 (+2.9%)	0.291
0.8	6369			1.329 (±0.0%)	1.308 (+3.7%)	1.308 (+3.7%)	2.944 (+3.7%)	0.291
0.9	7166			1.329 (±0.0%)	1.309 (+3.7%)	1.309 (+3.7%)	2.945 (+3.7%)	0.292
1.0	7962			1.329 (±0.0%)	1.316 (+4.3%)	1.316 (+4.3%)	2.962 (+4.3%)	0.291
0.3	2389	20	0.286	0.528 (±0.0%)	0.457 (−3.5%)	0.457 (−3.5%)	1.601 (−3.5%)	0.258
0.4	3185			0.528 (±0.0%)	0.459 (−3.0%)	0.460 (−3.1%)	1.608 (−3.1%)	0.254
0.5	3981			0.528 (Ref.)	0.474 (Ref.)	0.474 (Ref.)	1.659 (Ref.)	0.244
0.6	4777			0.528 (±0.0%)	0.487 (+2.8%)	0.487 (+2.8%)	1.706 (+2.8%)	0.246
0.7	5573			0.528 (±0.0%)	0.494 (+4.3%)	0.494 (+4.3%)	1.730 (+4.3%)	0.242
0.8	6369			0.528 (±0.0%)	0.516 (+9.0%)	0.516 (+9.0%)	1.808 (+9.0%)	0.235
0.9	7166			0.528 (±0.0%)	0.517 (+9.0%)	0.517 (+9.0%)	1.808 (+9.0%)	0.240
1.0	7962			0.528 (±0.0%)	0.523 (+10.3%)	0.523 (+10.3%)	1.830 (+10.3%)	0.236
0.3	2389	30	0.211	0.307 (±0.0%)	0.262 (−1.2%)	0.262 (−1.2%)	1.246 (−1.2%)	0.244
0.4	3185			0.307 (±0.0%)	0.268 (+1.0%)	0.268 (+1.0%)	1.274 (+1.0%)	0.240
0.5	3981			0.307 (Ref.)	0.266 (Ref.)	0.266 (Ref.)	1.262 (Ref.)	0.240
0.6	4777			0.307 (±0.0%)	0.272 (+2.3%)	0.272 (+2.3%)	1.291 (+2.3%)	0.232
0.7	5573			0.307 (±0.0%)	0.282 (+6.1%)	0.282 (+6.1%)	1.339 (+6.1%)	0.236
0.8	6369			0.307 (±0.0%)	0.291 (+9.4%)	0.291 (+9.4%)	1.381 (+9.4%)	0.233
0.9	7166			0.307 (±0.0%)	0.294 (+10.6%)	0.294 (+10.6%)	1.396 (+10.6%)	0.229
1.0	7962			0.307 (±0.0%)	0.304 (+14.5%)	0.304 (+14.5%)	1.445 (+14.5%)	0.229

References

1. Wahl, T.L. *Trash Control Structures and Equipment: A Literature Review and Survey of Bureau of Reclamation Experience*; U.S. Bureau of Reclamation: Denver, CO, USA, 1992.
2. Meusburger, H. *Energieverluste an Einlaufrechen von Flusskraftwerken*; ETH Zurich: Zurich, Switzerland, 2002. (In German) [CrossRef]
3. Giesecke, J.; Heimerl, S. *Wasserkraftanlagen: Planung, Bau und Betrieb*, 6th aktualisierte u. erw ed.; Springer: Berlin/Heidelberg, Germany, 2014; ISBN 978-3-642-53871-1. (In German)
4. Perham, R.E. *Floating Debris Control; A Literature Review*; US Army Cold Regions Research and Engineering Laboratory: Hanover, NH, USA, 1987.
5. Bradley, J.B.; Richards, D.L.; Bahner, C.D. *Debris Control Structures—Evaluation and Countermeasures*, 3rd ed.; Hydraulic Engineering Circular No. 9; U.S. Department of Transportation, Federal Highway Administration: Washington, DC, USA, 2005.
6. Schalko, I.; Schmocker, L.; Weitbrecht, V.; Boes, R.M. Backwater Rise Due to Large Wood Accumulations. *J. Hydraul. Eng.* **2018**, *144*, 04018056. [CrossRef]
7. Walczak, N. Operational Evaluation of a Small Hydropower Plant in the Context of Sustainable Development. *Water* **2018**, *10*, 1114. [CrossRef]
8. Hribernik, A. Evaluation of Clogged Hydropower Plant Trash Rack Losses. *SV-JME* **2020**, *66*, 142–152. [CrossRef]
9. Meister, J. *Fish Protection and Guidance at Water Intakes with Horizontal Bar Rack Bypass Systems*; ETH Zurich: Zurich, Switzerland, 2020. [CrossRef]

10. David, L.; Chatellier, L.; Courret, D.; Albayrak, I.; Boes, R.M. Fish Guidance Structures with Narrow Bar Spacing: Physical Barriers. In *Novel Developments for Sustainable Hydropower*; Rutschmann, P., Kampa, E., Wolter, C., Albayrak, I., David, L., Stoltz, U., Schletterer, M., Eds.; Springer International Publishing: Cham, Switzerland, 2022; pp. 91–98, ISBN 978-3-030-99137-1.
11. Čada, G.F.; Coutant, C.C.; Whitney, R.R. *Development of Biological Criteria for the Design of Advanced Hydropower Turbines*; U.S. Department of Energy, Idaho Operations Office: Idaho Falls, ID, USA, 1997.
12. Tsikata, J.M.; Tachie, M.F.; Katopodis, C. Open-Channel Turbulent Flow through Bar Racks. *J. Hydraul. Res.* **2014**, *52*, 630–643. [CrossRef]
13. Nielsen, N.; Szabo-Meszaros, M. *Hydropower and Fish—A Roadmap for Best Practice Management*; IEA: Paris, France, 2022. [CrossRef]
14. Ebel, G. *Fischschutz und Fischabstieg an Wasserkraftanlagen: Handbuch Rechen- und Bypasssysteme: Ingenieurbiologische Grundlagen, Modellierung und Prognose, Bemessung und Gestaltung; Mitteilungen aus dem Büro für Gewässerökologie und Fischereibiologie Dr. Ebel*, 3rd ed.; Büro für Gewässerökologie und Fischereibiologie Dr. Ebel: Halle, Germany, 2018; ISBN 978-3-00-039686-1. (In German)
15. Albayrak, I.; Maager, F.; Boes, R.M. An Experimental Investigation on Fish Guidance Structures with Horizontal Bars. *J. Hydraul. Res.* **2020**, *58*, 516–530. [CrossRef]
16. Larinier, M.; Travade, F. Downstream Migration: Problems and Facilities. *Bull. Fr. Pêche Piscic.* **2002**, 181–207. [CrossRef]
17. Williams, J.G.; Armstrong, G.; Katopodis, C.; Larinier, M.; Travade, F. Thinking like a Fish: A Key Ingredient for Development of Effective Fish Passage Facilities at River Obstructions: Fish Behaviour Related Fish Passage at Dams. *River Res. Applic.* **2012**, *28*, 407–417. [CrossRef]
18. Schwevers, U.; Adam, B. *Fish Protection Technologies and Fish Ways for Downstream Migration*; Springer International Publishing: Cham, Switzerland, 2020; ISBN 978-3-030-19241-9.
19. Meister, J.; Moldenhauer-Roth, A.; Beck, C.; Selz, O.M.; Peter, A.; Albayrak, I.; Boes, R.M. Protection and Guidance of Downstream Moving Fish with Electrified Horizontal Bar Rack Bypass Systems. *Water* **2021**, *13*, 2786. [CrossRef]
20. Tutzer, R.; Röck, S.; Walde, J.; Zeiringer, B.; Unfer, G.; Führer, S.; Brinkmeier, B.; Haug, J.; Aufleger, M. Ethohydraulic Experiments on the Fish Protection Potential of the Hybrid System FishProtector at Hydropower Plants. *Ecol. Eng.* **2021**, *171*, 106370. [CrossRef]
21. Haug, J.; Auer, S.; Frees, C.; Brinkmeier, B.; Tutzer, R.; Hayes, D.S.; Aufleger, M. Retrofitting of Existing Bar Racks with Electrodes for Fish Protection—An Experimental Study Assessing the Effectiveness for a Pilot Site. *Water* **2022**, *14*, 850. [CrossRef]
22. Haug, J.; Frees, C.; Brinkmeier, B.; Aufleger, M. Ethohydraulic Experiments Investigating Retention Rates of an Electrified Bar Rack. *Water* **2022**, *14*, 4036. [CrossRef]
23. Tutzer, R.; Röck, S.; Walde, J.; Haug, J.; Brinkmeier, B.; Aufleger, M.; Unfer, G.; Führer, S.; Zeiringer, B. A Physical and Behavioral Barrier for Enhancing Fish Downstream Migration at Hydropower Dams: The Flexible FishProtector. *Water* **2022**, *14*, 378. [CrossRef]
24. Kirschmer, O. *Untersuchungen über den Gefällsverlust an Rechen*; Mitteilungen des Hydraulischen Instituts der Technischen Hochschule München, Heft 1; Technische Hochschule München: Munich, Germany, 1926. (In German)
25. Raynal, S.; Courret, D.; Chatellier, L.; Larinier, M.; David, L. An Experimental Study on Fish-Friendly Trashracks—Part 1. Inclined Trashracks. *J. Hydraul. Res.* **2013**, *51*, 56–66. [CrossRef]
26. Raynal, S.; Chatellier, L.; Courret, D.; Larinier, M.; David, L. An Experimental Study on Fish-Friendly Trashracks—Part 2. Angled Trashracks. *J. Hydraul. Res.* **2013**, *51*, 67–75. [CrossRef]
27. Josiah, N.R.; Tissera, H.P.S.; Pathirana, K.P.P. An Experimental Investigation of Head Loss through Trash Racks in Conveyance Systems. *Engineer* **2016**, *49*, 1. [CrossRef]
28. Zayed, M.; El Molla, A.; Sallah, M. An Experimental Study on Angled Trash Screen in Open Channels. *Alex. Eng. J.* **2018**, *57*, 3067–3074. [CrossRef]
29. Böttcher, H.; Gabl, R.; Aufleger, M. Experimental Hydraulic Investigation of Angled Fish Protection Systems—Comparison of Circular Bars and Cables. *Water* **2019**, *11*, 1056. [CrossRef]
30. Beck, C.; Albayrak, I.; Meister, J.; Boes, R.M. Hydraulic Performance of Fish Guidance Structures with Curved Bars—Part 1: Head Loss Assessment. *J. Hydraul. Res.* **2020**, *58*, 807–818. [CrossRef]
31. Lučin, I.; Čarija, Z.; Grbčić, L.; Kranjčević, L. Assessment of Head Loss Coefficients for Water Turbine Intake Trash-Racks by Numerical Modeling. *J. Adv. Res.* **2020**, *21*, 109–119. [CrossRef]
32. Meister, J.; Fuchs, H.; Beck, C.; Albayrak, I.; Boes, R.M. Head Losses of Horizontal Bar Racks as Fish Guidance Structures. *Water* **2020**, *12*, 475. [CrossRef]
33. Baselt, I.; Malcherek, A. Determining the Flow Resistance of Racks and the Resulting Flow Dynamics in the Channel by Using the Saint-Venant Equations. *Water* **2022**, *14*, 2469. [CrossRef]
34. Latif, M.A.; Sarwar, M.K.; Farooq, R.; Shaukat, N.; Ali, S.; Hashmi, A.; Tariq, M.A.U.R. Estimating Energy Efficient Design Parameters for Trash Racks at Low Head Hydropower Stations. *Water* **2022**, *14*, 2609. [CrossRef]
35. Walczak, N.; Walczak, Z.; Tymiński, T. Laboratory Research on Hydraulic Losses on SHP Inlet Channel Trash Racks. *Energies* **2022**, *15*, 7602. [CrossRef]
36. Zayed, M.; El Molla, A.; Sallah, M. An Experimental Investigation of Head Loss through a Triangular "V- Shaped" Screen. *J. Adv. Res.* **2018**, *10*, 69–76. [CrossRef] [PubMed]
37. Spangler, J. *Untersuchungen Über den Verlust an Rechen Bei Schräger Zuströmung*; Mitteilungen des Hydraulischen Instituts der Technischen Hochschule München, Heft 2; Technische Hochschule München: Munich, Germany, 1928. (In German)
38. Berezinski, I. *Hydraulisches Handbuch*; Mostov, M.A.: Moscow, Russia, 1954. (In German)

39. United States Army Corps of Engineers (USACE). *Hydraulic Design Criteria*; USACE Waterways Experiment Station: Vicksburg, MS, USA, 1988.
40. Fellenius, W.; Lindquist, E.G.W. Experiments on the Head Loss Caused by Protecting Racks at Water-Power Plants. In *Hydraulic Laboratory Practice*; American Society of Mechanical Engineers (ASME): New York, NY, USA, 1929; pp. 533–538.
41. Escande, L. *Pertes de Charge a La Traversèe des Grilles*; Complements d'Hydraulique 1, Publications de l'Institut Electronique et de l'Institut de Mécanique des Fluides de l'Universté de Toulouse: Toulouse, France, 1947. (In French)
42. Orsborn, J.F. Rectangular-Bar Trashrack and Baffle Headlosses. *J. Power Div.* **1968**, *94*, 111–123. [CrossRef]
43. Zimmermann, J. *Widerstand Schräg Angeströmter Rechengitter*; Universität Fridericana Karlsruhe, Theodor-Rhebock-Flußbaulaboratorium, Mitteilungen Heft 157: Karlsruhe, Germany, 1969. (In German)
44. Bureau of Reclamation. *Design of Small Dams*, 3rd ed.; U.S. Department of the Interior: Denver, CO, USA, 1987.
45. Clark, S.P.; Tsikata, J.M.; Haresign, M. Experimental Study of Energy Loss through Submerged Trashracks. *J. Hydraul. Res.* **2010**, *48*, 113–118. [CrossRef]
46. Albayrak, I.; Kriewitz, C.R.; Hager, W.H.; Boes, R.M. An Experimental Investigation on Louvres and Angled Bar Racks. *J. Hydraul. Res.* **2018**, *56*, 59–75. [CrossRef]
47. Raynal, S.; Chatellier, L.; David, L.; Courret, D.; Larinier, M. Numerical Simulations of Fish-Friendly Angled Trashracks at Model and Real Scale. In Proceedings of the 35th IAHR World Congress, Chengdu, China, 8–13 September 2013.
48. Chatellier, L.; Wang, R.-W.; David, L.; Courret, D.; Larinier, M. Experimental Characterization of the Flow across Fish-Friendly Angled Trashrack Models. In Proceedings of the 34th IAHR World Congress, Brisbane, Australia, 26 June–1 July 2011; pp. 2776–2783.
49. Åkerstedt, H.O.; Eller, S.; Lundström, T.S. Numerical Investigation of Turbulent Flow through Rectangular and Biconvex Shaped Trash Racks. *Engineering* **2017**, *09*, 412–426. [CrossRef]
50. Leuch, C.; Beck, C.; Albayrak, I.; Vetsch, D.F.; Boes, R.M. Analysis and Optimization of Hydraulic Characteristics at Fish Guidance Structures Using CFD. In Proceedings of the 39th IAHR World Congress, Granada, Spain, 19–24 June 2022; pp. 3721–3728.
51. Pope, S.B. *Turbulent Flows*; Cambridge University Press: Cambridge, MA, USA; New York, NY, USA, 2000; ISBN 978-0-521-59125-6.
52. Keylock, C.J.; Hardy, R.J.; Parsons, D.R.; Ferguson, R.I.; Lane, S.N.; Richards, K.S. The Theoretical Foundations and Potential for Large-Eddy Simulation (LES) in Fluvial Geomorphic and Sedimentological Research. *Earth-Sci. Rev.* **2005**, *71*, 271–304. [CrossRef]
53. Rodi, W. Turbulence Modeling and Simulation in Hydraulics: A Historical Review. *J. Hydraul. Eng.* **2017**, *143*, 03117001. [CrossRef]
54. Nascimento, L.P.; Silva, J.B.C.; Di Giunta, V. Damage of Hydroelectric Power Plant Trash-Racks Due to Fluid-Dynamic Exciting Frequencies. *Lat. Am. J. Solids Struct.* **2006**, *3*, 223–243.
55. Ghamry, H.; Katopodis, C. Numerical Investigation of Turbulent Flow through Bar Racks in Closed Conduits. In Proceedings of the 9th International Symposium on Ecohydraulics, Vienna, Austria, 17–21 September 2012.
56. Paul, S.S.; Adaramola, M.S. Analysis of Turbulent Flow Past Bar-Racks. In Proceedings of the ASME 2014 International Mechanical Engineering Congress and Exposition, Montreal, QC, Canada, 14–20 November 2014.
57. Jeethulakshmi, G.; Sumam, K.S.; Sajikumar, N. Simulation of Head Loss in Trashrack—A Comparative Study. In Proceedings of the Flotek.g 2017—"Innovative Solutions in Flow Measurement and Control—Oil, Water and Gas", Palakkad, Kerala, India, 28 August 2017.
58. Čarija, Z.; Lučin, I.; Lučin, B.; Grbčić, L. Investigation of Numerical Simulation Parameters on Fluid Flow Around Trash-Racks. In Proceedings of the 29th DAAAM International Symposium, Vienna, Austria, 24–27 October 2018; Katalinic, B., Ed.; DAAAM International Vienna: Vienna, Austria, 2018; pp. 1046–1052, ISBN 978-3-902734-20-4.
59. Zhao, W.; Huhe, A. Large-Eddy Simulation of Three-Dimensional Turbulent Flow around a Circular Pier. *J. Hydrodyn.* **2006**, *18*, 765–772. [CrossRef]
60. Krajnović, S. Large Eddy Simulation of Flows around Ground Vehicles and Other Bluff Bodies. *Phil. Trans. R. Soc. A.* **2009**, *367*, 2917–2930. [CrossRef] [PubMed]
61. Gousseau, P.; Blocken, B.; Van Heijst, G.J.F. Quality Assessment of Large-Eddy Simulation of Wind Flow around a High-Rise Building: Validation and Solution Verification. *Comput. Fluids* **2013**, *79*, 120–133. [CrossRef]
62. Zdravkovich, M.M. *Flow around Circular Cylinders, Volume 1: Fundamentals*; Oxford Science Publications; Oxford University Press: Oxford, UK; New York, NY, USA, 1997.
63. Derakhshandeh, J.F.; Alam, M.M. A Review of Bluff Body Wakes. *Ocean Eng.* **2019**, *182*, 475–488. [CrossRef]
64. Khan, N.B.; Ibrahim, Z.; Bin Mohamad Badry, A.B.; Jameel, M.; Javed, M.F. Numerical Investigation of Flow around Cylinder at Reynolds Number = 3900 with Large Eddy Simulation Technique: Effect of Spanwise Length and Mesh Resolution. *Proc. Inst. Mech. Eng. Part M J. Eng. Marit. Environ.* **2019**, *233*, 417–427. [CrossRef]
65. Norberg, C. *Effect of Reynolds Number and a Low-Intensity Freestream Turbulence on the Flow around a Circular Cylinder*; Publikation Nr 87/2; Chalmers University of Technology, Department of Applied Thermodynamics and Fluid Mechanics: Gothenburg, Sweden, 1987.
66. Beaudan, P.; Moin, P. *Numerical Experiments on the Flow Past a Circular Cylinder at Sub-Critical Reynolds Number*; Report No. TF-62; Department of Mechanical Engineering, Stanford University: Stanford, CA, USA, 1994.
67. Williamson, C.H.K. Vortex Dynamics in the Cylinder Wake. *Annu. Rev. Fluid Mech.* **1996**, *28*, 477–539. [CrossRef]
68. Ong, L.; Wallace, J. The Velocity Field of the Turbulent Very near Wake of a Circular Cylinder. *Exp. Fluids* **1996**, *20*, 441–453. [CrossRef]

69. Kravchenko, A.G.; Moin, P. Numerical Studies of Flow over a Circular Cylinder at ReD = 3900. *Phys. Fluids* **2000**, *12*, 403–417. [CrossRef]
70. Franke, J.; Frank, W. Large Eddy Simulation of the Flow Past a Circular Cylinder at ReD = 3900. *J. Wind. Eng. Ind. Aerodyn.* **2002**, *90*, 1191–1206. [CrossRef]
71. Parnaudeau, P.; Carlier, J.; Heitz, D.; Lamballais, E. Experimental and Numerical Studies of the Flow over a Circular Cylinder at Reynolds Number 3900. *Phys. Fluids* **2008**, *20*, 085101. [CrossRef]
72. Lysenko, D.A.; Ertesvåg, I.S.; Rian, K.E. Large-Eddy Simulation of the Flow Over a Circular Cylinder at Reynolds Number 3900 Using the OpenFOAM Toolbox. *Flow Turbul. Combust* **2012**, *89*, 491–518. [CrossRef]
73. Ma, X.; Karamanos, G.-S.; Karniadakis, G.E. Dynamics and Low-Dimensionality of a Turbulent near Wake. *J. Fluid Mech.* **2000**, *410*, 29–65. [CrossRef]
74. Dong, S.; Karniadakis, G.E.; Ekmekci, A.; Rockwell, D. A Combined Direct Numerical Simulation–Particle Image Velocimetry Study of the Turbulent near Wake. *J. Fluid Mech.* **2006**, *569*, 185. [CrossRef]
75. Wissink, J.G.; Rodi, W. Numerical Study of the near Wake of a Circular Cylinder. *Int. J. Heat Fluid Flow* **2008**, *29*, 1060–1070. [CrossRef]
76. Xu, C.; Chen, L.; Lu, X. Large-Eddy and Detached-Eddy Simulations of the Separated Flow around a Circular Cylinder. *J. Hydrodyn.* **2007**, *19*, 559–563. [CrossRef]
77. D'Alessandro, V.; Montelpare, S.; Ricci, R. Detached–Eddy Simulations of the Flow over a Cylinder at Re = 3900 Using OpenFOAM. *Comput. Fluids* **2016**, *136*, 152–169. [CrossRef]
78. Young, M.E.; Ooi, A. Comparative Assessment of LES and URANS for Flow Over a Cylinder at a Reynolds Number of 3900. In Proceedings of the 16th Australasian Fluid Mechanics Conference, Crown Plaza, Gold Coast, Australia, 2–7 December 2007.
79. Deng, G.B.; Piquet, J.; Queutey, P.; Visonneau, M. Vortex-Shedding Flow Predictions with Eddy-Viscosity Models. In *Engineering Turbulence Modelling and Experiments*; Elsevier: Amsterdam, The Netherlands, 1993; pp. 143–152. ISBN 978-0-444-89802-9.
80. Rodi, W. Comparison of LES and RANS Calculations of the Flow around Bluff Bodies. *J. Wind. Eng. Ind. Aerodyn.* **1997**, *69–71*, 55–75. [CrossRef]
81. Lei, C.; Cheng, L.; Kavanagh, K. Spanwise Length Effects on Three-Dimensional Modelling of Flow over a Circular Cylinder. *Comput. Methods Appl. Mech. Eng.* **2001**, *190*, 2909–2923. [CrossRef]
82. Shao, J.; Zhang, C. Numerical Analysis of the Flow around a Circular Cylinder Using RANS and LES. *Int. J. Comput. Fluid Dyn.* **2006**, *20*, 301–307. [CrossRef]
83. Sumner, D. Two Circular Cylinders in Cross-Flow: A Review. *J. Fluids Struct.* **2010**, *26*, 849–899. [CrossRef]
84. Zhou, Y.; Mahbub Alam, M. Wake of Two Interacting Circular Cylinders: A Review. *Int. J. Heat Fluid Flow* **2016**, *62*, 510–537. [CrossRef]
85. Afgan, I.; Kahil, Y.; Benhamadouche, S.; Sagaut, P. Large Eddy Simulation of the Flow around Single and Two Side-by-Side Cylinders at Subcritical Reynolds Numbers. *Phys. Fluids* **2011**, *23*, 075101. [CrossRef]
86. Nguyen, Q.D.; Lu, W.; Chan, L.; Ooi, A.; Lei, C. A State-of-the-Art Review of Flows Past Confined Circular Cylinders. *Phys. Fluids* **2023**, *35*, 071301. [CrossRef]
87. Zovatto, L.; Pedrizzetti, G. Flow about a Circular Cylinder between Parallel Walls. *J. Fluid Mech.* **2001**, *440*, 1–25. [CrossRef]
88. Nguyen, Q.D.; Lei, C. Hydrodynamic Characteristics of a Confined Circular Cylinder in Cross-Flows. *Ocean Eng.* **2021**, *221*, 108567. [CrossRef]
89. Jiang, H.; Cheng, L. Large-Eddy Simulation of Flow Past a Circular Cylinder for Reynolds Numbers 400 to 3900. *Phys. Fluids* **2021**, *33*, 034119. [CrossRef]
90. Norberg, C. An Experimental Investigation of the Flow around a Circular Cylinder: Influence of Aspect Ratio. *J. Fluid Mech.* **1994**, *258*, 287–316. [CrossRef]
91. Ooi, A.; Lu, W.; Chan, L.; Cao, Y.; Leontini, J.; Skvortsov, A. Turbulent Flow over a Cylinder Confined in a Channel at Re = 3900. *Int. J. Heat Fluid Flow* **2022**, *96*, 108982. [CrossRef]
92. Celik, I.B.; Cehreli, Z.N.; Yavuz, I. Index of Resolution Quality for Large Eddy Simulations. *J. Fluids Eng.* **2005**, *127*, 949–958. [CrossRef]
93. Smagorinsky, J. General Circulation Experiments with the Primitive Equations: Part I. The Basic Experiment. *Mon. Wea. Rev.* **1963**, *91*, 99–164. [CrossRef]
94. ANSYS, Inc. *ANSYS Fluent Theory Guide*; Release 19.2; Southpointe: Canonsburg, PA, USA, 2018.
95. Menter, F.R. Two-Equation Eddy-Viscosity Turbulence Models for Engineering Applications. *AIAA J.* **1994**, *32*, 1598–1605. [CrossRef]
96. Idel'chik, I.E.; Ginevskiĭ, A.S. *Handbook of Hydraulic Resistance*, 4th ed. rev. and augmented; Begell House: Redding, CT, USA, 2007; ISBN 978-1-56700-251-5.
97. Gabl, R.; Gems, B.; Birkner, F.; Hofer, B.; Aufleger, M. Adaptation of an Existing Intake Structure Caused by Increased Sediment Level. *Water* **2018**, *10*, 1066. [CrossRef]
98. White, F.M. *Fluid Mechanics*, 6th ed.; Mcgraw-Hill Series in Mechanical Engineering; McGraw-Hill: New York, NY, USA, 2009; ISBN 978-0-07-352934-9.
99. Aigner, D.; Bollrich, G. *Handbuch der Hydraulik: Für Wasserbau und Wasserwirtschaft*, 1st ed.; Wissen Bauwesen; Beuth Verlag GmbH: Berlin, Germany; Vienna, Austria; Zürich, Seitzerland, 2015; ISBN 978-3-410-21346-8. (In German)

100. Lu, W.; Aljubaili, D.; Zahtila, T.; Chan, L.; Ooi, A. Asymmetric Wakes in Flows Past Circular Cylinders Confined in Channels. *J. Fluid Mech.* **2023**, *958*, A8. [CrossRef]
101. ANSYS, Inc. *ANSYS Fluent User's Guide*; Release 19.2; Southpointe: Canonsburg, PA, USA, 2018.
102. Wieselsberger, C. Neuere Feststellungen über die Gesetze des Flüssigkeits- und Luftwiderstandes. *Phys. Z.* **1921**, *22*, 321–328. (In German)
103. Blevins, R.D. *Applied Fluid Dynamics Handbook*; Van Nostrand Reinhold Co.: New York, NY, USA, 1984; ISBN 978-0-442-21296-4.
104. Singh, S.P.; Mittal, S. Flow Past a Cylinder: Shear Layer Instability and Drag Crisis. *Int. J. Numer. Meth. Fluids* **2005**, *47*, 75–98. [CrossRef]
105. Karabelas, S.J.; Koumroglou, B.C.; Argyropoulos, C.D.; Markatos, N.C. High Reynolds Number Turbulent Flow Past a Rotating Cylinder. *Appl. Math. Model.* **2012**, *36*, 379–398. [CrossRef]
106. Wen, P.; Qiu, W. Investigation of Drag Crisis Phenomenon Using CFD Methods. *Appl. Ocean. Res.* **2017**, *67*, 306–321. [CrossRef]
107. Lekkala, M.R.; Latheef, M.; Jung, J.H.; Coraddu, A.; Zhu, H.; Srinil, N.; Lee, B.-H.; Kim, D.K. Recent Advances in Understanding the Flow over Bluff Bodies with Different Geometries at Moderate Reynolds Numbers. *Ocean Eng.* **2022**, *261*, 111611. [CrossRef]
108. Naudascher, E.; Rockwell, D. *Flow-Induced Vibrations: An Engineering Guide*; Dover Publications: Mineola, NY, USA, 2005; ISBN 978-0-486-44282-2.
109. Kim, D.-H.; Yang, K.-S.; Senda, M. Large Eddy Simulation of Turbulent Flow Past a Square Cylinder Confined in a Channel. *Comput. Fluids* **2004**, *33*, 81–96. [CrossRef]
110. Bai, H.; Alam, M.M. Dependence of Square Cylinder Wake on Reynolds Number. *Phys. Fluids* **2018**, *30*, 015102. [CrossRef]
111. Davis, R.W.; Moore, E.F.; Purtell, L.P. A Numerical-Experimental Study of Confined Flow around Rectangular Cylinders. *Phys. Fluids* **1984**, *27*, 46. [CrossRef]
112. Alam, M.M.; Zhou, Y.; Wang, X.W. The Wake of Two Side-by-Side Square Cylinders. *J. Fluid Mech.* **2011**, *669*, 432–471. [CrossRef]
113. Zhao, F.; Wang, R.; Zhu, H.; Ping, H.; Bao, Y.; Zhou, D.; Cao, Y.; Cui, H. Large-Eddy Simulations of Flow Past a Circular Cylinder near a Free Surface. *Phys. Fluids* **2021**, *33*, 115108. [CrossRef]
114. Cao, Y.; Tamura, T. Large-Eddy Simulations of Flow Past a Square Cylinder Using Structured and Unstructured Grids. *Comput. Fluids* **2016**, *137*, 36–54. [CrossRef]
115. Lemkecher, F.; Chatellier, L.; Courret, D.; David, L. Contribution of Different Elements of Inclined Trash Racks to Head Losses Modeling. *Water* **2020**, *12*, 966. [CrossRef]
116. Feigenwinter, L.; Vetsch, D.; Kammerer, S.; Kriewitz, C.; Boes, R. Conceptual Approach for Positioning of Fish Guidance Structures Using CFD and Expert Knowledge. *Sustainability* **2019**, *11*, 1646. [CrossRef]
117. Maddahi, M.; Hagenbüchli, R.; Mendez, R.; Zaugg, C.; Boes, R.M.; Albayrak, I. Field Investigation of Hydraulics and Fish Guidance Efficiency of a Horizontal Bar Rack-Bypass System. *Water* **2022**, *14*, 776. [CrossRef]
118. Zöschg, H.; Dobler, W.; Aufleger, M.; Zeiringer, B. Evaluation of Hydraulics and Downstream Fish Migration at Run-of-River Hydropower Plants with Horizontal Bar Rack Bypass Systems by Using CFD. *Water* **2023**, *15*, 1042. [CrossRef]
119. Zhao, M.; Cheng, L.; Zhou, T. Direct Numerical Simulation of Three-Dimensional Flow Past a Yawed Circular Cylinder of Infinite Length. *J. Fluids Struct.* **2009**, *25*, 831–847. [CrossRef]
120. Liang, H.; Duan, R.-Q. Effect of Lateral End Plates on Flow Crossing a Yawed Circular Cylinder. *Appl. Sci.* **2019**, *9*, 1590. [CrossRef]
121. Zhou, B.; Wang, J.; Jin, G.; Zhou, Y.; Wang, X.; Gho, W.M. Large Eddy Simulation of Flow Past an Inclined Finite Cylinder. *Ocean. Eng.* **2022**, *258*, 111504. [CrossRef]
122. Kriewitz-Byun, C.R. *Leitrechen an Fischabstiegsanlagen: Hydraulik und fischbiologische Effizienz*; ETH Zurich: Zurich, Switzerland, 2015. (In German) [CrossRef]
123. Albayrak, I.; Boes, R.M. Fish Guidance Structure with Wide Bar Spacing: Mechanical Behavioural Barrier. In *Novel Developments for Sustainable Hydropower*; Rutschmann, P., Kampa, E., Wolter, C., Albayrak, I., David, L., Stoltz, U., Schletterer, M., Eds.; Springer International Publishing: Cham, Switzerland, 2022; pp. 99–104, ISBN 978-3-030-99137-1.
124. Heller, V. Scale Effects in Physical Hydraulic Engineering Models. *J. Hydraul. Res.* **2011**, *49*, 293–306. [CrossRef]
125. Haug, J.; Brinkmeier, B.; Tutzer, R.; Aufleger, M. Hybride Barrieren zur Optimierung von Stabrechen zum Fischschutz. *Wasserwirtsch* **2021**, *111*, 48–53. (In German) [CrossRef]
126. Khan, L.A.; Wicklein, E.A.; Rashid, M.; Ebner, L.L.; Richards, N.A. Computational Fluid Dynamics Modeling of Turbine Intake Hydraulics at a Hydropower Plant. *J. Hydraul. Res.* **2004**, *42*, 61–69. [CrossRef]
127. Waldy, M.; Gabl, R.; Seibl, J.; Aufleger, M. Alternative Methoden für die Implementierung von Rechenverlusten in die 3D-numerische Berechnung mit FLOW-3D. *Österr. Wasser-Abfallw.* **2015**, *67*, 64–69. (In German) [CrossRef]
128. Özbey, C.; Kucukali, S. A Numerical Investigation of Hydrodynamics of Fish-Friendly Fine Screens. In Proceedings of the 40th IAHR World Congress, Vienna, Austria, 21–25 August 2023; pp. 1935–1944. [CrossRef]
129. Wagner, F.; Keuneke, R.; Kampa, E.; Kemper, M.; Naumann, S. *Forum Fischschutz und Fischabstieg: Ergebnisse und Ausblick*; Umweltbundesamt: Dessau-Roßlau, Germany, 2023; p. 52. (In German)

fluids

MDPI

Article

Pressure Transient Solutions for Unbounded and Bounded Reservoirs Produced and/or Injected via Vertical Well Systems with Constant Bottomhole Pressure

Ruud Weijermars [1,2,*] and Clement Afagwu [1]

1 Department of Petroleum Engineering, College of Petroleum Engineering and Geosciences (CPG), King Fahd University of Petroleum & Minerals (KFUPM), Dhahran 31261, Saudi Arabia
2 Center for Integrative Petroleum Research (CIPR), College of Petroleum Engineering and Geosciences (CPG), King Fahd University of Petroleum & Minerals (KFUPM), Dhahran 31261, Saudi Arabia
* Correspondence: ruud.weijermars@kfupm.edu.sa

Abstract: Various analytical solutions for computing production and injection-induced pressure changes in aquifers and oil reservoirs have been derived over the past century. All prior solutions assumed a constant well rate as the boundary condition. However, in many practical situations, the fluid withdrawal from and/or injection into such subsurface reservoirs occurs with the aid of pump devices that maintain a constant bottomhole pressure in the well. Until now, how the well rate will decline over time, based on the pressure difference in the well relative to the initial reservoir pressure, could not be rapidly computed analytically (using the diffusivity as the key governing system parameter), because no concise expression had been derived with the boundary condition of a constant bottomhole pressure. The present study shows how the pressure diffusion equation can be readily solved for wells acting as sinks and sources with a constant bottomhole pressure condition. We consider both fractured and unfractured completions, as well as injection and production modes. The new solutions do not require an elaborate time-stepped pressure-matching procedure as in nodal analysis, the only other physics-based analytical method currently available to compute the well rate decline when a constant bottomhole pressure production system is used, which unlike our new method proposed here is limited to single well systems.

Keywords: pressure diffusion; bottomhole pressure; well rate; analytical solution; nodal analysis

1. Introduction

Natural resources such as water, hydrocarbons and geothermal heat occurring in subsurface reservoirs are massively exploited by humans for a multitude of industrial and agricultural purposes [1–3]. Similarly, depleted reservoirs are widely used for seasonal storage of natural gas, hydrogen and liquids, as well as for permanent sequestration of carbon dioxide, CO_2 [4,5]. The effective engineering of fluid withdrawal from, and plume movements of fluid injected into, porous rock formations requires practical expressions that describe the pressure changes in the reservoir [5,6]. This is needed, a.o., to prevent formation damage and potential leakage from occurring when the CO_2 pressure exceeds the seal failure pressure [7].

The prior scientific literature records a progressive development of insight, with solutions for fluid withdrawal rates via wells from subsurface reservoirs. The Fourier solution of the diffusivity equation for heat [8] was modified and scaled by Fick [9] for molecular diffusion, and then (in an indirect way) to pressure diffusion problems by Darcy [10]. From this general insight, specific solutions of well behavior for fluid withdrawal from porous reservoirs in the subsurface were formulated by Dupuit [11] and Thiem Sr. [12], and were tested with field data by Thiem Jr. [13]. Drawing analogy from heat diffusion solutions of Carlslaw [14], Theis [15] gave analytical solutions for the pressure drawdown profile around a well in an aquifer.

The prior work cited so far on the theoretical well performance was commonly focusing on wells sourcing water from a subsurface aquifer. Next, a modified Dupuit–Thiem equation was developed for application in hydrocarbon well test modeling by Raghavan [16] and Zimmerman [17,18], largely modeled after a textbook solution of heat diffusion equations by Jaeger and Carslaw [19]. The analytical solution of the diffusivity equation based on a constant well rate boundary condition is widely used in practical well tests assessing flow rates in newly drilled exploration wells [20,21]. The aim of well testing is to establish how fast the discovered hydrocarbons can be produced, and if the fluid production rate of the exploration well can meet the required economic threshold for additional wells to be drilled in order to validate the full-scale development of costly surface installations.

There have been recent advancements in both numerical and analytical frameworks aimed at elucidating the transport dynamics of hydrocarbon fluids [22–24] and delineating the movement of CO_2 plumes during both injection and post-injection phases in subsurface aquifer storage [5,25]. In the physical entrapment of CO_2, the dominant transport mechanisms include horizontal plume migration at rates controlled by viscous forces resulting from pressure gradients. Subsequently, buoyancy and capillary forces play crucial roles in post-injection trapping when the upward migration of CO_2 may occur [5]. Sbai and Azaroual [25] utilized numerical methods to predict potential formation damage caused by CO_2 injection. Weijermars [22,23] introduced a novel Gaussian pressure-transient-based solution to quantify the spatial and temporal progression of pressure and the associated fluid depletion, both in unfractured and fractured wells in unconventional reservoirs. Okoroafor et al. [24] utilized numerical methods to simulate hydrogen injection, storage and recovery from both depleted methane gas reservoirs and saline aquifers.

The present study derives new analytical solutions for fluid injection, storage and withdrawal via a vertical wellbore in a conventional reservoir from a horizontal, permeable reservoir, assuming a constant bottomhole pressure. Such a solution did not exist until now. The diffusivity equation was revisited, and the well rate equations were derived for a constant bottomhole pressure boundary condition; such a situation applies to many oil and gas wells produced with artificial lift systems (using a pump which imposes a certain constant bottomhole pressure). The new method does not require an elaborate time-stepped pressure-matching procedure as in nodal analysis [26–29], the only other physics-based analytical method currently available to compute the well rate decline when a constant bottomhole pressure system is used.

This research paper is structured as follows. Section 2 provides a historical review and details the workflow used in this study; after first tracing the evolution of analytical methods from modeling basic groundwater behavior to describing complex hydrocarbon flow behavior in oil and gas reservoirs. Section 3 revisits the governing diffusion equation [30] and presents pressure transient solutions applicable to vertical wells with artificial lift in reservoirs with no lateral boundaries. Wells produced with artificial lift have a constant bottomhole pressure and are comprehensively modeled in the present study. Section 4 presents pressure gradient solutions for such wells. Section 5 explores well rate solutions, including a complementary solution for a vertical well in a bounded reservoir, illustrated with case studies using real production data from two wells (UK offshore gas fields), which are history-matched with the new method using specific reservoir and well conditions.

2. Review of Prior Solutions

This section lays the groundwork for the present study. We begin by introducing foundational solutions for modeling how water table levels decline in aquifers (drawdown) in Section 2.1. We then transition to the pivotal pressure transient test equation used in oil and gas well analysis (Section 2.2). In Section 2.3, we present the crucial assumption of steady-state well rates that underpins the traditional well rate solution. Finally, considering the limitations of prior research, we restate the objective and workflow of this study in Section 2.4.

2.1. Hydraulic Head Drawdown

All prior solutions governing the injectivity and productivity of water and petroleum wells have assumed a constant well rate. Landmark solutions for practical use include the Thiem [12,13] solution, the Theis [15] solution and the pressure transient solution of the diffusion equation by Carslaw and Jaeger [19]. Thiem [13] provided a practical equation based on earlier work by Dupuit [11], which relates the constant well rate, q_{WH}, to the respective water table drawdown to heights (h_1,h_2) in nearby observation wells at radial distance (r_1,r_2) from a well in the origin with original water table height, h_0, in the aquifer of study as follows:

$$q_{WH} = -2\pi h_0 K \frac{h_1 - h_2}{\ln r_1 - \ln r_2}. \tag{1}$$

The parameter to be determined from the measurements in the observation well(s) is the hydraulic conductivity, K, of the aquifer. The solution of Equation (1) is based on comparing the mass balance of fluid flow using Darcy's Law [30–32].

Separately, Theis [15] gave a continuous solution for the drawdown profile of the fluid surface in aquifers, based on the 2D diffusion solution for heat dissipation as given in Carslaw [14]. The change in the hydraulic head, $\Delta h(r,t)$, in radial direction around the well in the origin is given by the exponential integral E_i [33] as follows:

$$\Delta h(r,t) = \frac{q_{WH}}{4\pi T} \int_{t=0}^{\infty} e^{\frac{r^2 S}{4Tt}} \frac{dt}{t} = \frac{q_{WH}}{4\pi T} E_i\left(\frac{r^2 S}{4Tt}\right). \tag{2}$$

The well rate in Equation (2) is again assumed constant, and the rate of drawdown of the hydraulic head is controlled by the aquifer transmissivity, T, and storativity, S. The practical purpose of Theis's [15] solution is that it solves for the transient change in hydraulic head due to a well pumping at a constant rate from a confined aquifer [34].

2.2. Well Test Equation

In petroleum engineering, new oil and gas reservoirs are tested in exploration wells over relatively short flow periods to measure what well rate can be achieved in response to a pressure differential between the reservoir and the wellbore. This solution is again based on Darcy's Law, and the assumed boundary condition is a steady flow in a porous media, at rate q_D:

$$q_D = -\frac{k}{\mu} \nabla P. \tag{3}$$

Equation (3) features the fluid viscosity, μ, and reservoir permeability, k; both are initially assumed to be constant and uniform. Note that q_D [L T^{-1}] in Equation (3) is used here for Darcy flux from the porous formation into the open space of the wellbore across a unit area of wellbore exposed; q_D should not be confused with the dimensionless well rate and is distinct from the volumetric flux into the well, q_{WH} [L^3 T^{-1}] in Equation (2). The actual velocity, v_R, in the porous reservoir is always greater than the Darcy flux and will be given by $v_R\ \phi = q_D$, with ϕ representing the connected porosity.

The associated steady-state pressure profile $P(r)$ for radial flow near the well producing at a constant rate, q_{WH}, with original reservoir pressure, P_0, is as follows [18]:

$$P(r) = P_0 - q_{WH} \frac{B}{2\pi h} \frac{\mu}{k} \ln\left(\frac{r}{r_w}\right). \tag{4}$$

where h is payzone thickness, and r_w is wellbore radius. The formation volume factor, B, accounts for the volume change in the fluid mostly due to drawdown pressure when the fluid withdrawn from the reservoir is transferred to the surface. The solution of Equation (4) itself does not account for friction losses in the wellbore's vertical section, the exclusion of which is a routine practice in well test equations. Such friction may be considered in a more comprehensive analysis of the later production system, which can account for the wellbore friction if critically needed. Provide the well flows, the solution of Equation (4) is correct, even if the wellbore fiction I left out of the equation, because the dynamics is computed at the reservoir level as controlled by Darcy's law, with ∇P specified for the reservoir level.

The radial flow solution of Equation (4) is termed a 'modified' Dupuit–Thiem equation for the steady-state pressure profile, because the original equations by Dupuit [11], Thiem Sr. [12] and (his son) Thiem Jr. [13] do not include the formation volume factor; for hydrological problems in shallow aquifers, B indeed can be safely assumed close to unity.

2.3. Steady-State Pressure Profile

The assumption of well production at steady state underlying Equation (4) is useful, because this way a steady-state pressure profile may exist, where the reservoir pressure increases logarithmically with distance to the well to converge to the original reservoir pressure P_0 at a certain distance from the well. The Dupuit–Thiem equation describes the steady-state pressure profile and still features centrally in today's well-testing literature [20], with small modifications such as the skin factor, s, a fidgeting factor which accounts for near-wellbore damage or stimulation of the reservoir:

$$P(r) = P_0 - q_{WH}\frac{B}{2\pi h}\frac{\mu}{k}\left[\ln\left(\frac{r}{r_w}\right) + s\right], \tag{5}$$

The steady well rate and constant pressure profile assumptions are only physically reasonable for reservoirs produced with pressure replenishment by balanced water injection [35]. When no fluid replenishment occurs, the pressure profile around the wells will be transient and a function of the hydraulic diffusivity [19]:

$$q_{WH} = [P_0 - P(r,t)]\frac{k}{\mu}\frac{2\pi h}{B}E_i\left(\frac{-r^2}{4D_h t}\right). \tag{6}$$

The purpose of most well tests is primarily to measure the rate of the well for an imposed pressure drop at the wellbore, where $P(r_w, 0) = P_{BH}$. Assuming the well rate q_{WH} stays constant, reservoir parameters can be estimated by coupling the Zimmerman [18] hydraulic diffusivity, D_h, expression (Equation (7)) into Equation (6).

$$D_h = \frac{k}{\mu\left[(1-\phi)c_R + \phi c_f\right]}. \tag{7}$$

D_h is simultaneously estimated by constraining one or more of the primary values that jointly control its value, such as the intrinsic properties of the reservoir (k,ϕ, c_R) and of the fluid (μ, c_f) in its pores of total compressibility, $c_t = \left[(1-\phi)c_R + \phi c_f\right]$, made up of rock compressibility, c_R, and fluid compressibility, c_f. Representative values of k, ϕ, c_R, μ, c_f can be obtained from lab tests on sampled reservoir fluids and core test data, as routinely conducted by oil companies during field appraisal programs. Alternatively, the value of D_h can be abstracted from history matching of well rates as demonstrated later in this study.

2.4. Study Objective and Research Workflow

The prior mathematical solutions cited above all assumed a vertical well pumping at a constant rate from a horizontal aquifer confined between an upper and a lower impervious boundary, with infinite lateral extent, uniform thickness, homogeneous conductivity, storativity and transmissivity. In the analyses by Thiem [13] and Theis [15], the diameter of the well was assumed to be negligibly small so that fluid storage in the well may be neglected.

In order to compute the well rate decline when a constant bottomhole pressure production system is used, one must currently use an elaborate time-stepped pressure-matching procedure called nodal analysis [26–29], which follows a history-matching procedure based on analytical solutions for transient and boundary dominated flow. The procedure is reasonably accurate but elaborate and technically complex due to the need for time-stepped history matching [36]. The benefits and limitations of the nodal analysis procedure are acknowledged in this present study, but the procedure is not further detailed.

Figure 1 visually summarizes the key steps of our research workflow. It begins with the governing partial differential equation (PDE) that incorporates an hydraulic diffusivity term. Next, we show how reservoir and well boundary conditions are applied to this equation. The present study focuses on vertical wells in the transient-flow regime of reservoirs without lateral boundaries (unbounded) and with boundary-dominated flow effect [37]. Figure 1 illustrates how subsequent steps lead to the development of the proposed analytical solutions for well performance, starting from the transient pressure field solutions of the diffusivity equation. The pressure transient solutions in laterally unbounded reservoirs are provided in Section 3 of the present study.

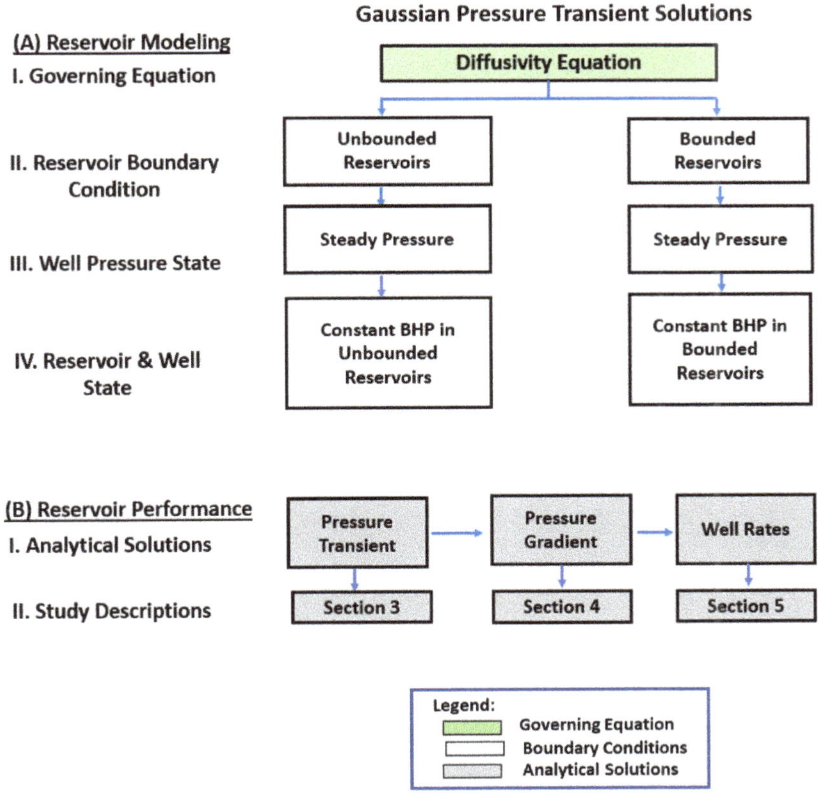

Figure 1. Summarized workflow for the mathematical model and application to either unbounded or bounded reservoirs produced with a vertical well.

3. Diffusion of Pressure Transients in Unbounded Reservoirs

This section focuses on pressure field solutions, which are the basis for the subsequent pressure gradient solutions' development in Section 4, and is structured as follows. The classical diffusivity equation is revisited in Section 3.1. Many oil and gas wells are produced with artificial lift systems (using a pump) which impose a certain constant bottomhole pressure. Thus, the pressure transient diffusion solution is first derived for a constant bottomhole pressure change in vertical well systems in laterally unbounded reservoirs (Section 3.2). one- and two-dimensional pressure diffusion solutions are provided. We further distinguish between producer wells and injection wells.

3.1. Governing Diffusion Equation

Consider the classical diffusion equation [38], adapted for pressure diffusion in three-dimensional space:

$$\frac{\partial P}{\partial t} = \nabla \cdot (D_h \nabla P). \quad (8)$$

The first Del operator ∇ in Equation (8) takes the dot product or divergence of $D_h \nabla P$ [MT^{-3}], and ∇P denotes the pressure gradient. If normalized pressure is initiated at initial time, and gravity gradients are negligible, the spatial changes in ∇P are governed by Equation (8) as would apply to a perfectly homogenous, isotropic porous medium with slow, laminar flow due to the pressure gradient (Figure 2a).

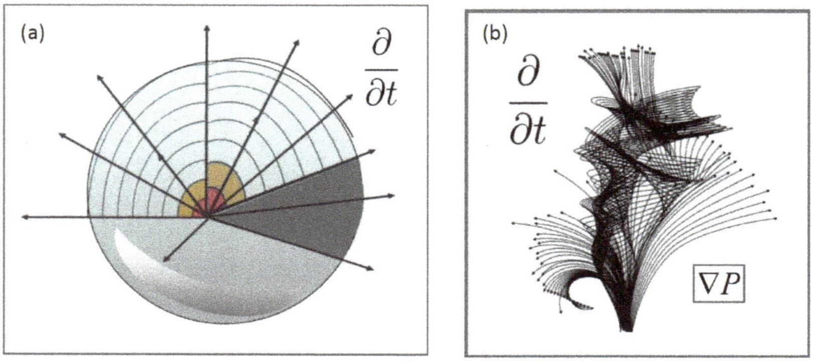

Figure 2. Principle sketch of (**a**) 3D diffusion from a point source in a homogenous isotropic porous medium. (**b**) General 3D pressure diffusion process with inertia and consequently complex pressure transients.

For a free flow of liquid media, inertia and gravity force terms will quickly dominate the spatial changes in the pressure field, which is why the pressure gradients of free flow with turbulence may quickly appear extremely complex (Figure 2b). The pressure diffusion Equation (8) does not account for such additional terms. Also, only diffusion mass transfer due to the pressure gradient is considered. If molecular diffusion is included as well, the relative importance of the convective mass transfer due to the pressure gradient and that due to molecular diffusion can be described by a Fokker–Planck (convection–diffusion) equation [23,39]. A Gaussian Péclet number has been proposed to compare the relative importance of the two processes for application in gas reservoir/well dynamics [40].

Next, assume the pressure diffusion process of interest occurs radially outward from a cylindrical well source placed vertically in a tabular homogeneous reservoir with an impervious upper and lower boundary (Figure 3) and at a certain initial pressure, P_0. The pressure changes $\Delta P(x, y, t)$ occur exclusively in the (x, y) plane, due to the no-flow condition at the upper and lower boundary. Specifically, 2D diffusion starting at the well is governed by:

$$\frac{\partial \Delta P(x,y,t)}{\partial t} = D_h \left(\frac{\partial^2 \Delta P(x,y,t)}{\partial x^2} + \frac{\partial^2 \Delta P(x,y,t)}{\partial y^2} \right). \quad (9)$$

where D_h is modeled according to Zimmerman's [18] hydraulic diffusivity equation (Equation (7)). The simplified PDE expression in Equation (9) is for an unbounded reservoir, meaning no lateral boundaries occur in the radial direction away from the well.

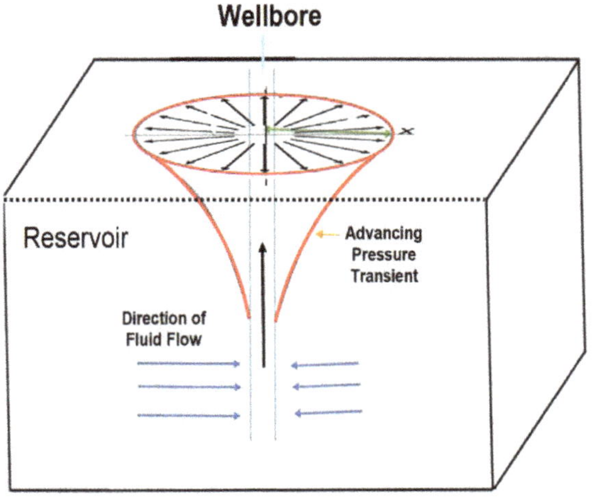

Figure 3. Pressure transient advance (red) for the 2D diffusion of pressure perturbation initiated from a vertical well in a reservoir confined between impervious upper and lower boundaries, but laterally unbounded.

3.2. Steady Pressure Transient Diffusion (1D) for Production and Injection Wells in an Unbounded Reservoir

The key differences between injection and production wells are first established prior to presenting the solutions for such possible applications. Figure 4 illustrates a schematic of both types alongside their bottomhole pressures relative to the initial reservoir pressure. This helps visualize the distinct well designs used for injection and production purposes.

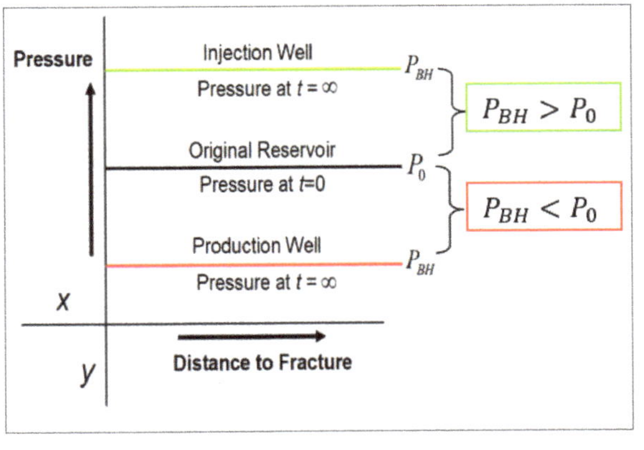

Figure 4. Initial pressure (black line for $t = 0$) and final pressure states (green and red lines for $t = \infty$) in reservoirs for injection and production wells, respectively.

Injection wells: These wells have a higher bottomhole pressure than the original reservoir pressure. This pushes fluids outward away from the well landing point, following the pressure gradient, when injected into the formation.

Production wells: In contrast, production wells have a lower bottomhole pressure compared to the initial reservoir pressure. This pressure difference draws fluids inward towards the wellbore, enabling their extraction, and the drainage patterns enlarges as time goes.

For injection wells, we strive for a boundary condition where the injection pressure is initiated at $(x,\ t) = (0,\ 0)$ with a value $\Delta P(0,0) = P_{BH} - P_0$ (Figure 4), and for $t > 0$, the source maintains a constant pressure differential $\Delta P(0,t) > 0$, given by:

$$\Delta P(0,t) = P_{BH} - P_0. \tag{10}$$

For production wells, the injection pressure is initiated at $(x,\ t) = (0,\ 0)$ with a value $\Delta P(0,0) = P_0 - P_{BH}$ (Figure 4), and for $t > 0$, the source maintains a constant pressure differential $\Delta P(0,t) > 0$, given by:

$$\Delta P(0,t) = P_0 - P_{BH}. \tag{11}$$

We also distinguish between well systems based on types of technical completion: wells drilled vertically and completed with a single longitudinal fracture (bi-wing fracture, Figure 5a) and those completed with simply a vertical, open-hole wellbore (Figure 5b). The bi-wing fractured wells are commonly used in low-permeability reservoirs; their pressure transient is governed by 1D diffusion from the fracture plane. The open wellbore without fracturing is used in high-permeability reservoirs; such wells act as 2D, cylindrical diffusion sources for the pressure transient. Studying the 1D and 2D cases jointly is instructive, in particular because we show in the present study how the 1D diffusion solution can be transformed to a 2D diffusion solution.

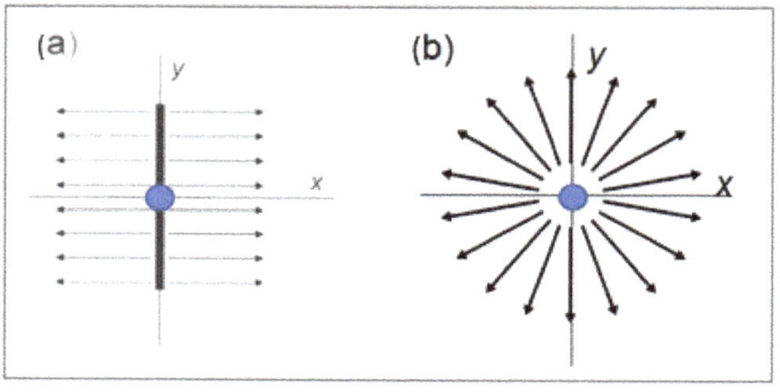

Figure 5. Well types in map view. (**a**) Vertical well with bi-wing fracture acting as a 1D diffusion source. (**b**) Vertical well without fractures acting as a 2D diffusion source from the wellbore of radius $r_w = \sqrt{x^2 + y^2}$.

3.3. *Fractured Well Solution (1D Diffusion)*

The pressure change component $\Delta P(x,t)$ in the x-direction due to a planar, 1D diffusion source along the y-axis (Figure 5a) is as follows:

$$\frac{\partial \Delta P(x,t)}{\partial t} = D_h \frac{\partial^2 \Delta P(x,t)}{\partial x^2}. \tag{12}$$

A valid 1D solution of Equation (12) with the boundary condition of Equation (10) was given by ([38], Equation 2.45):

$$\Delta P(x,t) = \Delta P_0(0,0)\left(1 - erf\frac{x}{2\sqrt{D_h t}}\right). \tag{13}$$

Outside the well system, the pressure transient will change the reservoir pressure according to:

$$P(x,t) = P_0 + \Delta P(x,t). \tag{14}$$

The time-dependent part $\Delta P(x,t)$ of Equation (14) is given by Equation (13), which after normalizing gives the non-dimensional pressure change function $\frac{\Delta P(x,t)}{|\Delta P_0|}$ [using ΔP_0 for $\Delta P(0,0)$] at any time and given position in the reservoir as follows.

For the production well case, the initial reservoir pressure at the well location is changed by the well according to a step function, which lowers the pressure of the reservoir at $(x,\ t) = (0,\ 0)$ with a value $\Delta P(0,0) = P_0 - P_{BH}$; for 1D diffusion from a production well with a single fracture. The equation for the normalized pressure transient that travels perpendicular to the vertical fracture plane acting as the pressure change source is (Figure 6a):

$$\frac{\Delta P(x,t)}{|\Delta P_0|} = erf\left(\frac{x}{2\sqrt{D_h t}}\right) - 1. \tag{15}$$

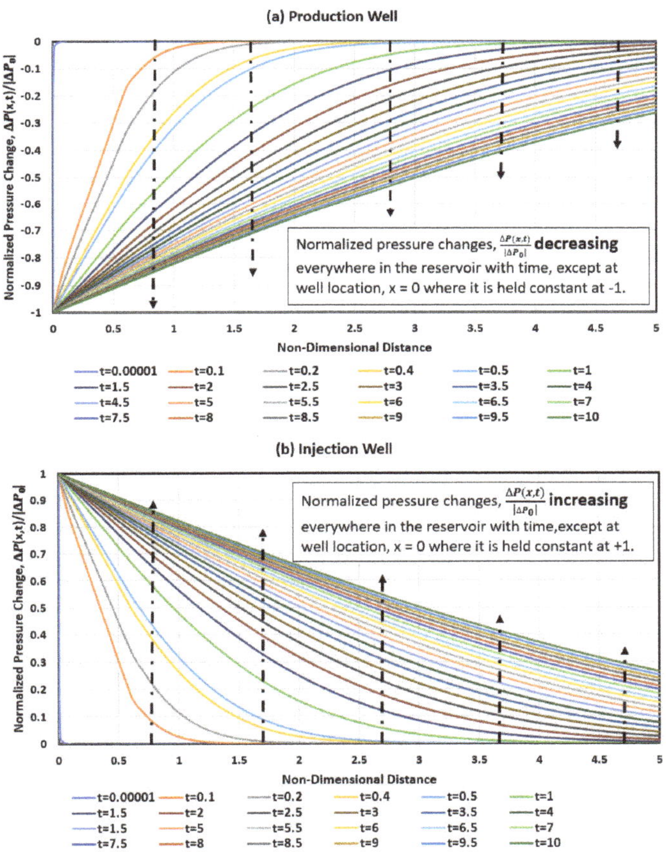

Figure 6. Normalized pressure changes $\frac{\Delta P(x,t)}{|\Delta P_0|}$ for (**a**) 1D (planar fracture) production source shows pressure drawdown over time everywhere in the reservoir except at the fracture location ($x = 0$), where normalized pressure change is fixed at -1. (**b**) Injection well with constant scaled pressure change at the wellbore; $\frac{\Delta P(x,t)}{|\Delta P_0|} = 1$ shows how the pressure increases everywhere in the reservoir (from $x = 0$ to $x = 5$) as time increases from $t = 0.00001$ to $t = 10$ units of time. Diffusivity used: $D_h = 1$.

For a 1D diffusion from an injection source, the step function of the initial pressure increase in the well advances in the lateral reservoir space (x-direction) as follows (Figure 6b):

$$\frac{\Delta P(x,t)}{|\Delta P_0|} = 1 - erf\left(\frac{x}{2\sqrt{D_h t}}\right). \tag{16}$$

Figure 6a shows curves of the pressure transient progressively advancing across the reservoir space for time steps separated by 10 time units for the 1D diffusion of a hydraulically fractured *production* well, using Equation (15). The top border of the graph (y = 0) is where the original reservoir pressure, P_0, remains unchanged. However, the non-dimensional pressure change initiated from the production well advances into the reservoir space and results in progressive pressure depletion associated with the fluid withdrawal process. Ultimately, after infinite time, the pressure everywhere in the reservoir will attain the bottomhole pressure, P_{BH}, which for the production case is given by the non-dimensional pressure change maximum at the bottom border of the graph (at $y = -1$) of Figure 6a.

Figure 6b shows the advance of the pressure change profiles at different times for the 1D diffusion of a fractured *injection* well, according to Equation (16). In this case, the bottom border of the graph (with y = 0) is where the original reservoir pressure, P_0, remains unchanged. However, at a particular time, say t = 0.1 (depicted with yellow line in Figure 6b), the reservoir pressure has changed close to the wellbore, but asymptotically, no pressure change has occurred (yet) further away from the well. For the injector case, the pressures near the well location are increased, and the imposed pressure changes dissipate into the reservoir space as graphed in Figure 6b. Ultimately, after infinite time, the pressure everywhere in the reservoir will reach the bottomhole pressure, P_{BH}, which is given by the normalized pressure change top border of the graph (at y = 1) of Figure 6b.

3.4. Cylindrical Well Solution (2D Diffusion)

The advancing pressure change and normalized pressure depletion profiles for unidirectional 1D diffusion were given in Figure 6a,b. If we were to obtain the 2D solution (Figure 7b) by an orthogonal superposition of 1D pressure change sources (Figure 7a), it would give radial pressure gradients useful for flow path mapping. However, the rate of advance of the 2D diffusion process would be inaccurately scaled by the 1D diffusion process (as in Figure 6a,b), due to which the divergence of the flow paths and the required conservation of mass during the mass transport process will not be satisfied.

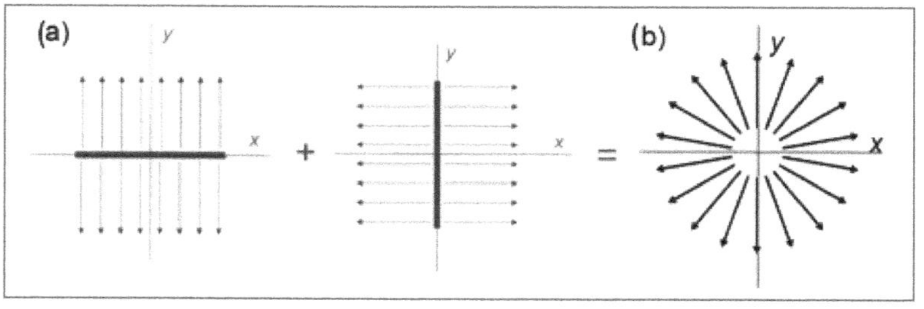

Figure 7. (**a**) Superposition of two 1D diffusion sources in the origin to approximate a 2D diffusion source in (**b**), assuming the 1D sources along x and y are short finite lengths, such that they fit into a wellbore section of radius $r_w = \sqrt{x^2 + y^2}$. It should be noted that such a superposition introduces a scaling error, as detailed in the main text. The conformal mapping of Figure 8 resolves this error.

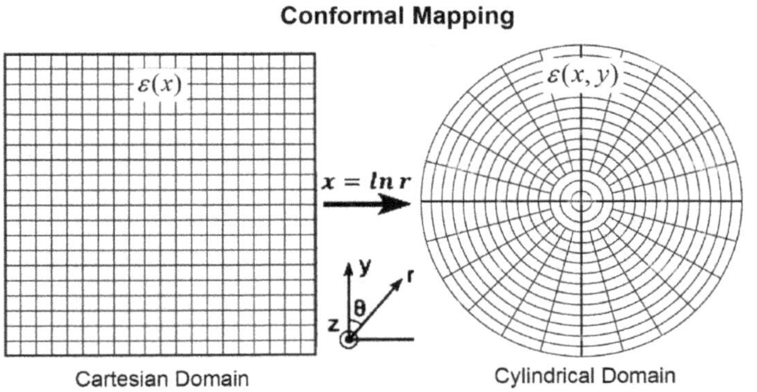

Figure 8. Conformal mapping of the 1D diffusion solution in Cartesian space to divergent diffusion in radial directions by conformal mapping of coordinates from the $\varepsilon(x)$ domain to the $\varepsilon(x,y)$ domain.

In order to correct for the divergence deficit, we apply a coordinate transformation of the 1D diffusion space to a 2D cylindrical diffusion space. The required transformation of the Cartesian reference system to a polar reference system can be achieved by conformal mapping (Figure 8), such that the normalized spatial coordinate x in domain $\varepsilon(x)$ is mapped to $\ln r$ in the domain $\varepsilon(x,y)$:

$$x \to \ln r, \text{ and } r = \sqrt{x^2 + y^2} \tag{17}$$

For the production case, the solution of Equation (15) becomes (Figure 9a):

$$\frac{\Delta P(r,t)}{\Delta P_0} = erf\left(\frac{(r+1)\ln(r+1)}{2\sqrt{D_h t}}\right) - 1 \tag{18}$$

For the injector case, the solution of Equation (16) becomes (Figure 9b):

$$\frac{\Delta P(r,t)}{\Delta P_0} = 1 - erf\left(\frac{(r+1)\ln(r+1)}{2\sqrt{D_h t}}\right) \tag{19}$$

It is important to note that the inclusion of "+1" in the numerator of Equations (18) and (19) serves to accurately represent the wellbore as a point source at $r = 0$. However, in practical wellbore applications, the spatial distance between the wellbore and the boundary typically precludes the necessity of this correction. To plot Figure 6a,b, no inputs are required other than normalized pressure change at the wellbore location and assuming a unit diffusivity. In Section 5, we will show how the well rate declines as a consequence of the lowered bottomhole pressure for a production well case with a specific dimensional hydraulic diffusivity. Unlike classical well test analysis, the well rate is not a boundary condition but an output of the applied bottomhole pressure, original reservoir pressure and hydraulic diffusivity (and well radius, which is assumed to be negligible in Figure 6a,b). The dimensional example of well rate computation and production forecasting using real field data from the early well life is included in Section 5.4.

Comparison of Figure 6a,b and Figure 9a,b reveals the slower advance of the pressure gradient due to the cylindrical source assumed as a consequence of the applied transformation.

Equations (18) and (19) are valid solutions of the general diffusivity equation in polar coordinates (Equation (20)):

$$\frac{1}{r}\frac{\partial}{\partial r}\left(r\frac{\partial P}{\partial r}\right) = \frac{-1}{D_h}\frac{\partial P}{\partial t}, \tag{20}$$

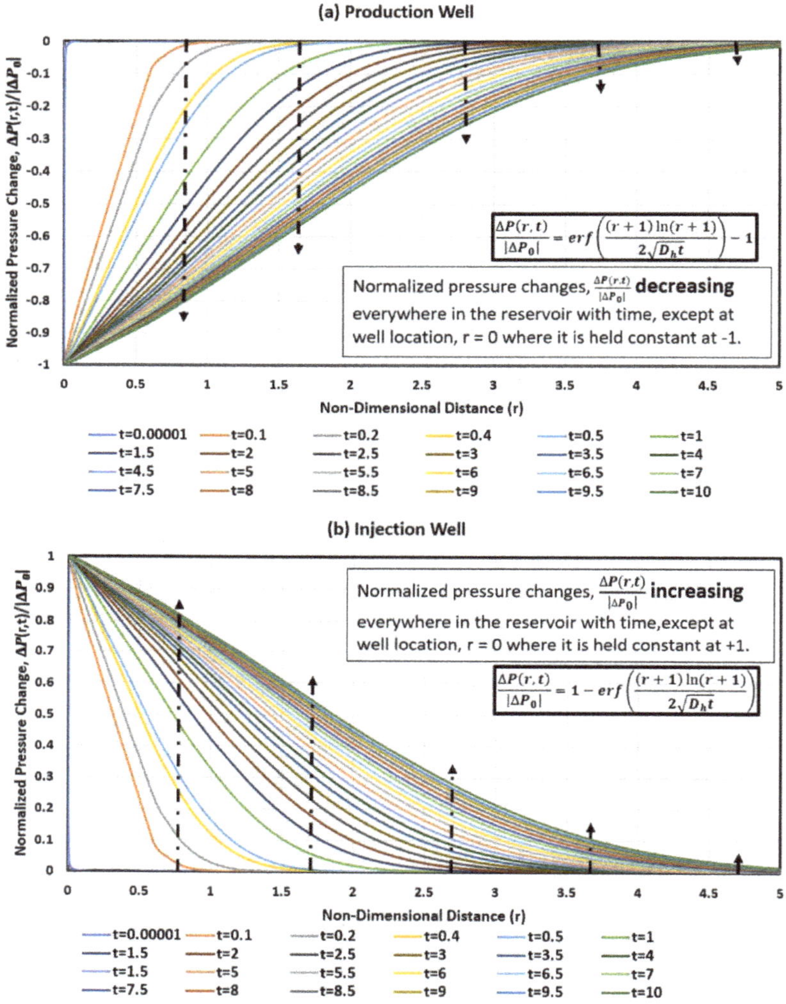

$$\frac{\Delta P(r,t)}{|\Delta P_0|} = erf\left(\frac{(r+1)\ln(r+1)}{2\sqrt{D_h t}}\right) - 1$$

$$\frac{\Delta P(r,t)}{|\Delta P_0|} = 1 - erf\left(\frac{(r+1)\ln(r+1)}{2\sqrt{D_h t}}\right)$$

Figure 9. Normalized pressure changes $\frac{\Delta P(r,t)}{\Delta P_0}$ for (**a**) cylindrical pressure change source (such as occurs in the case of a production well) shows pressure-drawdown over time everywhere in the reservoir except at well location ($x = 0$), where normalized pressure change is fixed at -1. (**b**) Injection well with normalized pressure change at the wellbore, $\frac{\Delta P(r,t)}{\Delta P_0} = 1$.

4. Pressure Gradient Solutions

In Section 3, we focused on solving the pressure transient diffusion process from vertical sources (both injection and production wells with constant bottomhole pressures). We provided illustrative examples of how pressure changes will advance into the reservoir space away from a specific well location over time.

This section develops the solutions for the pressure gradients, which are of practical value for (1) plotting flow paths in the reservoir space and (2) computing the well rate. The flow paths can be identified as the local maxima of the pressure gradient solution (Section 4). Section 5 will explore the concept of well rate. Our objective here is to first present the pressure gradients for the unbounded reservoir solutions.

Flow in reservoirs is in each point principally controlled by the convective movement of mass along the direction of the local maximum of the pressure gradient. The diffusive mass transfer due to Brownian motion in the absence of pressure gradients is neglected in what follows, but if both pressure diffusion and molecular diffusion are considered, a Gaussian Péclet number has been proposed to compare the relative importance of the convective mass transfer due to the pressure gradient and that due to molecular diffusion [40].

4.1. Pressure Gradient Solution for 1D Diffusion

The gradient of the pressure change field for a 1D source (Equation (13)) in an unbounded reservoir is [using ΔP_0 for $\Delta P(0,0)$]:

$$\frac{\partial}{\partial x}\Delta P(x,t) = \frac{\partial}{\partial x}\Delta P_0\left(1 - erf\frac{x}{2\sqrt{D_h t}}\right) = \frac{-\Delta P_0}{\sqrt{\pi D_h t}}\left[e^{\frac{-x^2}{4D_h t}}\right]. \tag{21}$$

The resulting pressure gradient in the radial direction away from a wellbore placed in the origin (depicted in the typical pressure diffusion configuration of sources Figure 5a,b), using a Cartesian grid for a 2D diffusion within the (x, y) plane, may be computed from the following equation, which gives the non-dimensional pressure gradient per length dimension [41]:

$$\frac{\nabla P(x,t)_{NEW}}{|\Delta P_0|} = \frac{1}{\sqrt{\pi D_h t}}\left[e^{\frac{-x^2}{4D_h t}}\right]. \tag{22}$$

4.2. Pressure Gradient Solution for 2D Diffusion

To move from the pressure change solution of Equations (18) and (19) to the pressure gradient solution of the radial well for the injection case, we apply the derivative (Equation (23)):

$$\frac{\partial}{\partial r}\frac{\Delta P(r,t)}{|\Delta P_0|} = \frac{\partial}{\partial r}\left[1 - erf\left(\frac{(r+1)\ln(r+1)}{2\sqrt{D_h t}}\right)\right] = \frac{-\ln(r+1)+1}{\sqrt{\pi D_h t}}e^{\frac{-(r+1)^2\ln^2(r+1)}{4D_h t}} \tag{23}$$

Subsequently, the non-dimensional pressure gradient per unit length is (Equation (24)):

$$\frac{\nabla P(r,t)_{NEW}}{\Delta P_0} = \frac{-\ln(r+1)+1}{\sqrt{\pi D_h t}}e^{-\frac{((r+1)\ln(r+1))^2}{4D_h t}}. \tag{24}$$

4.3. Comparison of 1D and 2D Pressure Gradient Solutions

Consider an injection well located at the center of the field of view, at $(x, y) = (0, 0)$. Equations (22) and (24) describe the dimensionless pressure gradient solution per unit length of a 1D planar and a 2D cylindrical pressure source, respectively. The 1D pressure change solution of Equation (22) is here used to compare the pressure change advance from a well with two perpendicular sets of bi-wing fractures (short, with fraction of unit length) as in Figure 7b with the solution of the 2D-diffusion-driven flow around the cylindrical well as in Figure 8.

The resulting pressure gradient in the radial direction away from a wellbore placed in the origin (using a conjugate set of 1D planar sources as in Figure 5a,b), represented in a Cartesian grid for the resulting apparent 2D diffusion within the (x, y) plane, may be computed from [41]:

$$\nabla P(x,y,t)_{NEW} = \sqrt{\left(\frac{\partial\Delta P(x,t)}{\partial x}\right)^2 + \left(\frac{\partial\Delta P(y,t)}{\partial y}\right)^2}. \tag{25}$$

Substituting the 1D diffusion solution of Equation (22) into Equation (25) gives a Gaussian pressure transient solution for a vertical wellbore with two perpendicular, planar sources, using Cartesian coordinates (Equation (26)):

$$\nabla P(x,y,t)_{NEW} = \sqrt{\left(\frac{-\Delta P_0}{\sqrt{\pi D_h t}}\left[e^{\frac{-x^2}{4D_h t}}\right]\right)^2 + \left(\frac{-\Delta P_0}{\sqrt{\pi D_h t}}\left[e^{\frac{-y^2}{4D_h t}}\right]\right)^2} = \frac{\Delta P_0}{\sqrt{\pi D_h t}}\left[e^{\frac{-(x^2+y^2)}{4D_h t}}\right]. \quad (26)$$

The non-dimensional pressure gradient per length dimension is (Equation (27)):

$$\frac{\nabla P(x,y,t)_{NEW}}{\Delta P_0} = \frac{1}{\sqrt{\pi D_h t}}\left[e^{\frac{-(x^2+y^2)}{4D_h t}}\right]. \quad (27)$$

Figure 10a–f illustrates the early time development of the pressure fields, pressure gradients and flow paths for the 1D and 2D diffusion cases. The difference between the two cases for t = 0.1 remains modest.

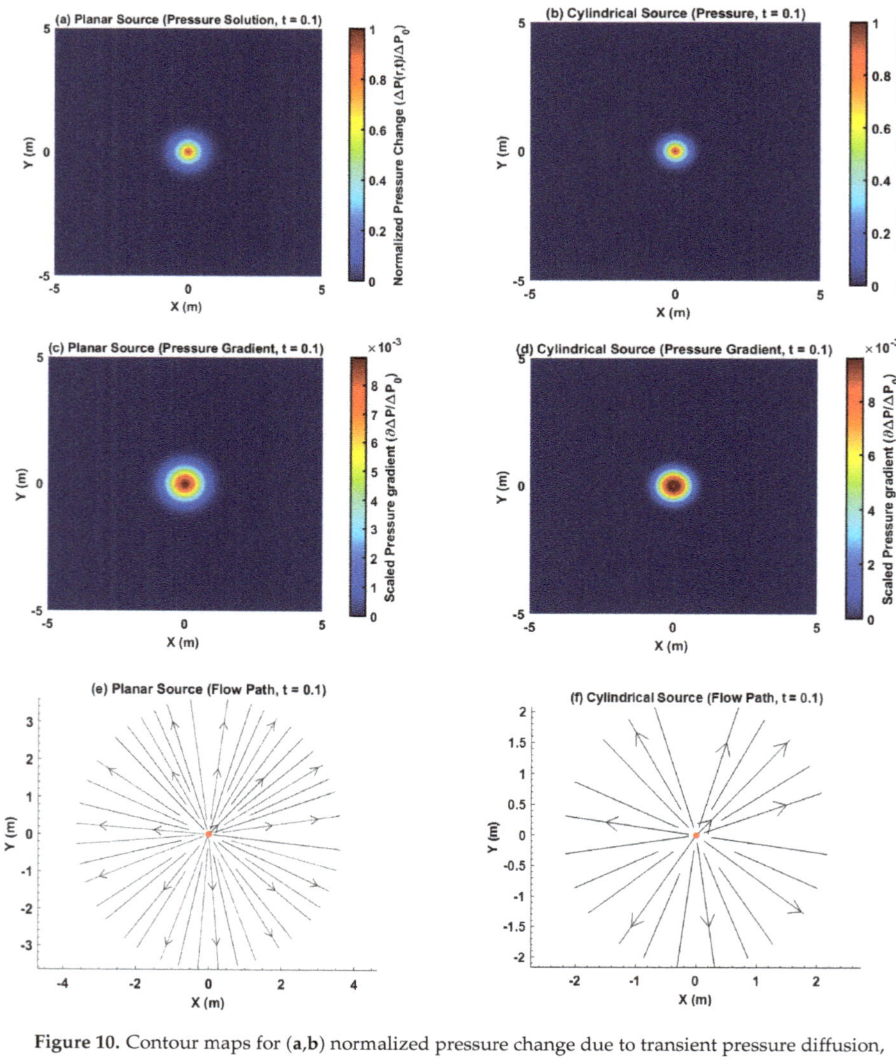

Figure 10. Contour maps for (**a**,**b**) normalized pressure change due to transient pressure diffusion, (**c**,**d**) pressure gradients and (**e**,**f**) flow paths, all at t = 0.1. The left column is for two orthogonal planar sources (**a**,**c**,**e**). The right column is for a cylindrical source (**b**,**d**,**f**).

We next re-examine the same two cases later in the well-life (Figure 11a–f), after considering $t = 100$, keeping input parameters the same, except that the field view dimension is increased from 5 to 50 units of spatial distance from the well location at $r = 0$. The difference between the two cases now becomes more pronounced. The arrows in Figure 10e,f and Figure 11e,f indicate the direction of fluid flow, which is radially outward for the injection well.

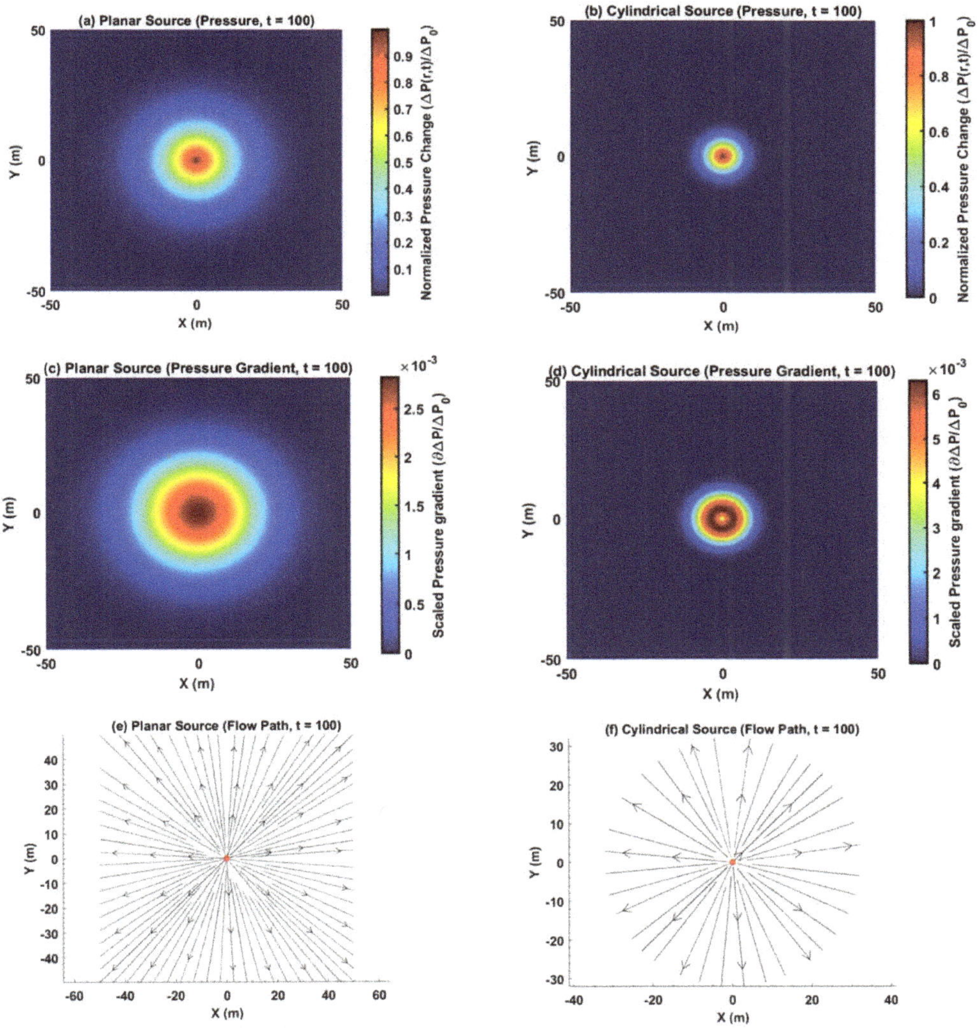

Figure 11. Contour maps for normalized pressure change, pressure gradient and flow paths for two orthogonal planar sources (left column: **a**,**c**,**e**) and a cylindrical source (right column: **b**,**d**,**f**) at $t = 100$.

The purpose of computing the superposition of two orthogonal 1D planar diffusion sources (Figure 7 configuration) and the cylindrical 2D diffusion source obtained by the transformation of Figure 8 is to demonstrate the different outcomes. The correct procedure for a pressure transient initiated by a pressure disturbance at a cylindrical wellbore with a constant bottomhole pressure is to use the solution of Equation (24) not (27).

5. Well Rate Solutions

Building upon the foundation of pressure gradients laid out in Section 4, this section seeks to establish analytical solutions for well rates applicable to both injection and production wells.

5.1. Well Rate Solutions for Single Bi-Wing Fractured Well

The petroleum industry has shifted to massive use of horizontal wells that withdraw the fluid from ultra-low-permeability reservoirs via sets of transverse hydraulic fractures. Gaussian solutions for the 1D diffusion from such multi-stage fractured wells have been derived in recent companion studies [22,23,41–45]. The well rate solution for a fractured vertical well in a 2D Cartesian coordinate system is obtained using the time-dependent pressure gradient solution of Equation (21) for steady-state pressure diffusion in an unbounded reservoir in combination with Darcy's Law. The pressure gradient accounts for the volumetric flow rate per unit cross-sectional area entering the fracture system of the well, using the relationship between flow rate and volumetric flow rate per unit fracture area, $q_{WH}(t) = 4lh.\ q_D(t)/B$ and $q_D(t)$, from Equation (3) to give the wellhead flow rate:

$$q_{WH}(t) = \frac{4lhk}{\mu B} \frac{(P_0 - P_{BH})}{\sqrt{\pi D_h t}} e^{\frac{-x_0^2}{4D_h t}}. \tag{28}$$

The factor 4 in the solution of Equation (28) stems from use of the fracture half-length and the fact that flow into the bi-wing fracture occurs from the reservoir space at either side of the fracture. The parameter B accounts for the volume changes when fluid is lifted from reservoir space to the surface and is determined in specialized laboratories of sampled reservoir fluid in so-called PVT-analysis.

5.2. Well Rate Solutions for Cylindrical Wells in Laterally Unbounded Reservoirs

The solution for the rate of a vertical well is obtained using the time-dependent pressure gradient solution for steady-state pressure diffusion in unbounded reservoirs (Equation (23)). The volumetric flow rate per unit wellbore area, $q_{WH}(t) = 2\pi r_w h.\ q_D(t)/B$, was used to obtain the well rate, replacing $r + 1$ by r_w; assuming $r_w > 0$:

$$q_{WH}(t) = \frac{2\sqrt{\pi}(hk)(r_w \ln(r_w) + r_w)}{\mu B} \frac{(P_0 - P_{BH})}{\sqrt{D_h t}} e^{\frac{-r_w^2 \ln^2 r_w}{4D_h t}}. \tag{29}$$

The solution of Equation (29) differs from traditional well-testing equations of Equation (6). The latter assumes a constant well rate, which offers advantages for reservoir characterization using various well test rates at a constant imposed bottomhole pressure over brief periods of time, justifying treatment of both the bottomhole pressure and well rate as being approximately constant. However, it is possible to adapt our new solution for well test analysis [37]. Additionally, the future well rate can be predicted based on the pressure differential between the reservoir and the wellbore applied by the artificial lift system used.

5.3. Well Rate Solutions for Vertical Wells in Laterally Bounded Reservoirs

The solutions in the present paper so far are for vertical wells in transient flow due to a pressure drop originating from the well advancing into the ambient reservoir. The 2D diffusion is described with Gaussian pressure transients. While the research focus was initially on the production analysis of fractured wells used in so-called unconventional reservoirs (ultra-low-permeability rocks), the solutions offered in the present study are of equal practical significance for vertical injection wells and for vertical production wells in conventional formations. When the reservoir is at the lower end of the permeability spectrum, the transient flow period may last for much of the economic well life. Anisotropic diffusivity can be accounted for by using a hydraulic diffusivity tensor rather than scalar values. Examples have been given in Alotaibi et al. [35].

However, for conventional reservoirs with high permeability, the transient flow period will be limited to the early well life (first few months of production). To be of more practical value, the unbounded solution is expanded here in a solution for bounded reservoirs with an assumed cylindrical container shape of finite radius approximating the confined reservoir. Given that the Gaussian distribution relies on probabilistic concepts, a novel probability concept described in a prior study [46] was employed to formulate the innovative pressure diffusion model tailored for reservoirs dominated by bounded flow. To simulate the cylindrical reservoir boundary, the pressure front from the wellbore was superimposed, with a virtual pressure front originating from a cylindrical pressure source located at twice the distance between the well location, r_w, and the effective reservoir boundary, r_e (as elucidated in Figure A2, Appendix A of Afagwu and Weijermars [46]). The normalized pressure transient diffusion solution for laterally bounded cylindrical reservoirs for a production case is given as follows (Equation (30)):

$$\frac{\Delta P(r,t)}{|\Delta P_0|} = erf\left(\frac{(r+1)ln(r+1)}{2\sqrt{D_h t}}\right) erf\left(\frac{(2r_e - r + 1)ln(2r_e - r + 1)}{2\sqrt{D_h t}}\right) \tag{30}$$

The validity of the normalized pressure solution (Equation (30)) is demonstrated in Figure 12 using the same approach originally described in our prior publication [46]. The tracked time-dependent pressure changes between the scaled wellbore location and real reservoir boundary confirmed that the boundary conditions were honored. The normalized pressure drawdown is at its maximum at the wellbore throughout the production period because of the constant bottomhole pressure boundary condition imposed at the wellbore. At the beginning, t = 0.06, there is no pressure change around the normalized real reservoir boundary, r = 1, (Figure 12a). However, from t = 0.1 units of time, there is gradual pressure change, suggesting that the pressure transient originating from the wellbore has reached the boundary. During late production time depicted by t = 1000 units of time, the pressure is nearly depleted everywhere in the reservoir (Figure 12b). Furthermore, the reduction in the steepness of the slope of spatial pressure changes indicates a corresponding decrease in gradient flux between the start of production at t = 0.06 units of time to late time, t = 1000.

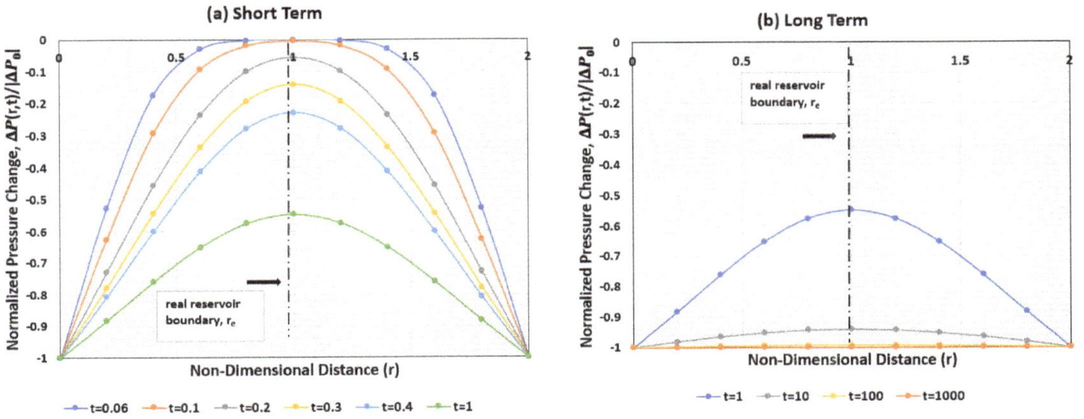

Figure 12. Time-dependent normalized pressure change solution (Equation (30)) from the scaled wellbore position to the reservoir boundary extent, estimated with unit hydraulic diffusivity, $D_h = 1$. (**a**) Short unit of time, t = 0 to 1 unit; (**b**) longer period of production, from 1 to 1000 units of time.

The well rate expression of Equation (29) for the unbounded vertical well can now be modified as follows to account for the bounded reservoir case. The derivation of the wellbore flow rate solution for laterally bounded reservoirs follows the three-step procedure as follows: (1) Compute the time-independent pressure gradient solution for steady-state pressure diffusion in bounded reservoirs (Equation (30)). (2) Darcy's Law integration: the pressure gradient is then integrated across the wellbore radius using Darcy's Law (Equation (3)) to relate the pressure gradient to the volumetric flow rate per unit cross-sectional area entering the wellbore. (3) Wellbore flow rate relationship: the resulting expression is substituted into the relationship between flow rate and volumetric flow rate per unit area, $q_{WH}(t) = (2\pi r_w h.\ q_D)/B$, to give the wellbore flow rate model as follows (Equation (31)), where $C_1 = 0.00632829\ \left(\left(\text{cp.ft}^2\right)/(\text{mD.psi.day})\right)$ is the conversion factor from metric to field unit.

$$q_{WH}(t) = \frac{2C_1\sqrt{\pi}(hk)}{\mu B_0}\frac{(P_0-P_{BH})}{\sqrt{D_h t}}[(r_w \ln(r_w) + r_w).e^{\frac{-r_w^2 \ln^2 r_w}{4D_h t}}.erf\left(\frac{(2r_e-r_w)\ln(2r_e-r_w)}{2\sqrt{D_h t}}\right) - (r_w \ln(2r_e - r_w) + r_w).e^{\frac{-(2r_e-r_w)^2\ r_w^2 \ln^2(2r_e-r_w)}{4D_h t}}.erf\left(\frac{r_w \ln(r_w)}{2\sqrt{D_h t}}\right)] \quad (31)$$

5.4. Application to Real Well Data

In this section, we give examples of history matching the production data from real wells with the GPT-based production forecasting method. In prior studies, partial solutions were compared with results from numerical simulators with excellent matches [37,41]. In the present study, we use real production data to demonstrate the value of our new solutions in practical applications. The production data for our chosen study wells come from decommissioned UK offshore fields, which have been made publicly available by the UK North Sea Transition Authority, as was detailed in [47].

The daily production data and cumulative production data of study wells from two UK offshore gas fields (Tethys and Kelvin) are plotted in Figure 13a–d. The Tethys field occurs in rocks of Permian age: the Leman Sandstone Formation of the Lower Permian Rotliegendes Group [47]. The Kelvin field is hosted in fluvial sandstones of the Upper Devonian–Carboniferous Fairway. When the full field data are used, it appeared in our prior study that use of the empirical decline curve analysis (DCA) method by Arps to historically match the production data can give curve matches that closely match the actual well behavior (Figure 13a–d).

However, for production forecasting of newly drilled wells and establishing the estimated ultimate recovery (EUR) volumes for reserve estimation, only a few months' production data would typically be available. We therefore tested how well Arps forecast the actual well behavior, if we used only the first 6 months of bootstrapped daily production data [47]. It then appeared that the Arps method gave rather inaccurate forward prediction of the well rates and either overestimated (Tethys) or underestimated (Kelvin) the ultimate recovery (EUR) volumes (Figure 14a–d).

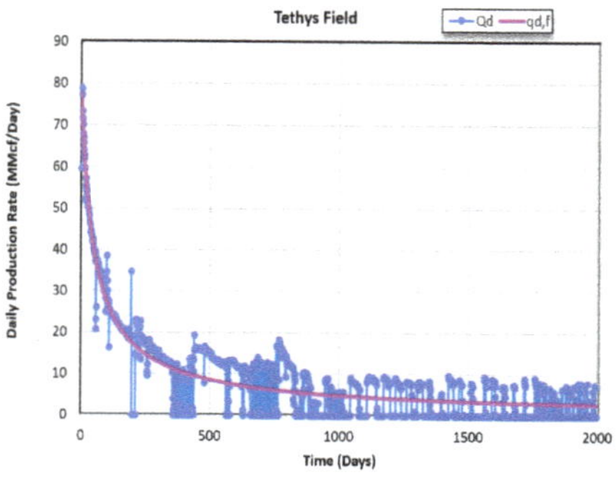

(a)

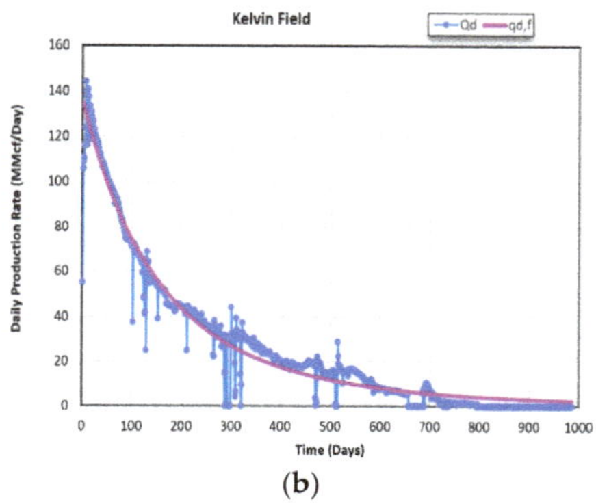

(b)

Figure 13. *Cont.*

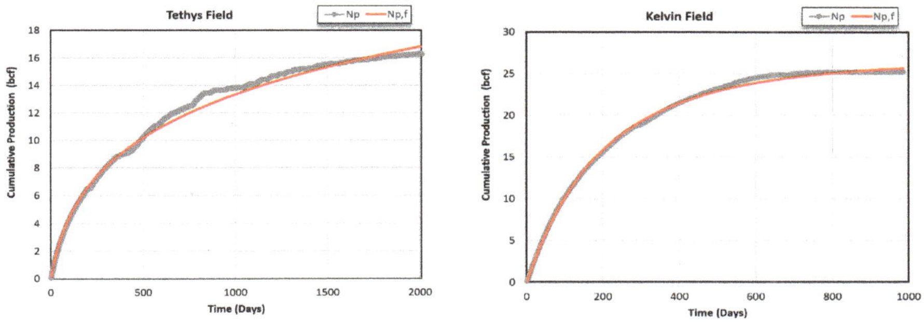

(c) (d)

Figure 13. Daily production rates (top row, (**a**,**b**)) and cumulative production (bottom row, (**c**,**d**)) of the Tethys and Kelvin Wells were fitted with Arps DCA curves [47] and provided excellent history matches throughout the well life. The detailed fitting parameters are given in [47].

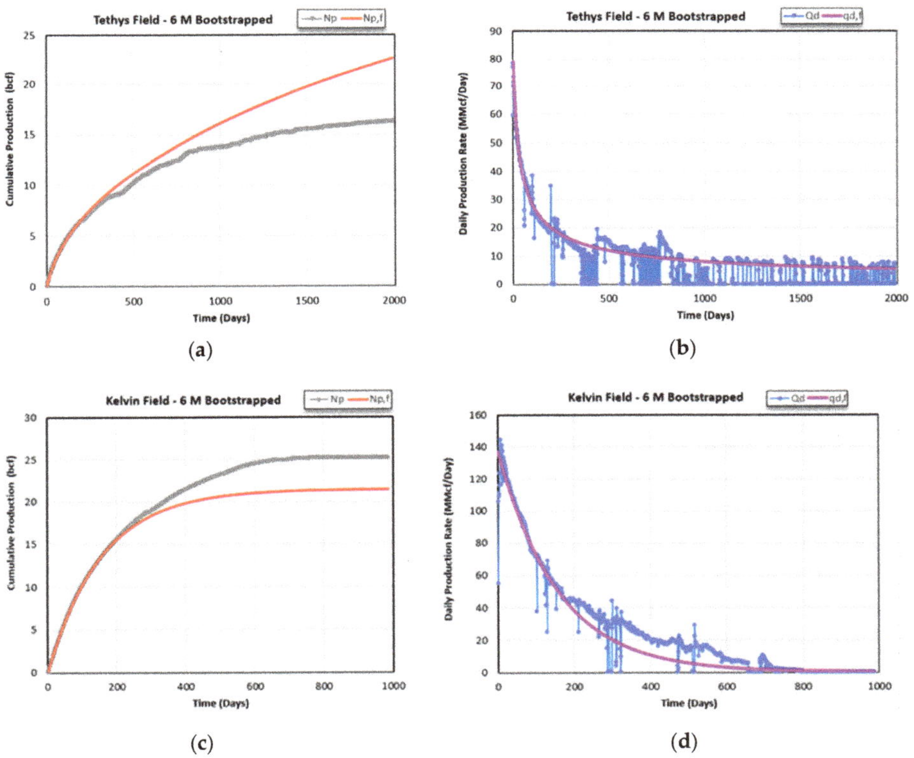

Figure 14. History matching of full well-life cumulative production and daily rates using only the first 6 months of daily production [47] gave poor history matches for the post-6-month production behavior for both the Tethys Well (top (**a**,**b**)) and the Kelvin Well (bottom (**c**,**d**)). The detailed fitting parameters are given in [47].

The conclusion was that the empirical Arps method becomes increasingly inaccurate when only a few months of production data are available. Remedying this limitation was one of the motivations to develop our physics-based GPT history-matching solution. The bounded solution of Equation (31) was incorporated in a spreadsheet for history matching the early production data, together with the traditional Arps equation described in [47], to check how well the physics-based GPT history match will predict the future well performance and EUR if again only the first 6 months of the historic production data were used. Figure 15a–d gives the outcome of the least square error history matches for both the Arps DCA and the GPT methods. Our conclusion is that using the physics-based GPT solution for a bounded reservoir gives excellent history matches. The detailed fitting parameters used in Equation (31) are given in Table 1.

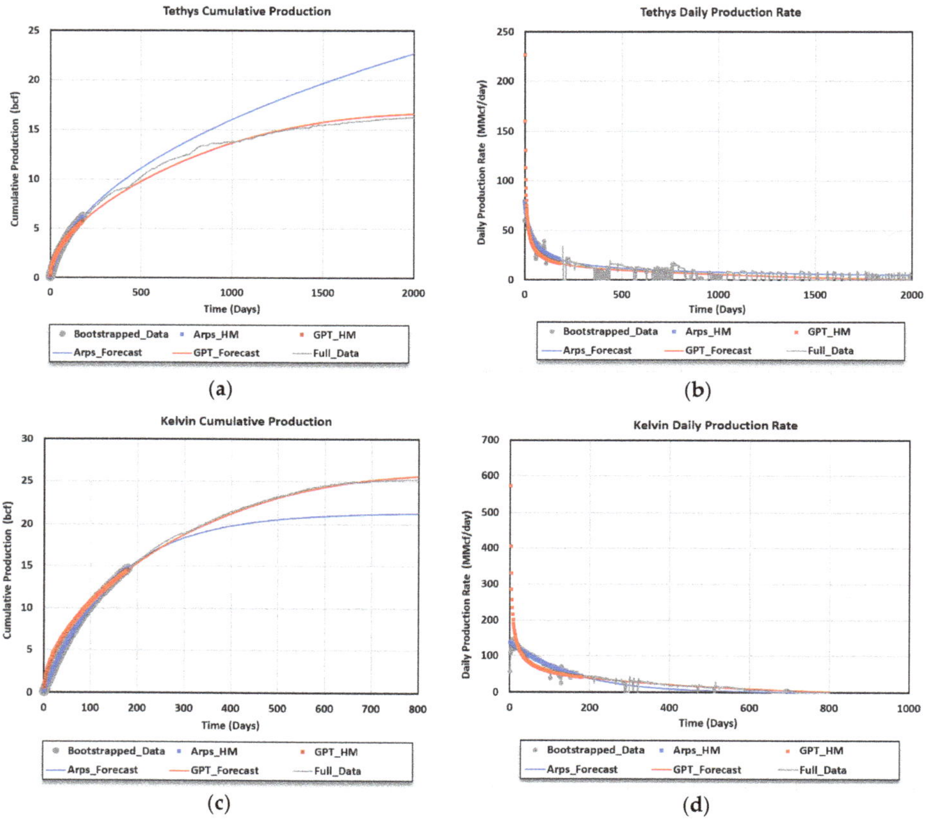

Figure 15. History matching of full well life cumulative production and daily rates using only the first 6 months of daily production [using the GPT solution-method, Equation (31) (red curves) gave excellent history matches for the post-6-month production behavior (gray curves) for both the Tethys Well (**top** (**a**,**b**)) and the Kelvin Well (**bottom** (**c**,**d**))]. The detailed fitting parameters are given in Table 1.

Apart from the excellent history matches obtained with GPT, an additional strength is its diagnostic value for reservoir characterization. For example, the effective radius of the reservoir is obtained from the history-matching process (last row in Table 1). Once the effective radius has been determined in the history match, changing the reservoir radius parameter shows how the transition from transient flow to boundary dominated flow would shift. For example, if the radius of Kelvin reservoir reduces from 5500 to 2750 ft, the boundary-dominated flow occurs much earlier, and the reservoir would effectively be depleted after 200 days (Figure 16a). If we use a 4000 ft radius, the reservoir would deplete after 450 days (Figure 16b). Because the well is known to produce for 800 days, the physics-based GPT model correctly shows a larger reservoir radius of 5500 ft.

Table 1. GPT fitting parameters used for Figure 15 results.

Physical Attribute	Unit	Tethys Field Well	Kelvin Field Well
Age trap system		Permian	Carboniferous
Effective thickness	ft	1034	50
Bottomhole pressure	psi	1800	3386
Original reservoir pressure	psi	3800	6310
Formation volume factor	rcf/scf	0.0029	0.0029
Viscosity	cPoise	0.02242	0.022422
Rock compressibility	1/psi	0.0000434	0.0000434
Fluid compressibility	1/psi	0.000033	0.000033
Permeability	mD	10	400
Porosity	-	0.11	0.11
Hydraulic diffusivity	ft^2/day	66,792	2,671,675
Effective well radius	ft	3.6	13
Reservoir boundary (r_e)	ft	1750	5500

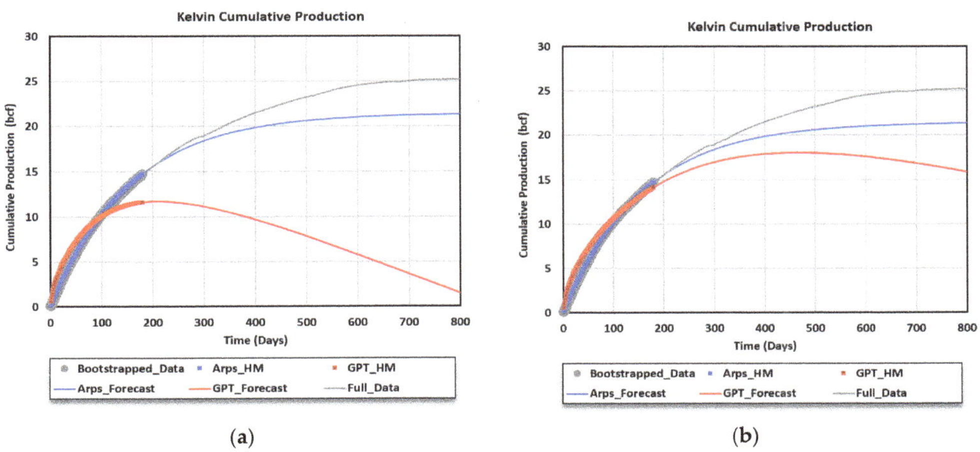

Figure 16. Sensitivity of GPT solution, Equation (31), to reservoir radius. (**a**) Radius reduced to half of the history-matched value of 5500 ft (Table 1); the EUR would reduce from 25.8 to 12.2 bcf. (**b**) Radius reduced from 5500 ft to 4000 ft shrinks the productive well life from 800 to 450 days; the EUR would reduce from 25.8 to 18 bcf.

5.5. Further Applications of Pressure-Transient-Based Well Rate Solutions

The pressure-transient-based well rate solutions can be used in a variety of practical applications:

1. *Pressure transient simulation*: These solutions can simulate the pressure transients caused by fluid flow towards a well system in unbounded and boundary-dominated flow reservoirs. Such solutions allow for analyzing the dynamic behavior of pressure during production or injection (see Figures 6 and 9).
2. *Reservoir petrophysical characterization:* By analyzing well test data using these solutions, reservoir properties can be characterized [37], such as:

a. *Permeability (k):* A key parameter controlling fluid flow in the reservoir.
b. *Skin factor:* A dimensionless parameter that quantifies wellbore damage (positive skin) or stimulation (negative skin).
c. *Effective wellbore radius* (which includes skin factor): This reflects the actual flow area around the wellbore, potentially influenced by formation damage or stimulation.

3. *Field development planning:* By integrating the information obtained from well test analysis (permeability, skin factor and transient distance), these solutions can guide investment decisions for field development. This can help ensure the reservoir meets stakeholder expectations by selecting an appropriate development plan based on exploratory and appraisal well data.

6. Discussion

6.1. Importance of Well Rate Forecasting Methods

Accurate production forecasts and reserve estimations are particularly important for the global petroleum industry in order for individual companies to estimate their remaining hydrocarbon reserves which underpin the book value of their fixed assets in the balance sheet and serves as collateral for credit lines. Reporting of the proved reserves (P90 certainty) is mandatory as per the guidelines of the US Security and Exchange Commission (SEC) and applies to all petroleum companies listed on the New York Stock Exchange (NYSE).

Two types of practical tools for well rate forecasting are widely used.

One type of tool is the commercial reservoir simulators (GMC, Kappa, Eclipse, etc.), of which many different varieties are offered by service companies, which provide slick interfaces with ready-to-use input boxes. The actual computations are invariably based on numerical, finite difference computational solution methods. While the results are reasonably accurate, the user licenses for the commercial simulators are costly, the methods and assumptions involved are non-transparent and filling out the detailed well data in the user interface requires quite a bit of practical experience. Also, the runtime to obtain output may take several hours, especially when probabilistic determination of reserve categories is required.

The other principal type of tools popular for determining hydrocarbon well performance and related reserves reporting is based on empirical decline curve analysis. Numerous decline curve analysis (DCA) formulas have been proposed in the past, as reviewed in considerable detail elsewhere [48,49]. These tools are not physics-based and largely phenomenological and use real well data to match decline curve formulas and then project the future well performance based on the historic trend. Analogy and type-curve well assumptions are extensively used to book reserves based on still undrilled but vacant well locations in the acreage leased by the companies.

The solutions for well rate forecasting for fractured wells (Equation (28)) and cylindrical wells (Equation (29)) in unbounded reservoirs and vertical wells in bounded reservoirs (Equation (31)) provide physics-based analytical solutions to predict well performance. The key unknown parameter for new reservoirs to be drilled is the hydraulic diffusivity. The diffusivity can be computed either using Equation (7) with primary parameter inputs, or by history matching early pilot well data, as demonstrated in [50].

Table 2 lists the strengths and weaknesses of the existing types of tools for well analysis, production forecasting and reserve estimation. Because nodal analysis is rather technical and requires specialist knowledge, companies apply the tool only sparsely. Strengths and weaknesses of the new Gaussian pressure transient solution are also included. A major strength of the Gaussian method is that historic well data are not needed to predict the performance of a new well, provided the hydraulic diffusivity is already known from applying Zimmerman's Equation (7).

Table 2. Strengths and weaknesses of various production analysis tools.

Method	Strengths	Weaknesses
Reservoir Simulators	• Reliable outputs (for multi-phase) production forecasting of (unfractured wells)	• Fractured well models involve many simplifying assumptions • Proppant transport algorithms are inaccurate, leading to overestimation of propped fracture half-lengths • Costly licenses; interface use requires training
Decline Curve Analysis	• Fast and easy to use • Rapid Monte Carlo simulation possible for reporting reserve categories	• Empirical solution method • Requires historic well data (preferably at least 18 months)
Nodal Analysis	• Can use the average reservoir pressure and production system outlet pressure. • Not limited to assumption of a fixed bottomhole pressure, because the pressure balance in the nodal analysis will generate the bottomhole pressure as an output • Can model well rates and pressure when the well is on natural flow without artificial lift support	• Involves elaborate time-stepping to match the pressures at the nodes • Detailed friction parameters needed • Multi-phase flow effects in wellbore are hard to quantify correctly, but needed for nodal matches
Gaussian Pressure Transient Solutions	• Fast and easy to use • Rapid Monte Carlo simulation possible for reporting reserve categories	• Physics-based solution methods • Does not require historic well-data

6.2. Relation to Prior Work

In previous research, various analytical solutions have been devised for interpreting well testing data and calculating well rates resulting from pressure changes induced by production in aquifers and oil reservoirs [17,18,20,21]. However, these earlier solutions typically assumed a constant well rate as the boundary condition. This assumption does not align with real-world scenarios where fluid withdrawal or injection is managed by pump devices that maintain a constant bottomhole pressure, a common practice in oil and gas wells equipped with artificial lift systems. Consequently, the ability to rapidly compute the decline and associated changes in the well rate over time based on the pressure differential applied at the well by the lift system (pump) relative to initial reservoir conditions has been constrained by the absence of concise expressions accommodating the constant bottomhole pressure boundary conditions.

The present study refined and expanded upon the analytical solutions established in recent works [22,42]. The earlier contributions provided valuable insights into fluid dynamics within subsurface reservoirs and introduced initial techniques for analytical modeling to compute well rates based on hydraulic diffusivity derived from real field data and production-induced pressure changes, with a particular emphasis on unconventional reservoirs.

By introducing a new analytical solution that accommodates constant bottomhole pressure boundary conditions and incorporates pressure interference from neighboring wells or external boundaries through the superposition of a cylindrical image source, the present study represents a significant leap forward from prior work. This advancement bridges the gap between theoretical models and practical applications by offering a practical and efficient approach for analyzing well testing data and computing well rate changes under constant bottomhole pressure conditions. Moreover, the new solution eliminates the need for time-intensive pressure-matching procedures, such as nodal analysis, for determining well rate declines in systems operating under constant bottomhole pressures. Through a reassessment of the diffusivity equation and the derivation of tailored well rate equations, this study provides a practical and efficient means to compute well rate changes based on fundamental physical parameters, while also accounting for pressure interference effects on diffusion and offering insights into the sensitivity of engineering parameters to outcomes.

The following practical recommendations can be formulated based on the insights from this study:

1. Analyze real-world historical reservoir pressure and production measurements with instantaneous and steady pressure analytical solutions. Refer to well test paper success in providing valuable estimates of parameters relevant for modeling real-world reservoir behavior.
2. Expanding analytical models to incorporate multi-phase flow phenomena, such as gas–liquid interactions or phase behavior effects, would advance our understanding of complex reservoir systems and enable more accurate predictions of production rates.
3. Explore methods for incorporating uncertainty analysis into analytical models, allowing for probabilistic assessments of reservoir performance and providing decision-makers with more robust risk assessments.
4. Explore the application of analytical models to emerging reservoir types, such as geothermal reservoirs, to expand the scope of their applicability and address pressing challenges in energy production.
5. Conduct comparative studies to validate the evolving analytical models and benchmark their performance against empirical data to enhance confidence in their predictive capabilities and facilitate their adoption in practical engineering applications.

These recommendations represent potential avenues for further research aimed at advancing our understanding of reservoir behavior and improving the effectiveness of reservoir management practices. By addressing these research areas, we can meaningfully contribute to the sustainable development of energy resources and the optimization of reservoir performance.

7. Conclusions

This study introduces novel analytical solutions for computing well rates in aquifers and oil reservoirs under the condition of constant bottomhole pressure, a scenario often encountered in practical applications involving pump devices, addressing a significant gap in the prior literature. Previous solutions assumed a constant well rate. The new solutions presented in this study enhance the understanding of system dynamics and ensure the precision of production metric estimations. This represents a significant step forward in the field of analytical modeling for well testing and reservoir management. This new approach enables the rapid computation of well rate changes based on pressure differences, reservoir permeability, diffusivity and fluid properties. This study derives a solution for the pressure diffusion equation with a constant bottomhole pressure boundary condition, allowing for the computation of well rate changes without the need for elaborate time-stepped procedures.

Our research builds upon the foundation laid by previous works, as evidenced by the contributions of the Weijermars research group [22,23,42]. The emphasis in the present study is on applying the diffusion equation to a vertical wellbore in a horizontal, permeable reservoir, assuming a constant bottomhole pressure. Furthermore, our approach provides a comprehensive framework for understanding reservoir dynamics and optimizing production strategies.

The discussion section highlights the advantages of the proposed solution over nodal analysis, a traditional method for computing declining well rates. The new solution eliminates the need for time-stepped pressure matching and provides a more straightforward and rapid approach for forecasting well rates. The paper also compares the new solution to other forecasting tools, such as reservoir simulators, decline curve analysis and nodal analysis, outlining the strengths and weaknesses of each approach. The derived Gaussian pressure transient solutions offer practical benefits, particularly for wells produced with artificial lift systems, and provide a valuable alternative and complement to existing methods used in the petroleum industry. In conclusion, our study represents a significant advancement in the field of reservoir engineering, offering a robust analytical toolkit for estimating well rates and paving the way for more accurate and efficient reservoir management in the oil and gas industry.

Author Contributions: Conceptualization, R.W.; methodology, C.A.; software, C.A.; validation, R.W. and C.A.; formal analysis, R.W. and C.A.; investigation, R.W. and C.A.; data curation, R.W.; writing—original draft, R.W.; writing—review and editing, C.A.; visualization, R.W. and C.A. All authors have read and agreed to the published version of the manuscript.

Funding: The APC was funded by the College of Petroleum Engineering & Geosciences (CPG) at King Fahd University of Petroleum & Minerals (KFUPM).

Data Availability Statement: The data presented in this study are available from the UK North Sea Transition Authority.

Acknowledgments: The authors acknowledge the generous support provided by the College of Petroleum Engineering & Geosciences (CPG) at King Fahd University of Petroleum & Minerals (KFUPM).

Conflicts of Interest: The authors declare no conflict of interest.

Nomenclature

Nomenclature is provided with units in metric and field units in parentheses.

B	Formation volume factor	m^3/sm^3(cf/scf)
c_f	Fluid compressibility	1/Pa (1/psi)
c_R	Rock compressibility	1/Pa (1/psi)
c_t	Total compressibility	1/Pa (1/psi)
D_h	Hydraulic diffusivity	m^2/s (ft^2/day)
E_i	Exponential integral function	
h_0	Original water table height	m (ft)
h_1, h_2	Drawdown heights from two observation wells	m (ft)
h	Reservoir thickness	m (ft)
k	Reservoir permeability	m^2 (mD)
K	Hydraulic conductivity of the aquifer	m/s (ft/day)
P	Reservoir pressure	Pa (psi)
P_0	Original reservoir pressure	Pa (psi)
P_{BH}	Bottomhole flowing pressure	Pa (psi)
q_D	Darcy flux per unit area of wellbore	m/s (ft/day)

q_{WH}	Wellhead production rate	m^3/s (scf/day)
r_1, r_2	Radial distances of the two observation wells from the origin	m (ft)
r_e	Reservoir extent or distance from wellbore to real reservoir boundary	m (ft)
r_w	Wellbore radius	m (ft)
r	Radial or cylindrical coordinate distance from well location	m (ft)
s	Dimensionless skin factor	
S	Aquifer storativity, dimensionless	
t	Time	s (day)
T	Aquifer transmissivity	m^2/s (ft^2/day)
v_R	Actual velocity	m/s (ft/day)
x, y	Radial 2D Cartesian coordinate distance from well location	m (ft)
$\Delta h(r,t)$	Change in hydraulic head	m (ft)
∇P	Reservoir pressure gradient	Pa/m (psi/ft)
ϕ	Connecting porosity	
μ	Fluid viscosity	Pa.s (cPoise)
∇	Del operator	1/m (1/ft)
∂	Partial differential operator	

References

1. Matos, C.R.; Carneiro, J.F.; Silva, P.P. Overview of Large-Scale Underground Energy Storage Technologies for Integration of Renewable Energies and Criteria for Reservoir Identification. *J. Energy Storage* **2019**, *21*, 241–258. [CrossRef]
2. Stober, I.; Bucher, K. *Geothermal Energy*; Springer International Publishing: Cham, Switzerland, 2021. [CrossRef]
3. Olah, G.A.; Goeppert, A.; Prakash, G.K.S. *Beyond Oil and Gas: The Methanol Economy*; Wiley: Hoboken, NJ, USA, 2009. [CrossRef]
4. Quaschning, V. *Renewable Energy and Climate Change*; Wiley: Hoboken, NJ, USA, 2010. [CrossRef]
5. Ajayi, T.; Gomes, J.S.; Bera, A. A review of CO_2 storage in geological formations emphasizing modeling, monitoring and capacity estimation approaches. *Pet. Sci.* **2019**, *16*, 1028–1063. [CrossRef]
6. Berkowitz, B. Characterizing flow and transport in fractured geological media: A review. *Adv. Water Resour.* **2002**, *25*, 861–884. [CrossRef]
7. Hesse, M.A.; Tchelepi, H.A.; Orr, F.M. Scaling Analysis of the Migration of CO_2 in Saline Aquifers. In Proceedings of the SPE Annual Technical Conference and Exhibition, San Antonio, TX, USA, 17–20 September 2006. [CrossRef]
8. Fourier, J.B.J. *Theorie Analytique de la Chaleur*; Didot: Paris, France, 1822; pp. 499–508.
9. Fick, A.V. On liquid diffusion. London, Edinburgh, Dublin Philos. *Mag. J. Sci.* **1855**, *10*, 30–39. [CrossRef]
10. Darcy, H. *Les Fontaines Publiques de la ville de Dijon: Exposition et Application des Principes à Suivre et des Formules à Employer dans les Questions de Distribution d'eau*; Recherche: Paris, France, 1856.
11. Dupuit, J. *Etudes Théoriques et Pratiques sur le Mouvement des eaux dans les Canaux Découverts et à Travers les Terrains Perméables [Theoretical and Practical Studies on the Movement of Water in Open Channels and Permeable Ground]*; Dunod: Paris, France, 1863.
12. Thiem, A. Verfahress fur Naturlicher Grundwassergeschwindegkiten (Movement of natural groundwater flow). *Polytechnisches Notizblatt* **1887**, *42*, 229.
13. Thiem, G. *Hydrologische Methoden*; J. M. Gebhardt: Leipzig, Germany, 1906; Volume 56. (In German)
14. Carslaw, H.S. *Introduction to Themothematical Theory of the Conduction of Heat in Solids*, 2nd ed.; Macmillan Inc.: New York, NY, USA, 1921.
15. Theis, C.V. The relation between the lowering of the Piezometric surface and the rate and duration of discharge of a well using ground-water storage. *Trans. Am. Geophys. Union* **1935**, *16*, 519–524. [CrossRef]
16. Raghavan, R. Well Test Analysis: Wells Producing by Solution Gas Drive. *Soc. Pet. Eng. J.* **1976**, *16*, 196–208. [CrossRef]
17. Zimmerman, R.W.; Bodvarsson, G.S. Hydraulic conductivity of rock fractures. *Transp. Porous Media* **1996**, *23*, 1–30. [CrossRef]
18. Zimmerman, R.W. Pressure Diffusion Equation for Fluid Flow in Porous Rocks. In *The Imperial College Lectures in Petroleum Engineering*; World Scientific: Singapore, 2018. [CrossRef]
19. Jaeger, J.J.C.; Carslaw, H.S. *Conduction of Heat in Solids*, 2nd ed.; Oxford University Press: New York, NY, USA, 1959.
20. Kikani, J.; Horne, R.N. Modeling Pressure-Transient Behavior of Sectionally Homogeneous Reservoirs by the Boundary-Element Method. *SPE Form. Eval.* **1993**, *8*, 145–152. [CrossRef]
21. Ehlig-Economides, C.A.; Valko, P.; Dyashev, I.S.; Economides, M.J. Pressure transient and production data analysis for hydraulic fracture treatment evaluation. *SPE Russ. Oil Gas Tech. Conf. Exhib.* **2006**, *1*, 286–292.
22. Weijermars, R. Production rate of multi-fractured wells modeled with Gaussian pressure transients. *J. Pet. Sci. Eng.* **2022**, *210*, 110027. [CrossRef]
23. Weijermars, R. Diffusive Mass Transfer and Gaussian Pressure Transient Solutions for Porous Media. *Fluids* **2021**, *6*, 379. [CrossRef]
24. Okoroafor, E.R.; Saltzer, S.D.; Kovscek, A.R. Toward underground hydrogen storage in porous media: Reservoir engineering insights. *Int. J. Hydrogen Energy* **2022**, *47*, 33781–33802. [CrossRef]
25. Sbai, M.; Azaroual, M. Numerical modeling of formation damage by two-phase particulate transport processes during CO_2 injection in deep heterogeneous porous media. *Adv. Water Resour.* **2011**, *34*, 62–82. [CrossRef]

26. Vogel, J. Inflow Performance Relationships for Solution-Gas Drive Wells. *J. Pet. Technol.* **1968**, *20*, 83–92. [CrossRef]
27. Brown, K.E.; Lea, J.F. Nodal Systems Analysis of Oil and Gas Wells. *J. Pet. Technol.* **1985**, *37*, 1751–1763. [CrossRef]
28. Beggs, H.D. *Production Optimization Using Nodal (TM) Analysis*; Oil & Gas Consultants International Inc.: Tulsa, OK, USA, 2003.
29. Jansen, J.D. *Nodal Analysis of Oil and Gas Production Systems*; Society of Petroleum Engineers (SPE): Richardson, TX, USA, 2017.
30. Wenzel, L.K. *The Thiem Method for Determining Permeability of Water-Bearing Materials*; US Department of the Interior: Washington, DC, USA, 1935.
31. Tügel, F.; Houben, G.J.; Graf, T. How appropriate is the Thiem equation for describing groundwater flow to actual wells? *Hydrogeol. J.* **2016**, *24*, 2093–2101. [CrossRef]
32. Henry, J.C.; Neville, C.J.; Olson, A.N. Olson, Estimation of Aquifer Transmissivity from Analysis of Long-Term Monitoring With the Thiem Solution. *Groundwater* **2023**, *62*, 295–302. [CrossRef] [PubMed]
33. Flores, L.; Bailey, R.T. Review: Revisiting the Theis solution derivation to enhance understanding and application. *Hydrogeol. J.* **2019**, *27*, 55–60. [CrossRef]
34. Barry, D.; Parlange, J.-Y.; Li, L. Approximation for the exponential integral (Theis well function). *J. Hydrol.* **2000**, *227*, 287–291. [CrossRef]
35. Alotaibi, M.; Alotaibi, S.; Weijermars, R. Stream and Potential Functions for Transient Flow Simulations in Porous Media with Pressure-Controlled Well Systems. *Fluids* **2023**, *8*, 160. [CrossRef]
36. Weijermars, R.; van Harmelen, A.; Zuo, L.; Nascentes, I.A.; Yu, W. High-Resolution Visualization of Flow Interference Between Frac Clusters (Part 1): Model Verification and Basic Cases. In Proceedings of the 5th Unconventional Resources Technology Conference, American Association of Petroleum Geologists, Tulsa, OK, USA, 2 August 2017. [CrossRef]
37. Afagwu, C.; Weijermars, R. Rapid Well-Test Analysis based on Gaussian Pressure-Transients. *Geoenergy Sci. Eng.* **2024**, *241*, 213168. [CrossRef]
38. Crank, J.J. *The Mathematics of Diffusion*, 1st ed.; Clarendon Press: Oxford, UK, 1956.
39. Weijermars, R.; Afagwu, C. Hydraulic diffusivity estimations for US shale gas reservoirs with Gaussian method: Implications for pore-scale diffusion processes in underground repositories. *J. Nat. Gas Sci. Eng.* **2022**, *106*, 104682. [CrossRef]
40. Afagwu, C.; Alafnan, S.; Weijermars, R.; Mahmoud, M. Multiscale and multiphysics production forecasts of shale gas reservoirs: New simulation scheme based on Gaussian pressure transients. *Fuel* **2023**, *336*, 127142. [CrossRef]
41. Afagwu, C.; Alafnan, S.; Abdalla, M.; Weijermars, R. Gaussian Pressure Transients: A Toolkit for Production Forecasts and Optimization of Multi-Fractured Well Systems in Shale Formations. *Arab. J. Sci. Eng.* **2024**, *49*, 8895–8918. [CrossRef]
42. Weijermars, R. Gaussian Decline Curve Analysis of Hydraulically Fractured Wells in Shale Plays: Examples from HFTS-1 (Hydraulic Fracture Test Site-1, Midland Basin, West Texas). *Energies* **2022**, *15*, 6433. [CrossRef]
43. Ibrahim, A.F.; Weijermars, R. Estimation of fracture half-length with fast Gaussian pressure transient and RTA methods: Wolfcamp shale formation case study. *J. Pet. Explor. Prod. Technol.* **2023**, *14*, 239–253. [CrossRef]
44. Alvayed, D.; Khalid, M.S.A.; Dafaalla, M.; Ali, A.; Ibrahim, A.F.; Weijermars, R. Probabilistic Estimation of Hydraulic Fracture Half-Lengths: Validating the Gaussian Pressure-Transient Method with the Traditional RTA-Method (Wolfcamp Case Study). *J. Pet. Explor. Prod. Technol.* **2023**, *13*, 2475–2489. [CrossRef]
45. Weijermars, R. Potential Production Gains of Multi-Stage Fractured Wells in Shale Plays: Sensitivity of Well Performance to Changes in Design Parameters Assessed with Fast and Accurate Gaussian Solutions. *First Break.* **2023**, *41*, 63–70. [CrossRef]
46. Fetkovich, M.J. Decline Curve Analysis Using Type Curves. *J. Pet. Technol.* **1980**, *32*, 1065–1077. [CrossRef]
47. Afagwu, C.; Weijermars, R. Production Analysis of Offshore Gas Wells (UK North Sea) using Bootstrapped Arps Method: Reserves Estimation without Reservoir Model. In Proceedings of the International Petroleum Technology Conference, Dhahran, Saudi Arabia, 18–20 February 2024.
48. Tan, L.; Zuo, L.; Wang, B. Methods of Decline Curve Analysis for Shale Gas Reservoirs. *Energies* **2018**, *11*, 552. [CrossRef]
49. Yehia, T.; Abdelhafiz, M.M.; Hegazy, G.M.; Elnekhaily, S.A.; Mahmoud, O. A comprehensive review of deterministic decline curve analysis for oil and gas reservoirs. *Geoenergy Sci. Eng.* **2023**, *226*, 211775. [CrossRef]
50. Pratama, M.A.; Al Qoroni, O.; Rahmatullah, I.K.; Jameel, M.F.; Weijermars, R. Probabilistic Production Forecasting and Reserves Estimation: Benchmarking Gaussian Decline Curve Analysis against the Traditional Arps Method (Wolfcamp Shale Case Study). *Geoenergy Sci. Eng.* **2024**, *232*, 212373. [CrossRef]

water

MDPI

Article

Optimizing Flow Conditions and Fish Passage Success in Vertical Slot Fishways: Lessons from Fish Behavior Observations

Damien Calluaud [1,2,*], Vincent Cornu [3], Philippe Baran [3], Gérard Pineau [1,2], Pierre Sagnes [2,4] and Laurent David [1,2]

1 Institut Pprime, CNRS-Université de Poitiers-Isae Ensma, 86000 Poitiers, France; gerard.pineau@univ-poitiers.fr (G.P.); laurent.david@univ-poitiers.fr (L.D.)
2 Pôle R&D écohydraulique, OFB-IMFT-PPRIME, 31400 Toulouse, France; pierre.sagnes@ofb.gouv.fr
3 Bureau d'études ECOGEA, 31600 Muret, France; vincent.cornu@ecogea.fr (V.C.); philippe.baran@ecogea.fr (P.B.)
4 Direction de la Recherche et de l'Appui Scientifique, Office Français de la Biodiversité (OFB), 31000 Toulouse, France
* Correspondence: damien.calluaud@univ-poitiers.fr

Abstract: This study investigates the behavior of chubs (*Squalius cephalus*) of mid-body length (9.7–15.6 cm) with respect to turbulent flow conditions in a pool representing an experimental vertical slot fishway. Velocity and turbulence were characterized using PIV data. The influence of turbulent flow on fish behavior was assessed through the number of successful fish passage attempts, the associated passage times, and the spatial distribution of fish in the pool. Turbulence conditions were modified by the addition of one or three vertical rigid cylinders inside the pool. The results show that these adaptations may facilitate the passage of chubs. Results provide valuable insights and information to understand the relationship between fish behavior and hydraulic conditions, especially in the context of improving the design of fishways.

Keywords: fish trajectories; hydraulic parameters; turbulent kinetic energy reduction; fishway efficiency

1. Introduction

Habitat fragmentation is known to be one of the main impacting factors on biodiversity. This is especially true in networked river ecosystems, where a lot of obstacles have been constructed over time and sometimes induce habitat modifications and/or connection losses between functional habitats. One of the solutions to restore some kind of ecological continuity, at least for fish species, is to provide devices that allow or facilitate upstream and downstream fish movements at the level of impacting obstacles. Among the various types of fishways, vertical slot fishways (VSFs) are commonly used to improve fish passage. Historically, lots of VSFs were designed and sized by considering the passage of "good swimmer" fish species, particularly Salmonid species. A current consideration is that, once they have found the entrance of the fishway, other species (e.g., those with small body sizes) may encounter difficulties in progressing along the fishway due to inappropriate flow conditions in the successive pools. One aim of this study was to better understand the relationships between flow characteristics in VSFs and fish behavior, with the objective to investigate a "low-cost" way to modify flow characteristics in pools in order to improve the passage efficiency of small species in existing VSFs.

Fish can withstand turbulent flows if their stabilization aptitudes are sufficient with respect to the hydrodynamic environment, but they tend to avoid flows that interfere with their swimming trajectories [1–3]. Unfortunately, accurate links between hydraulic conditions in VSFs and fish swimming capabilities cannot be correctly anticipated from the hydraulic criteria commonly used to design such fishways (i.e., the maximum head difference between successive pools, Δh, and the maximum volumetric dissipated power,

Pv). As a consequence, a better understanding of the relationships between fish behavior and positioning and accurate hydraulic conditions in the pools of VSFs is necessary to optimize these design criteria, as mentioned by [2].

In the past few decades, important turbulent descriptors with respect to fish swimming behavior have been identified: turbulent kinetic energy, Reynolds shear stress, turbulence intensity, strain, eddy length scale, flow orientation, and vorticity [4–10]. Moreover, numerous studies detailed the turbulent structures in several designs of VSFs and provided major knowledge for subsequent improvements to facilitate fish passage [8,9,11–20]. For example, Wang et al. [19] have explored the influence of slope, the shape ratio of pools and discharge on VSFs flow characteristics. They have defined two typical flow patterns, depending on the slope and the relative width of the pools. In the first flow pattern (FP1), generally observed in pools with large widths and low slope conditions, the principal flow enters the pools as a curved jet, which opens out before converging again towards the next (downstream) slot. This type of jet creates a large recirculation zone occupying roughly half of the pool surface and presents an unsteady beating movement near the large baffle. In the second flow pattern (FP2), generally observed in pools of small widths with high slope conditions, the entering jet has a high curve form and impacts the opposite side wall. Two large contra-rotating swirls are generated in the upstream corner of the pool in the convex part of the jet and a smaller one occurs near the large baffle. In FP2 conditions, the topology, spatial distributions of the velocity, turbulence kinetic energy, vorticity and non-stationary dynamics seem to be less appropriate for fish passage because VSF were previously designed using inadequate criteria [18,21,22].

The primary objective of the present study is to investigate the effects of turbulent flow conditions on the behavior of mid-body length chubs in a VSF pool. Fish behavior and their trajectories have been detailed and related to the four characteristics of turbulent flow suggested in the IPOS framework [23]: turbulence intensity (I), flow periodicity (P), flow orientation (O), and scale (S) of the significant eddies. For this purpose, the turbulence intensity, defined by turbulent kinetic energy, dynamics of the flow, instantaneous and 3D topologies, and the size of vortices were considered. Consequently, flow features were modified within a VSF pool through the addition of rigid, vertical cylinders, to better investigate the relationships between hydrodynamic conditions and fish behavior, particularly in terms of passage efficiency.

2. Materials and Methods

2.1. VSF Model and Cylinders

In the present paper, only a brief description of the VSF model and the metrology are given; further details can be found in [20]. The VSF model used in the present study (Figure 1) consisted of five similar pools (length L = 0.75 m, height H = 0.55 m, width B = 0.50 m). The width of the slots was b = 0.075 m, corresponding to a ratio of B/b = 6.67. The slope was set to 10% and the discharge to 23 $L{\cdot}s^{-1}$, resulting in a maximum velocity on a slot of 1.21 $m{\cdot}s^{-1}$. These settings induced a mean water depth in the pool of about 0.3 m, and a mean height drop between pools of about 0.075 m. The measurements of velocity were determined using particle image velocimetry (PIV) in a plane located at Z/b = 2 above the center of the bottom of the pool (which corresponds to a plane located at the mid-depth of the pool). All measurements were taken in the third pool of the model, to ensure uniform flow as the gravity is balanced by upstream dissipation. In total, 3000 instantaneous velocity fields were acquired throughout the pool, with a spatial resolution of 8 mm^2 at an acquisition frequency rate of 10 Hz. To modify flow (especially turbulence) conditions, one cylinder (diameter equal to b or diameter equal to b/2) or three cylinders (diameter equal to b/2) were placed vertically in each pool, near the upstream slot (Figure 2). The choice to introduce cylinders is based on the principle of dividing the entering jet into two parts (1 cylinder) and then dividing the secondary jet again into two parts (3 cylinders, in a triangle configuration). Cylinder positioning was optimized by calculation [24].

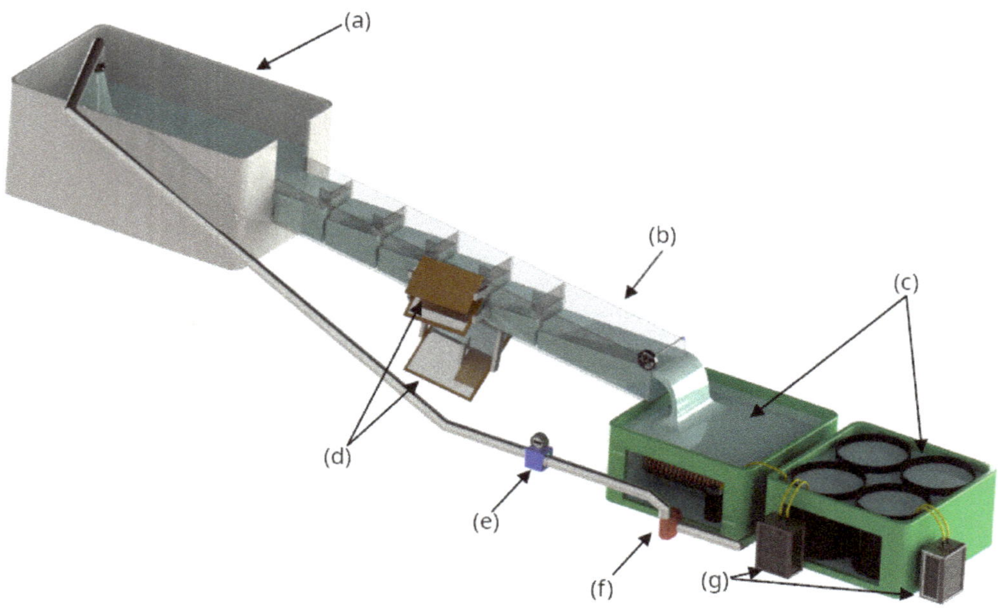

Figure 1. Experimental device: the 5 pools of the vertical-slot fish pass model, (a) upper tank, (b) lower tank, (c) fish-tank with 4 compartments, (d) mirrors for fish observation and image acquisition, (e) flowmeter, (f) pump, (g) filter pump.

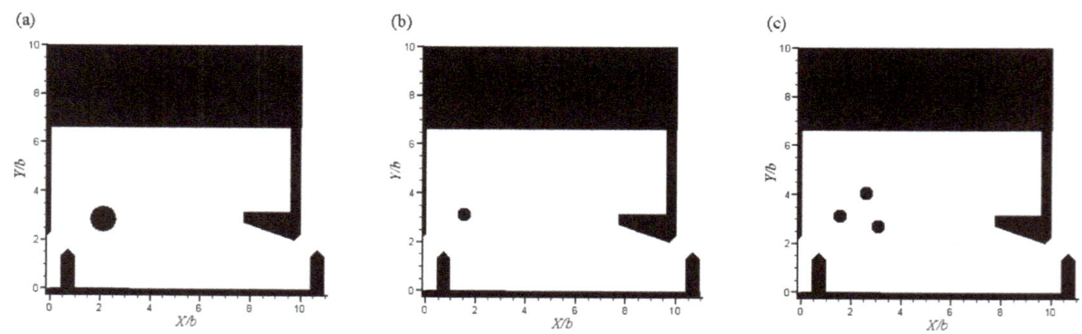

Figure 2. Experimental settings of the pool with additional cylinder(s). (**a**) one cylinder (diameter equal to b), (**b**) one cylinder (diameter equal to b/2), (**c**) three cylinders (diameters equal to b/2) (top views; flow direction: from left to right).

2.2. Fish Manipulation and Observation

Fish sampling was conducted in the "Petite Blourde" stream, a tributary of the Vienne River, in central France, using an electrofishing technique (HERON® model, DREAM Électronique Company, Pessac, France). Ninety chubs [min total length (TL) = 9.7 cm; max TL = 15.6 cm; average TL ± SD = 13.1 ± 1.9 cm] were collected and subsequently introduced, in the laboratory, into a fish tank that provided resting conditions with water characteristics similar to those during the experiments. Fish were kept in the fish tank (Figure 1) for one day before the beginning of the experiments, without food. After starting the pumps and waiting for flow stabilization in the pools, a group of 20–25 individuals, randomly selected from the fish tank, were transferred to the downstream pool (pool #5), separated from pool #4 by a mesh screen. The fish were held in pool #5 for 15 min before each trial, to allow them to adapt to flow conditions. The screen was then removed and the experiment began for a duration of 90 min (i.e., each 90 min period corresponds to one "trial"). A total of 18 trials were performed. At the end of each trial, the specimens used were transferred to a specific compartment in the fish tank, in order not to mix them with others and to use each individual only once per day at a maximum. Differences in water temperature among all trials were limited to <0.3 °C (mean water temperature during the experimental period = 16 °C). The water temperature during the experiments was very close to the optimum used by this species during its movements and in its breeding behavior [25]. The test room was illuminated with projectors at a constant brightness between 08.00 a.m. and 18.00 p.m. The flume walls were papered black, except the bottom of pool #3, to not disturb fish during observations and place them under light conditions as close as possible to real ones (walls of real fishways are opaque). We used young-of-the-year fish to carry out the tests. The chub (*Squalius cephalus* L.) is a very generalist species, colonising many rivers. It is highly adaptable to water temperature conditions and habitat characteristics. Its migratory behaviour is less marked than other species such as brown trout (*Salmo trutta* L.), but individuals regularly move between habitats.

A mirror positioned at a 45° angle under pool #3 enabled the observation of fish movements inside the pool (Figure 1), allowing a lateral view (to get the vertical fish position) and an underside view (to get the horizontal fish position) simultaneously. During each trial, these positions of fish and their behavior were continually recorded by a video camera (HDR-HC9E model, Sony Corporation, Tokyo, Japan). Frosted plexiglass plates were positioned above the pools to obtain video images without reflected light. The video recording device was surrounded by a tarp to avoid any stray light on the monitored pool. Each 90 min sequence was analyzed to (i) track fish movements between pools #2, 3 and 4, (ii) measure transit times through pool #3, (iii) track fish movements among and between previously defined hydraulic zones within pool #3 (see below) and (iv) measure the "resting-holding" time of individuals in each hydraulic zone. The 18 experimental trials performed represented 27 h of video.

There are few precise data available about the swimming capacity of chubs. In a study about the swimming capabilities of juvenile cyprinid species, Garner [26] gave a relation between the critical swimming speed (U_{crit}, in $cm{\cdot}s^{-1}$) of chubs, their body length (TL, in mm) and water temperature (T, in °C):

$$U_{crit} = 0.45 + 0.23{\cdot}T + 0.55{\cdot}TL \tag{1}$$

According to Equation (1), the U_{crit} of chub individuals used in our study ranged between 57 and 90 $cm{\cdot}s^{-1}$. We did not have information about the maximum swimming speed (U_{max}) of chubs, potentially used to overcome high flow velocities, but U_{max} is higher than U_{crit} in fish (e.g., up to 57% higher for rainbow trout, *Oncorhynchus mykiss* [27]). Therefore, considering the maximum velocity of V_{th} = 1.21 $m{\cdot}s^{-1}$ on the slot, chubs were placed in constraining flow conditions in our experimental VSF model.

3. Results

3.1. Flow Description

As previously mentioned (and see [19]), in the configuration studied without a cylinder, the jet had a notable curved form and hit the opposite side wall (Figure 3a). Two contra-rotating swirls were generated in the upstream part of the pool (one at each side of the jet) and another one occurred in the downstream part of the pool, near the large baffle (at its side opposite to the downstream slot).

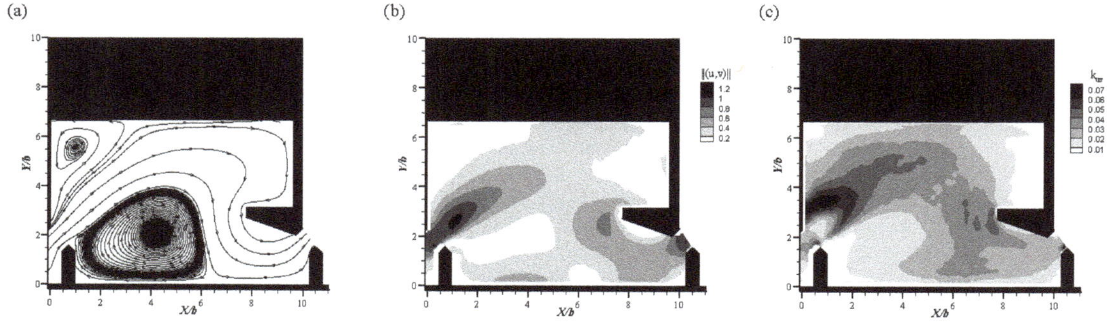

Figure 3. Mean flow features in pool #3 of the vertical-slot fishway model. (**a**) Streamlines, (**b**) velocity amplitude ($m \cdot s^{-1}$), (**c**) in-plane turbulent kinetic energy($m^2 \cdot s^{-2}$) (flow direction: from left to right).

The three zones containing the swirls can be assimilated as potential resting zones for fish. A layer with the mapping of these different zones and the boundaries of the pool has been added during the treatment of video-sequences, in order to facilitate image analyses (Figure 4).

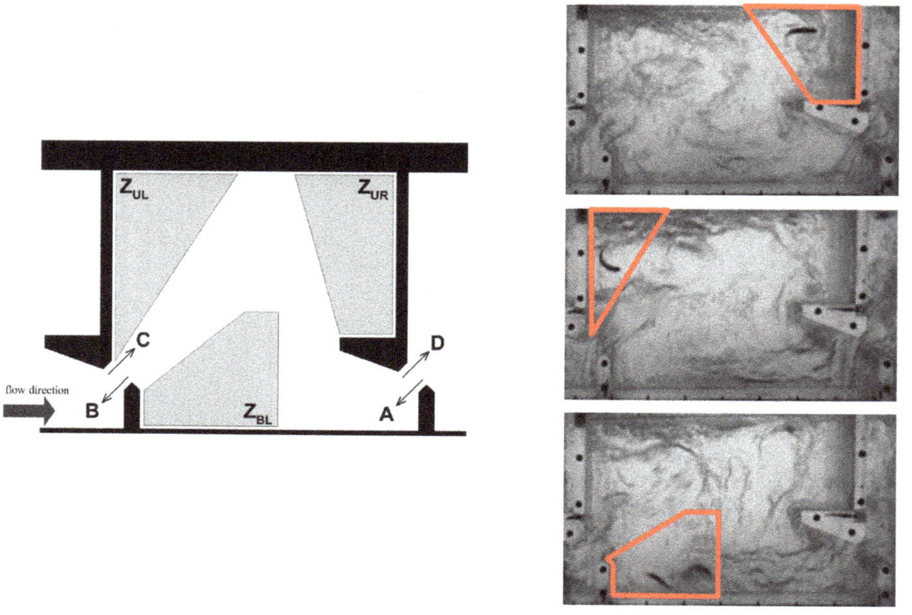

Figure 4. Bottom view of the third pool, recorded by a video camera, with an overlay of the 3 red zones containing the swirls (flow direction: from left to right). One fish is visible in each zone.

In the pool, high velocities were particularly present (i) at the upstream slot level, due to the head drop and flow constriction, and (ii) close to the large downstream baffle (Figure 3b). The large baffle locally generated high-velocity fluctuations that could stop or delay fish passage. At the upstream slot, the maximum velocity amplitude was equal to 93% of the theoretical maximum velocity in the slot, V_{th}, and the space-averaged mean flow velocity in the pool represented 28.2% of V_{th}. The average turbulence intensity, defined by the ratio between average turbulent kinetic energy (Figure 3c) and velocity (Figure 3b), may be of importance to understand fish behavior. The turbulence pattern in the pool was mainly due to the presence of the large baffle, which generated jet oscillations ("beating movements of the jet" defined in [20]), a complex kinematic process described in detail in [20,21]. The beating has a significant influence on the fluid motion in the pool, with a destabilization by the jet of the small region of recirculation located downstream from the small baffle (i.e., zone 2, Figure 4). Some velocity bursts, generated at the level of the upstream slot, also influenced the flow topology. Those bursts produced perturbations on the fluid motions and increased the average velocity magnitude in the pool [22]. The spatial distribution of the average turbulent kinetic energy (Figure 3c) was used to estimate the mean turbulent kinetic energy in the pool, which was equal to $0.017.V_{th}^2$. A total of 62% of the pool surface was characterized by a turbulence kinetic energy value lower than $0.017.V_{th}^2$.

3.2. Fish Trajectories in the Pool and Relationships with Turbulent Flow and IPOS Parameters

The displacement of fish within a pool is composed of upstream movements, downstream movements and stops or breaks in preferential zones. Video analyses showed three main scenarios of motion between pools:

- scenario A: crossing success, fish enter pool #3 from pool #4 and manage to enter pool #2,
- scenario B: crossing failures, fish enter pool #3 from pool #4 but are washed away in pool #4 later,
- scenario C: wash downstream, fish enter pool #3 after being washed downstream from pool #2.

No individual trackings were performed. As consequence, the same fish can be included in more than one scenario, for example, wash downstream and crossing success. In total, 111 situations were recorded during the 18 trials. Non-parametric statistical analysis (Kruskal–Wallis and Mann–Whitney tests) were used to compare (i) fish transit time through the pool, (ii) fish trajectories between the three different zones and (iii) the fish passage efficiency. We also used a homogeneity X^2 test with Yates' continuity correction to compare the distribution of fish associated with different patterns of using the zones inside the pool.

Scenario A, B and C represented 63.1%, 19.8% and 17.1% of the situations, respectively. Finally, the overall fish passage efficiency was calculated as the ratio between the number of success scenarios and failures and wash back scenarios accounted for 26.2%. The average time a successful fish spent in pool #3 was 6.4 s.

A more detailed investigation of fish trajectories inside the monitored pool was also performed (Figure 5). Upstream and downstream movements were observed. These movements can be continuous or discontinuous with stops in the pool. A detailed study of the trajectories reveals a very marked movement pattern. The most frequent trajectories are identified by thicker arrows. A total of 98.9% of the fish entered the pool directly into the ZBL. From the ZBL zone, 81.9% of the fish directly crossed the upstream slot and entered the upper pool without having entered any other zone of the pool. Fish movement patterns differ according to whether or not individuals have stopped in one of the zones of the basin. The movements of the fish that stopped for a moment in the pool were more diversified: they crossed the basin making a maximum of 16 movements between zones, whereas the fish that did not stop made a maximum of 5 movements across the pool. A total of 396 fish trajectories were recorded during the 18 trials. An average of 1.28 movements per fish viewed in the observed pool was acquired. Globally, fish located in zone ZBL orientated their head towards the upstream slot, whereas they orientated their tail towards this slot when they were located in zones Z_{UL} and Z_{UR}. Most of the fish entering the pool from the downstream slot sought to leave the jet by swimming toward zone Z_{BL} (98.9% of fish movements). Then, a little more than the half of the individuals located inside zone Z_{BL} left the pool via the upstream slot. A total of 23.9% of the remaining individuals swam to zone Z_{UL}, from which 27.3% of them successfully migrated upstream. Zone Z_{UR} was characterized by a significant rate of downstream migration (20.8% of fish). In summary, zones Z_{BL} and Z_{UL} were preferentially employed by fish before successfully migrating upstream.

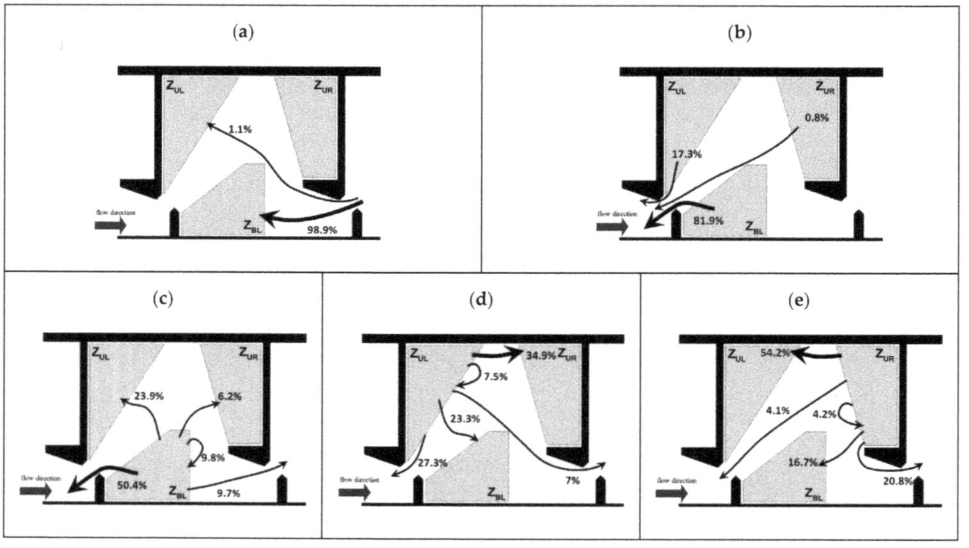

Figure 5. Chub trajectories to and from three potential resting zones (Z_{BL}, Z_{UL} and Z_{UR}, see Figure 4) in pool #3 of the VSF model (downstream = right-hand): (**a**) destination of individuals entering the pool; (**b**) origin of the individuals leaving the pool; (**c**) destination of the individuals leaving zone Z_{BL}; (**d**) destination of the individuals leaving zone Z_{UL}; (**e**) destination of the individuals leaving zone Z_{UR}.

Studying the relationships between fish behavior (e.g., trajectories and their use of resting zones) and the turbulence conditions in the pool could help to determine if the Intensity-Periodicity-Orientation-Scale (IPOS) approach [23] should be considered for improving fish passage through VSFs.

First, in terms of Intensity, the aim was to identify areas in the pool where intensities could be quite low and stable enough to be predictable and sustainable for fish. Four areas were considered (Figure 6), with respect to two threshold values for space-averaged velocity norm and turbulence kinetic energy, which could be considered as criteria to define the hydrodynamic environment of fish, see [21]. The choice of thresholds is only based on these hydraulic criteria. The aim is to identify locations in a pool where the flow could be considered as a "barrier" for fish. These values were set at $0.282 \cdot V_{th}$ (velocity) and $0.017.V_{th}^2$ (turbulence kinetic energy), respectively. In F11 areas (Figure 6), velocity and turbulence kinetic energy values were both over their respective threshold values. Inside these areas, the flow might be unpredictable for fish. In F00 areas, velocity and kinetic energy values were both under their respective threshold values. These areas might be considered as suitable for resting fish and/or for initiating an upstream movement and, as seen before, correspond to the preferential areas used by fish during the experimental trials. In F10 areas (compared to F01), the velocity value (relative to the turbulence kinetic energy value) was the only value higher than its threshold value.

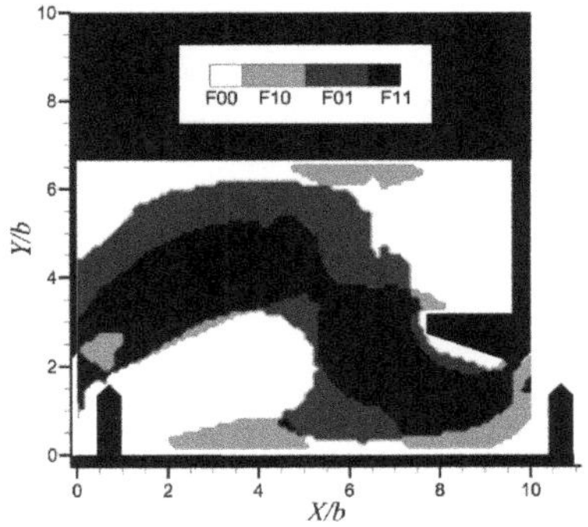

Figure 6. Representation of four levels of flow intensity, on average, in the studied pool (see text; flow direction: from left to right).

Second, in terms of Periodicity, non-stationary dynamics of the flow produced an oscillation of the F00 surface area (Figure 7). For example, velocity burst dynamics sometimes decreased or increased the total F00 area, in comparison to the average situation (e.g., Figures 7a and 7b, respectively). The small size vortices visualized were characterized by high rotation rates and were localized in the shearing region of the jet. Based on observations of fish videos, the assumption that velocity burst dynamics is the source of fish destabilization can be proposed: velocity bursts altered, during a short time, the stability of fish while they were initially resting in an F00 zone; subsequently, fish moved to another preferential zone or migrated upstream or downstream. Moreover, when velocity bursts decreased velocity magnitudes inside the slot, upstream migration could be facilitated (as fish should spend less energy moving) during short time sequences.

Third, in terms of Orientation, the topology and velocity field of the flow inside the VSF were mainly two-dimensional [2,19,20,26] and composed of vertical axis vortices. Based on our observations, the orientation of the large eddies with high vorticity values did not alter chub swimming behavior. Fish have not lost their postural alignment because of these eddies dominating the turbulent flow field. The orientation component of the IPOS framework (see [23]) seems therefore not to be a main parameter to consider to enhance fish passage success in VSFs when velocity fields are mainly 2D-oriented (i.e., upstream/downstream and laterally).

An example of the location of the small vortices is presented in Figure 7d,e. The small size vortices visualized are defined by significant rotation rates and are localized in the shearing region of the jet. The small size coherent structures seem to affect the swimming behavior of fish, as mentioned in the "Scales" parameter of IPOS framework. The small size vortices were used by fish and may allow fish to minimize swimming energy costs by taking advantage of the energy associated with eddies for propulsion. We studied the distribution of the Reynolds shear stresses (Figure 8a) that are estimated using a statistical analysis of turbulence based on time fluctuations. High-value areas are mainly located near the upstream slot, due to the mixing layer of the jet. To understand the physical barrier these conditions can oppose to fish displacement, the swirling structures must be precisely described in space and not only in time. We focused our analysis on the vorticity based on the mean flow (Figure 8b), which allowed us to quantify areas in the pool where shearing, stretching and rotation phenomena are significant, as in [20]. This detailed study of the vorticity allowed us to define the spatial distribution of coherent structures, which are not defined by the Reynolds shear stress data analysis. Significant rates of rotation, shearing and stretching phenomenon, marked by substantial vorticity intensity, are parameters to minimize in order to increase the biological efficiency of a VSF.

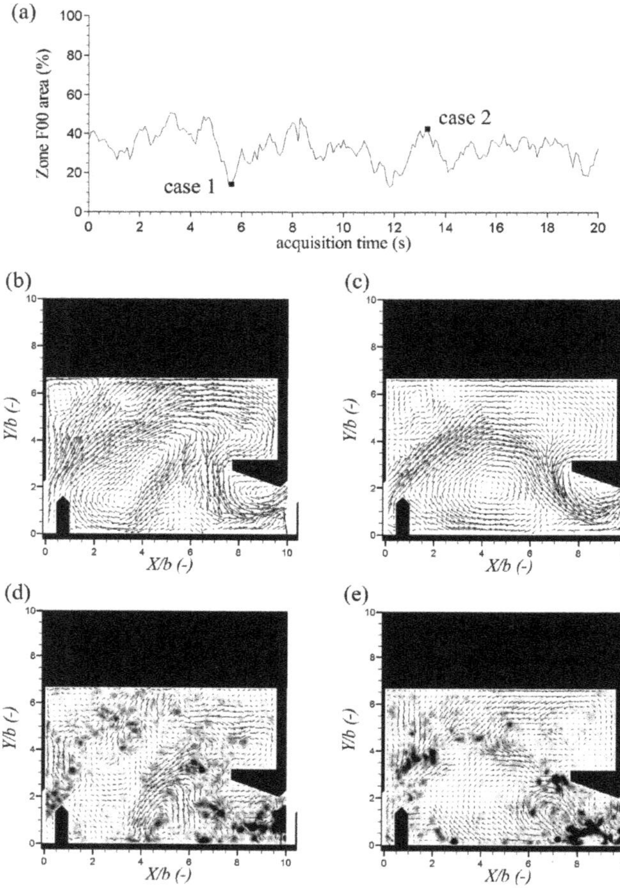

Figure 7. (**a**) Time evolution of the F00 area. (**b**) instantaneous velocity field, case 1. (**c**) instantaneous velocity field, case 2. (**d**) instantaneous location of the small size coherent structures, case 1. (**e**) instantaneous location of the small size coherent structures, case 2 (flow direction: from left to right).

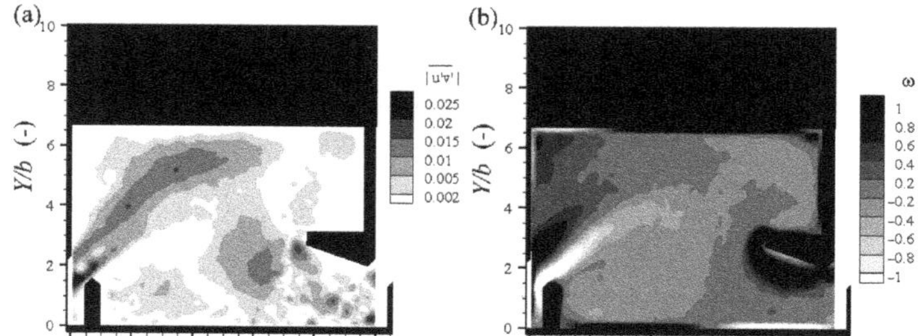

Figure 8. Mean flow features in the studied pool: (**a**) Reynolds shear stresses; (**b**) out-of-plane vorticity component (flow direction: from left to right).

3.3. Fish Passage Improvement in Existing VSFs

The previous results provided some explanations and hypotheses about the relationships between flow characteristics in VSF pools, fish behavior inside them and, subsequently, fish passage success. The following step was to try to improve these hydraulic characteristics in a simple (i.e., neither destructive nor expensive, with respect to existing VSFs) way to facilitate fish passage, especially for "small body species". Based on previous studies [21,22], we changed the turbulence conditions in the pools, by installing vertical cylinders inside (see Figure 2). Three different configurations were tested: (i) pools equipped with one cylinder of diameter D = b, (ii) pools equipped with one cylinder of diameter D = b/2 and (iii) pools equipped with three cylinders of diameter D = b/2 arranged in a triangle (Figure 9).

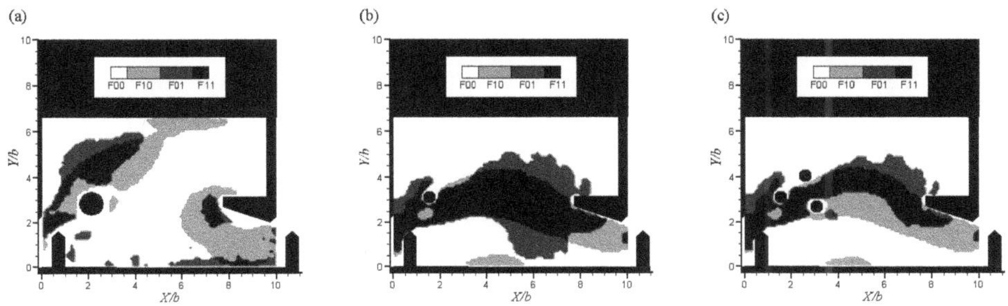

Figure 9. Representation of four levels of flow intensity (from F00 to F11, see text; flow direction: from left to right) for 3 VSF configurations: (**a**) with 1 cylinder of diameter D = b, (**b**) with 1 cylinder of D = b/2, (**c**) with 3 cylinders of D = b/2 arranged in a triangle.

The previous results helped us to understand the positive relationship between an increase in the proportion of low velocity and turbulence kinetic energy conditions in the pools, and the biological efficiency of VSFs. The effects of one or three cylinder(s) on the spatial distribution of these flow characteristics should therefore modify fish passage success in VSFs. In general, migrating fish are attracted to turbulent flows, but they have to maintain their stability in such hydrodynamic environments, especially when they try to pass over an obstacle. Consequently, they tend to avoid flows that have unpredictable velocity and turbulence components [3]. In this context, a quantification of "global estimators", in relation to the local hydrodynamic parameters, could help (i) to predict fish trajectories inside VSFs and (ii) to quantify the hydrodynamic effects of a cylinder, which could not be anticipated by the commonly used criteria, PV and Δh, during the dimensioning process of fishways. For that purpose, we have defined the hydrodynamic environment in the pool by using the averages, over space, of the following target flow quantities: velocity, turbulent kinetic energy and vorticity.

The adding of cylinder(s) in the VSF model globally reduced the proportion of areas where velocity, turbulent kinetic energy, rotation, shear and stress intensities are high, and reduced both average flow velocity and turbulent kinetic energy inside the pool (Figure 9, to be compared to Figure 6; Table 1). The presence of a cylinder with a diameter D = b significantly increased the average vorticity. The values of flow features that should affect fish passage success are poorly reduced by the use of one cylinder with a diameter D = b/2 (Figure 9b, to be compared to Figure 6). The use of three cylinders influenced the flow inside the VSF, in comparison with the experimental setting without a cylinder, by decreasing velocity, turbulence kinetic energy and vorticity average values by 28.8%, 40.0% and 17.3%, respectively (Table 1). The presence of three cylinders modified the flow inside the VSF by changing the jet shape from a curved pattern (Figure 6) to a more linear one, straightened towards the downstream slot (Figure 9c).

Table 1. Averages values over space of three flow characteristics in a VSF pool, for four experimental configurations (with and without cylinders, see Figure 2). D = cylinder(s) diameter, b = slot width.

Configuration	Velocity Norm	Turbulent Kinetic Energy	Absolute Value of Vorticity
No cylinder	$0.282 \cdot V_{th}$	$0.017 \cdot {V_{th}}^2$	$0.307 \cdot V_{th}/b$
1 cylinder D = b	$0.183 \cdot V_{th}$	$0.011 \cdot {V_{th}}^2$	$0.354 \cdot V_{th}/b$
1 cylinder D = b/2	$0.216 \cdot V_{th}$	$0.015 \cdot {V_{th}}^2$	$0.303 \cdot V_{th}/b$
3 cylinders	$0.263 \cdot V_{th}$	$0.013 \cdot {V_{th}}^2$	$0.254 \cdot V_{th}/b$

As hypothesized, lowering the average values of velocity, turbulent kinetic energy and vorticity in the pool by adding cylinders increased the proportion of fish passage success, from 26% without any cylinder to 47.4% with three cylinders (Table 2). The increase in biological efficiency (i.e., passage success) was also significant after adding one cylinder with a diameter D = b/2, shifting from 26.2% to 42.3%. There are statistically significant differences in the use of the three zones identified for each configuration studied (χ^2, $p < 0.05$). Nevertheless, note that the median transit time increased from 6.4 s in the configuration without any cylinder to 9.8 s in the experimental configuration with three cylinders (Kruskall–Wallis, $p = 0.12$).

Table 2. Fish passage efficiency and proportion of F00 areas in the pool (see text and Figures 6 and 9), with respect to different experimental configurations (with and without cylinders, see Figure 2). D = cylinder(s) diameter, b = slot width.

Configuration	Fish Passage Efficiency	Proportion of F00 Zone
No cylinder	26.2%	34.6%
1 cylinder D = b	28.5%	58.6%
1 cylinder D = b/2	42.3%	54.6%
3 cylinders	47.4%	60.9%

These results underline the importance of considering "global flow estimators" (e.g., levels of flow intensity) during the VSF design process to improve their biological efficiency in terms of passage success and transit time.

4. Conclusions

The relationship between flow characteristics and fish behavior inside a VSF pool was investigated. The preferential zones used by fish in the pool before attempting to pass upstream are characterized by relatively low turbulence intensities, quantified in this study as the ratio between average turbulent kinetic energy and velocity values.

Flow patterns in the pool could be changed and some hydrodynamic characteristics unfavorable to chub passage were reduced by adding cylinders in the VSF. Such devices modified the spatial intensity distribution of turbulent quantities (velocity, turbulent kinetic energy and vorticity) and stabilized the unsteady flow dynamics into a low disturbance flow state, making it less constraining for fish migration success (at least for chub, in our experimental conditions). The hydraulic conditions created by the addition of three cylinders resulted in a significantly higher rate of chub passage (47.4%) when compared to the VSF configuration without any cylinders (26.2%).

These results confirm the importance of optimizing flow characteristics in VSF pools in order to maximize fish passage success, for example, by modifying the length-to-width ratios of vertical slot passes [19], or by introducing energy dissipating devices like vertical cylinders [21].

The joint investigation of fish behavior (i.e., trajectories, time they spend in particular areas) and of the spatial distribution of some hydrodynamic characteristics in a fishway provides valuable insights to help engineers and biologists to improve design criteria for fishways. These results now have to be confirmed by in situ studies and for other fish species. Retrofitting of existing VSFs, which were initially designed for good swimmers, could be applied by adding cylindrical structures inside the pool for small fish species.

Author Contributions: Methodology, V.C., P.B. and L.D.; Validation, L.D.; Investigation, D.C. and V.C.; Writing—original draft, D.C.; Writing—review & editing, G.P., P.S. and L.D. All authors have read and agreed to the published version of the manuscript.

Funding: This project received funding from the European Union's Horizon 2020 research and innovation program FITHydro (www.fithydro.eu), under grant agreement [No 727830].

Data Availability Statement: The data presented in this study are available on request from the corresponding author.

Acknowledgments: The authors would like to acknowledge the Nouvelle Aquitaine CPER and the Environmental Hydrodynamic Platform of the University of Poitiers and CNRS for their facilities and equipment. The local OFB staff is also thanked for its assistance in catching the fish.

Conflicts of Interest: Vincent Cornu and Philippe Baran were employed by Bureau d'études ECO-GEA. The remaining authors declare that the research was conducted in the absence of any commercial or financial relationships that could be construed as a potential conflict of interest.

References

1. Castro-Santos, T. Optimal swim speeds for traversing velocity barriers: An analysis of volitional high-speed swimming behavior of migratory fishes. *J. Exp. Biol.* **2005**, *208*, 421–432. [CrossRef]
2. Katopodis, C.; Cai, L.; Johnson, D. Sturgeon survival: The role of swimming performance and fish passage research. *Fish. Res.* **2019**, *212*, 162–171. [CrossRef]
3. Liao, J.C. A review of fish swimming mechanics and behaviour in altered flows. *Philos. Trans. R. Soc. B Biol. Sci.* **2007**, *362*, 1973–1993. [CrossRef]
4. Pavlov, D.S.; Lupandin, A.I.; Skorobogatov, M.A. The effects of flow turbulence on the behavior and distribution of fish. *J. Ichthyol.* **2000**, *40*, 232–261.
5. Hotchkiss, R. *Turbulence Investigation and Reproduction for Assisting Downstream Migrating Juvenile Salmonids. Part I*; BPA Report DOE/BP-00004633-I; Bonneville Power Administration: Portland, OR, USA, 2002.
6. Odeh, M.; Noreika, J.F.; Haro, A.; Maynard, A.; Castro-Santos, T.; Cada, G.F. *Evaluation of the Effects of Turbulence on the Behavior of Migratory Fish. Final Report 2002*; Bonneville Power Administration: Portland, OR, USA, 2002.
7. Tritico, H.M.; Cotel, A.J. The effects of turbulent eddies on the stability and critical swimming speed of creek chub (*Semotilus atromaculatus*). *J. Exp. Biol.* **2010**, *213*, 2284–2293. [CrossRef] [PubMed]
8. Romão, F.; Quaresma, A.L.; Branco, P.; Santos, J.M.; Amaral, S.; Ferreira, M.T.; Katopodis, C.; Pinheiro, A.N. Passage performance of two cyprinids with different ecological traits in a fishway with distinct vertical slot configurations. *Ecol. Eng.* **2017**, *105*, 180–188. [CrossRef]
9. Romão, F.; Quaresma, A.L.; Simão, J.; Bravo-Córdoba, F.J.; Viseu, T.; Santos, J.M.; Sanz-Ronda, F.J.; Pinheiro, A.N. Debating the Rules: An Experimental Approach to Assess Cyprinid Passage Performance Thresholds in Vertical Slot Fishways. *Water* **2024**, *16*, 439. [CrossRef]
10. Silva, A.T.; Katopodis, C.; Santos, J.M.; Ferreira, M.T.; Pinheiro, A.N. Cyprinid swimming behaviour in reponse to turbulent flow. *Ecol. Eng.* **2012**, *44*, 314–328. [CrossRef]
11. Silva, A.T.; Santos, J.M.; Ferreira, M.T.; Pinheiro, A.N.; Katopodis, C. Effects of water velocity and turbulence on the behaviour of Iberian barbel (*Luciobarbus bocagei*, Steindachner 1864) in an experimental pool-type fishway. *River Res. Appl.* **2011**, *27*, 360–373. [CrossRef]
12. Rajaratnam, N.; Katopodis, C.; Solanki, S. New designs for vertical slot fishways. *Can. J. Civ. Eng.* **1992**, *19*, 402–414. [CrossRef]
13. Wu, S.; Rajaratnam, N.; Katopodis, C. Structure of flow in vertical slot fishway. *J. Hydraul. Eng.* **1999**, *125*, 351–360. [CrossRef]
14. Puertas, J.; Pena, L.; Teijeiro, T. An experimental approach to the hydraulics of vertical slot fishways. *J. Hydraul. Eng.* **2004**, *130*, 10–23. [CrossRef]
15. Liu, M. Turbulence Structure in Hydraulic Jumps and Vertical Slot Fishways. Ph.D. Thesis, Department of Civil Engineering, University of Alberta, Edmonton, AB, Canada, 2004.
16. Liu, M.; Rajaratnam, N.; Zhu, D.Z. Mean flow and turbulence structure in vertical slot fishways. *J. Hydraul. Eng.* **2006**, *132*, 765–777. [CrossRef]
17. Peake, S.; McKinley, R.S.; Scruton, D.A. Swimming performance of various freshwater Newfoundland salmonids relative to habitat selection and fishway design. *J. Fish Biol.* **1997**, *51*, 710–723. [CrossRef]
18. Tarrade, L.; Texier, A.; David, L.; Larinier, M. Topologies and measurements of turbulent flow in vertical slot fishways. *Hydrobiologia* **2008**, *609*, 177–188. [CrossRef]
19. Wang, R.; David, L.; Larinier, M. Contribution of experimental fluid mechanics to the design of vertical slot fish passes. *Knowl. Manag. Aquat. Ecosyst.* **2010**, *396*, 1–21. [CrossRef]

20. Tarrade, L.; Pineau, G.; Calluaud, D.; Texier, A.; David, L.; Larinier, M. Detailed experimental study of hydrodynamic turbulent flows generated in vertical slot fishways. *Environ. Fluid Mech.* **2011**, *11*, 1–21. [CrossRef]
21. Calluaud, D.; Pineau, G.; Texier, A.; David, L. Modification of vertical slot fishway flow with a supplementary cylinder. *J. Hydraul. Res.* **2014**, *52*, 614–629. [CrossRef]
22. Calluaud, D.; Pineau, G.; Texier, A.; David, L. Estimation of the turbulent features of flow in vertical slot fishway. Improvements on fishway design criteria. In Proceedings of the 3rd IAHR European Congress, Porto, Portugal, 14–16 April 2014.
23. Lacey RW, J.; Neary, V.S.; Liao, J.C.; Enders, E.C.; Tritico, H.M. The IPOS framework: Linking fish swimming performance in altered flows from laboratory experiments to rivers. *River Res. Appl.* **2012**, *28*, 429–443. [CrossRef]
24. Bourtal, B.; Pineau, G.; Calluaud, D.; Texier, A.; David, L. Design optimization of existing vertical slot fishways by cylinder adjunction. In Proceedings of the 2nd IAHR European Congress, Munich, Germany, 27–29 June 2012.
25. Keith, P.; Poulet, N.; Denys, G.; Changeux, T.; Feunteun, E.; Persat, H. *Les Poissons d'Eau Douce de France*, 2nd ed.; Collection Inventaires et Biodiversité; Biotope Editions, Mèze; Muséum National d'Histoire Naturelle: Paris, France, 2020; 704p.
26. Garner, P. Swimming ability and differential use of velocity patches by O+ cyprinids. *Ecol. Freshw. Fish* **1999**, *8*, 55–58. [CrossRef]
27. Farrel, A.P. Comparisons of swimming performance in rainbow trout using constant acceleration and critical swimming speed tests. *J. Fish Biol.* **2008**, *72*, 693–710. [CrossRef]

water

MDPI

Article

A Framework for Quantifying Reach-Scale Hydraulic Roughness in Mountain Headwater Streams

Tae-Hyun Kim [1], Jeman Lee [1], Taehyun Kim [1], Hyung Tae Choi [2] and Sangjun Im [1,3,*]

1 Department of Agriculture, Forestry and Bioresources, Seoul National University, Seoul 08826, Republic of Korea; thkim96@snu.ac.kr (T.-H.K.); jemahn@snu.ac.kr (J.L.); kta40@snu.ac.kr (T.K.)
2 Urban Forests Division, Forest Environment and Conservation Department, National Institute of Forest Science, Seoul 02455, Republic of Korea; choiht@korea.kr
3 Research Institute of Agriculture and Life Sciences, Seoul National University, Seoul 08826, Republic of Korea
* Correspondence: junie@snu.ac.kr

Abstract: The presence of macroroughness elements directly affects the flow velocity in mountain headwater streams. Hydraulic roughness is the dominant resistance to flow caused by objects protruding into the water, but it is not measurable in the field. This study quantified the reach-average hydraulic roughness based on the channel morphology in two mountain streams. The average flow velocities of the reaches were measured using the dye-tracing method. The magnitude of the hydraulic roughness was derived from the grain size of the streambed materials (D_{50} and D_{84}) and the cross-sectional/longitudinal bed roughness. The observations for low flows (0.04–0.43 m^2/s discharge per unit width) indicated that the longitudinal 90% inter-percentile range (IPR_{90}^{L}) seemed to have considerable merit in examining the influences of large roughness elements on flow conveyance. A dimensionless hydraulic geometry relation that can reflect the field measurements over a limited range of hydraulic characteristics was also developed for estimating the reach-average flow velocity in steep and rough streams. Thus, the research framework used appears to provide a reliable method of quantifying reach-average hydraulic roughness from local data in mountain headwater streams.

Keywords: reach-average velocity; roughness height; mountain headwater stream; dimensionless hydraulic geometry

1. Introduction

Mountain headwater streams are the smallest reaches at the highest end of a mountain landscape and are often referred to as zero-order rather than first-order streams. Headwater streams strongly influence downstream river networks. In South Korea, headwater streams play a vital role in mitigating floods and debris flow disasters, and they represent a significant portion of the total channel length in mountainous regions. Mountain headwater streams typically refer to small, undulating channels in mountainous forests with relatively shallow flow depths and varying flow velocities. Headwater streams differ from large rivers since they exhibit small depth-to-width ratios, large relative roughness, steep channel gradients, and different channel morphology [1,2]. Although differences in the process characteristics between mountain streams and larger flat rivers have been extensively studied, relatively little is known about the hydraulic and morphological processes of mountain headwater streams.

Determining the flow velocity in a stream has been the main focus of many hydrological and geomorphological studies [2]. Different approaches exist for quantifying the flow characteristics of torrent channels. The most widely used method for measuring the flow velocity in natural channels is the point-velocity method using a current meter [3]. The direct measurement of the instantaneous flow velocity at the location of interest has several advantages over other measurement methods. These include ease of use and high accuracy when using a relatively small number of measurements in regular and steady flows. However, the point-velocity technique can be impractical or yield significant errors in torrent channels owing to various bedforms, steep slopes, and isolated large particles [4,5].

An alternative method for obtaining flow velocity is the water-tracing approach. This approach has been increasingly used in mountain hydrology over the past two decades, as it enables the measurement of the reach-averaged velocity for low-depth flows and/or rapidly varying flows in mountain gravel-bed channels [6]. The average velocity is derived by measuring the travel time of solutes in water over a channel section of a certain length using several tracers, including salt, buoyant particles, fluorescent dyes, and thermal tracers. Salt and dye tracers are commonly used in water-tracing due to their ease of use and other desirable features [4,7]. Salt tracers tend to be more widely used for flow measurements than dye tracers, but their use under high-flow conditions is limited due to measurement uncertainty [4,8]. Dye tracers, which are more expensive than salt ones, enable the measurement of higher flows even in small quantities because the dye can be detected at much lower concentrations than salt [4]. Rhodamine WT is preferred in most dye fluorometry applications because of its ease of use, relatively low cost, low absorptive tendency, strong fluorescence, and chemical stability [7,9]. Compared with other ambient water substances, Rhodamine WT has distinct spectral features, making it suitable for use as a water tracer [7,8].

The flow resistance is an essential determinant of the hydraulic and sedimentary processes in mountain channels. It reflects the processes by which the physical shape and bed roughness of a channel influence the flow depth and mean flow velocity in steep channels [10]. Flow resistance caused by objects protruding into the water is quantified by the hydraulic roughness [11]. The hydraulic roughness in mountain streams with large rough elements is particularly difficult to quantify because the large grains in the bed typically protruding above the water surface at low flows can have heights similar to the flow depth during high flows [12,13]. The hydraulic roughness at channel beds can be explicitly associated with the presence of bedform roughness, which is empirically expressed as a log-law formula structure, such as the Manning formula. This formula performs well under deep and uniform flow conditions with relatively small roughness, such as in plain rivers, but its use in mountain streams is still limited owing to the shallow water depth, steep slopes, wide grain size distribution, different types of bed structures, and sharp variations in the flow path direction [13,14].

Mountain torrents typically feature large grains randomly distributed in channels. The bed materials in mountain channels are very coarse; therefore, they protrude well into or completely through the flow. The hydraulic roughness in mountain channels is significantly increased by the presence of larger bed materials such as boulders or rocks [1], while the relative roughness, roughness shape, size distribution, and spacing of its elements may influence the resistance flow [12]. Ferguson [15] and Rickenmann and Recking [16] selected the characteristic grain size (D_{84}) to describe the channel roughness, whereas Aberle and Smart [5] and Lee and Ferguson [17] argued that the grain size might not be an appropriate roughness measure for steep streams. As an alternative, they derived roughness measures from the standard deviation of the bed elevation.

Efficient and accurate field measurements enable all researchers to explore the hydraulic relation between flow resistance and streambed characteristics. This could lead to a better understanding of the influences of bed gradient and roughness on flow behavior in mountain channels. The presence of macrorough elements, both across and down the reach, makes it difficult to obtain a reliable measure of hydraulic variables.

Currently, there is no universal consensus on quantifying the hydraulic roughness in shallow mountain channels. This is mainly due to limited data on the flow and associated roughness measurements in the field, making accurate field measurements essential for elucidating the relationship between the flow characteristics and bed microtopography at a reach scale. The aim of this study was to propose a framework for quantifying the reach-average hydraulic roughness from limited local data in mountain headwater streams.

To achieve this objective, dye fluorometry was used to estimate the reach-average flow velocity under various hydraulic conditions. Between-site hydraulic parameters were obtained from LiDAR surveys and field measurements.

2. Materials and Methods

2.1. Study Site

Data were collected from two mountain headwater (zero-order) streams in the Republic of Korea (Figure 1). The general characteristics of the mountain headwater streams are listed in Table 1. The Gwanak headstream (GA) is the headwater of the Samsung Stream (first-order) situated in the Seoul National University Arboretum (SNUA). The SNUA is located in a temperate zone between 37°26′–28′ N latitude and 126°55′–58′ E longitude, with an elevation ranging from 110 to 620 m a.s.l. The GA site includes 266.7 ha of natural coniferous forest. The mean annual precipitation and temperature are 1294.9 mm and 13.1 °C, respectively [18]. The GA consists of a cascade stream with an average slope of 2–6%. The Baekun (BU) headwater stream drains into the Donggok stream in the upper portion of Mt. Baekun (N 35°01′–20′, E 127°30′–34′). The BU is characterized by a cascade of steep slopes varying from 3 to 20%, while its elevation ranges from 20 to 1218 m a.s.l. The annual precipitation measured at the Gwangyang station ranges from 835 to 2497 mm, with an annual mean temperature of 14.7 °C [18].

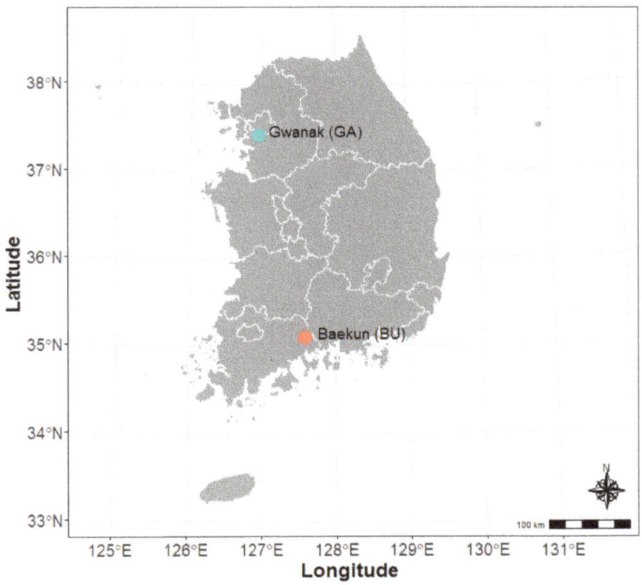

Figure 1. Location maps of the study sites.

Table 1. Characteristics of the study sites.

Location	Coordinates (°)	Average Slope (%)	Morphology	Basin Area (ha)
Gwanak	37°26′–28′ N 126°55′–58′ E	6	Cascade	266.7
Baekun	35°01′–20′ N 127°30′–34′ E	11	Cascade	284.7

Field measurements were performed at the reaches of the mountain headwater streams in 2022–2023. The selected streams are mainly located at the head of forest watersheds at an average altitude of 320 m. The streams are composed of a series of undulations, such as small-scale pools and riffles, which form heterogeneous channel structures as a result of the flow conditions. The streambed has a relatively thin layer of gravel and boulders, but large boulders are randomly spaced throughout the channels. Photographs illustrating the stream characteristics of the selected headwaters are shown in Figure 2.

Figure 2. Photos of selected reaches in the study sites: (**a**) Gwanak, and (**b**) Baekun.

In mountain streams where roughness elements have a spatial extent exceeding the water's depth of flow, the flow resistance caused by flow obstructions is affected by the relative magnitude of submergence. In this study, the flow discharge rates were automatically measured at the GA and BU sites (Figure 3). The flow measuring rectangular weirs have been operated since 2021 to monitor the water levels in the two sites continuously. The flow rate corresponding to the water level was estimated from the predefined head–discharge relation of the weir. The field experiments were arbitrarily conducted at the upstream and downstream reaches of the weir structures over the study period. The flow discharge was considered constant along the target reach, as no substantial lateral flow contribution to the reach was observed. The spatial and temporal variations in the flow discharge can be neglected in this study, because averaged hydraulic parameters over a reach were only considered.

Figure 3. Flow discharge measurements in the study sites: (**a**) Gwanak, and (**b**) Baekun.

Various reach sections in the headwater streams were selected for field measurements. These reaches were relatively uniform in terms of the slope, channel geometry, and bed material along their lengths. No wake flows or local inflow, which could delay or separate the dye from the flow, existed between the injection and measurement points. Prior to field measurements using tracers, the distance required for complete mixing was experimentally determined by comparing the variations in the trace concentrations at multiple points across the section. This was performed by choosing certain sections at different distances from the release point and then measuring the concentration variations at many points in each of them. The selected reach length ranged from 3.9 to 46.8 times the wetted width of the water surface, which is enough to achieve the complete mixing of the dye solution [4]. The reach slopes ranged from 8.2% to 26.0%, with an average of 14.5%. Measurements of the flow discharge and velocity were performed immediately after summer rainfall events.

2.2. Dye Trace Method

The dye trace method using slug injection was selected in this study to measure the flow velocity and discharge of the mountain river flows, where relatively short mixing lengths of lateral and vertical dispersions of injected solutes are observed. This method requires the instantaneous release of a certain quantity of dye solution into the stream and the subsequent measurement of the variations in the dye trace concentration at a point where it is fully mixed with the flow. Although several dyes have been utilized in many fields, Rhodamine WT is preferred for most dye fluorometry hydrologic studies due to its ease of use, chemical stability, and low absorptivity [7,9].

The flow discharge can be calculated from the dilution of Rhodamine WT (A13572 Rhodamine B, Alfa Aesar, Haverhill, MA, USA) by equating the changes in the stream tracer mass to the mass being injected using the tracer mass conservation equation applied in steady flows:

$$Q = \frac{M}{\int [c(x_m, t) - c_0(x_m, t)]dt} \tag{1}$$

where Q is the flow discharge, M is the injected tracer mass, c is the measured tracer concentration at x_m and c_0 is the background concentration.

The average velocity through a reach is defined as the ratio of the distance between two points of traveling time [4]. The time of travel refers to the movement of water or waterborne solutes from point to point in a stream and is inversely proportional to the velocity in a stream.

To acquire the reach-average velocity, the travel time of the dye solution is typically calculated using either the time between peaks, the time between centroids, or the harmonic mean of the measurements [19]. Previous studies have demonstrated that the harmonic mean is the best statistic for computing the reach-average velocity [6,20]. The harmonic mean is the reciprocal of the expected value of the reciprocal of a random variable.

The average reach velocity ($\overline{V}$) over reach length (x_L) is expressed as:

$$\overline{V} = \frac{x_L}{t_L} \tag{2}$$

where t_L is the travel time of the Rhodamine WT solution, which is replaced by the harmonic mean travel time. The harmonic mean time (t_{HM}) of the dye tracers passing through the two locations is expressed as the probability density function of the travel time for each tracer [6]:

$$t_{HM} = \frac{1}{\int_{t=0}^{\infty} \frac{1}{t} p_x(t)dt} \tag{3}$$

where $p(x)$ is the temporal probability function defined as

$$p_x(t) = \frac{1}{I_x} c(x, t) \tag{4}$$

where $c(x,t)$ denotes the tracer concentration. I_x is defined as the sum of the concentration over time at a fixed point x and is calculated as follows:

$$I_x = \int_{t=0}^{\infty} c(x,t)dt \tag{5}$$

Consequently, the reach-average flow velocity can be estimated using the dye-tracing method, with three replicates for each reach.

2.3. Determining Hydraulic Parameters from Field Measurements

Hydraulic roughness is a fundamental hydraulic property used to quantify the flow resistance of the protruding elements. Mountain headwater streams incorporate a wide range of morphological structures, resulting in significant variations in the substrate material, water depth, and local flow velocity. Therefore, being related to the particular roughness topography, quantifying the hydraulic roughness in mountain streams is challenging [21]. In this study, the hydraulic roughness height was quantified using the surface roughness caused by bedform friction and bedform material and microtopography-associated drag [22].

Longitudinal and cross-sectional profiles were surveyed with a LiDAR scanner (L-Leplica-BL, Mobiltech, Seoul, South Korea) along the entire reach length under low-flow conditions. Laser scanning seems to be a suitable tool for collecting surface topography data as it quickly acquires a large amount of accurate point data. At each study reach, geometric data such as the cross-sectional profile, longitudinal gradient, and bedform morphology were extracted from the LiDAR point-cloud data using the CloudCompare software (v2.12.4, GNU). All the reaches were scanned with an average point-cloud density of 0.0318 points/cm^2, with a mean absolute registration error of 0.11 mm. Figure 3a represents the gradient norm scale of the GA site.

The surface roughness of each reach was quantified using the standard parameters, the 90% inter-percentile range (IPR_{90}) and the standard deviation of the bed elevation (STD_z) from the detrended elevations. The IPR_{90} is defined as the 95th percentile minus the 5th percentile of all the detrended roughness, and STD_z reflects the macroroughness of the bed structure at the reach or cross-section scale [22]. These surface roughness variables were indirectly derived from the LiDAR-based bed profile by applying a similar procedure to Nitsche et al. [1] (Figure 4). Geomorphological effects, introduced by the channel slope or large bedforms, can be removed by fitting a trend surface to the longitudinal and cross-sectional profiles (Figure 4b). In this study, locally weighted scatterplot smoothing (lowess) was employed to calculate the trend value for the data sequence (Figure 4c,d). The lowess is a non-parametric and polynomial smoothing technique for fitting a trend curve to data points [23]. Subsequently, the detrended profile was produced by subtracting the trend values from the LiDAR data. The longitudinal IPR_{90}^{L} and STD_z^{L} were derived from the detrended profile of the entire reach (Figure 4c). Since the channel geometry and roughness parameters may vary considerably within a given channel, the cross-sectional hydraulic variable was sampled on regularly spaced cross-sections along the reach. The equidistant points along the reach were selected for the sampling to include at least 10 cross-sections. Once these fixed spacings were defined, the cross-sectional hydraulic parameters for a specified reach were extracted from the cross-sectional profile data. The cross-sectional IPR_{90}^{X} and STD_z^{X} were then estimated at 10 different sections over the entire reach and then averaged for each reach, where the cross-section was confined by the water depth during the experimental period (Figure 4d).

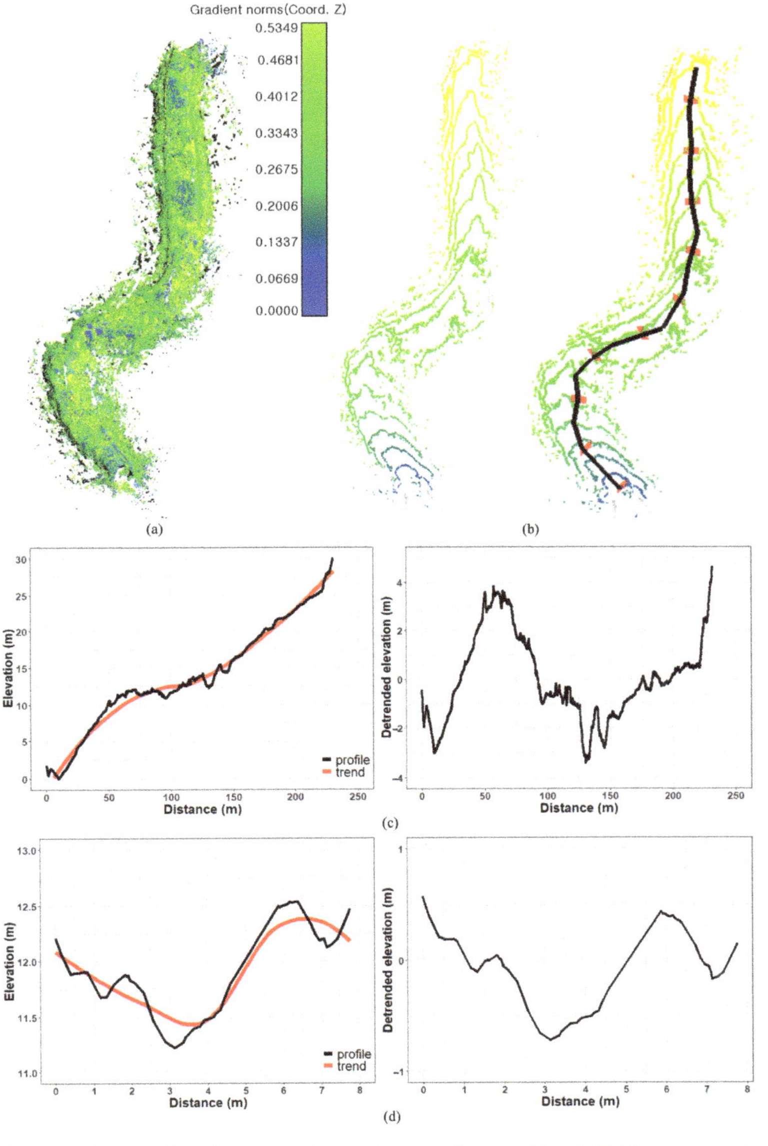

Figure 4. A procedure for measuring surface roughness with a LiDAR survey at the Gwanak site: (**a**) LiDAR point-cloud data, (**b**) gradient norm scaling topography, (**c**) detrended longitudinal profile, and (**d**), detrended cross-sectional profile.

The grain size distribution of the streambed materials was monitored using the pebble-count technique [24]. Grain size samples were taken from the entire reach in short reaches, while in longer reaches, more than 300 counts of individual particles were taken from the upper, middle, and lower parts of the sampled reach, which is considered adequate for representing the actual grain size distribution [25]. The intermediate axes of the sampled individual stones were manually measured to obtain the cumulative size distribution curve for each reach. The particle sizes were averaged to obtain the average grain size statistics. Based on the grain size distribution, the values of D_{50} (median diameter) and D_{84} (diameter below which 84% of the samples lie) for each reach were used as grain roughness parameters.

2.4. Hydraulic Geometry Relation

The mean velocity is a crucial determinant of the hydraulic process in mountain streams. The flow velocity varies between sites depending on the magnitude of the flow and the nature of the hydraulic geometry on the flow conveyance surface [13]. Therefore, in the absence of a direct measurement, it is possible to predict the flow velocity using the hydraulic geometry relationships with a particular discharge.

Several approaches have been used to estimate the flow velocity association with a given flow discharge [13,15,26]. These approaches are essentially empirical in origin and are based on direct laboratory measurements [26] and field studies [13]. Some researchers have studied commonly used formulas, such as the Manning, Chezy, and Darcy–Weisbach coefficients, whereas others have utilized hydraulic geometry approaches. Whether the traditional approach or the dimensional hydraulic geometry approach is more appropriate is still under debate.

Recently, a number of researchers have attempted to revise the traditional approach to accurately reflect the hydraulic processes occurring in mountain streams. Rickenmann [26] and others [5,15,27] have proposed dimensionless hydraulic relations to represent the flow characteristics in a range of rough streams. Such relationships are expressed in the following equations:

$$v_* = c\left(q_*^{\alpha} S^{\beta}\right) \tag{6}$$

$$v_* = \frac{v}{\sqrt{gk}}, \quad q_* = \frac{q}{\sqrt{gk^3}} \tag{7}$$

where v_*, q_* are the dimensionless velocity and discharge, respectively; k is the hydraulic roughness that can be expressed by the grain size or streambed geomorphology; S is the bed slope; and g is the gravitational acceleration. c is the discharge coefficient, and α, β are independent exponents in the velocity–discharge hydraulic geometry relations. These quantitative relationships generally provide better flow velocity estimates in steep streams.

When selecting the best parameter for flow resistance in mountain channels, the dimensionless hydraulic relations in Equation (7) were tested with different roughness measures. Instead of the point velocity, the reach-average velocity measured with the dye-tracing technique was used in its dimensionless form. Various hydraulic roughness heights used to characterize flow resistance were incorporated to represent the dimensionless velocity and discharge in Equation (7). In the final analysis, the hydraulic relation was proposed to demonstrate the influence of streambed roughness on the flow velocity in mountain headwater streams.

3. Results

3.1. Reach-Average Hydraulic Roughness

Twenty-two reach-average hydraulic geometry data series were used, including a range of bed roughness values, grain sizes, and flow characteristics. The between-site hydraulic geometry influences the flow conveyance over the reach and explains the variation in the flow velocity between sites at a given discharge. Therefore, differences between sites probably arose from differences in the grain size, microtopography, and surface structures in streams. All the reaches have sufficient distance to achieve the complete mixing of the tracer.

Locally, the reach gradients were approximately 0.02–0.07 m/m at the GA site and 0.05–0.20 m/m at the BU site. The longitudinal profile contributed to the presence of large coarse materials in the streambed. Table 2 shows the reliability of the bed gradient and D_{50} relationship. Notably, the D_{84} values between the two sites did not differ significantly, which was probably a consequence of fluvial sediment transport. The presence of large boulders is uncommon in small, forested headwater streams in South Korea. Wide variations in microtopography (IPR^L_{90}, STD^L_z) along the reach length were observed at both sites; however, small differences in cross-sectional profiles were observed.

Table 2. Average reach velocity, weir discharge and geometric characteristics.

Site	V^* (m/s)	Q^* (m^3/s)	L^* (m)	Slope (m/m)	D_{50} (m)	D_{84} (m)	IPR^L_{90}	STD^L_z	IPR^X_{90}	STD^X_z
GA1	0.57	2.02	102.2	0.04	0.17	0.48	3.59	1.12	0.77	0.23
GA2	0.96	1.87	84.4	0.05	0.18	0.54	3.59	1.12	0.92	0.28
GA3	0.43	1.76	82.6	0.03	0.18	0.54	3.49	1.06	0.92	0.28
GA4	0.23	0.94	113.2	0.07	0.16	0.39	5.29	1.78.	0.84	0.26
GA5	0.46	1.59	113.2	0.07	0.16	0.39	5.29	1.78	0.84	0.26
GA6	0.66	1.96	164.8	0.05	0.16	0.39	6.20	1.79	0.82	0.25
GA7	0.42	1.64	42.2	0.07	0.18	0.54	2.15	0.62	0.60	0.19
GA8	0.43	1.23	32	0.04	0.18	0.54	1.24	0.42	0.53	0.17
GA9	0.34	1.23	51.8	0.06	0.17	0.48	2.32	0.65	0.82	0.25
GA10	0.28	1.23	30.4	0.02	0.18	0.54	2.19	0.67	0.95	0.29
GA11	0.33	1.76	176.9	0.04	0.16	0.39	6.04	1.74	0.78	0.24
GA12	0.24	1.31	90.5	0.02	0.17	0.48	4.16	1.30	0.84	0.26
GA13	0.22	1.31	163.9	0.04	0.16	0.39	6.22	1.79	0.82	0.25
BU1	0.63	2.01	51.2	0.20	0.20	0.40	1.63	0.54	0.63	0.21
BU2	0.48	1.76	60.4	0.15	0.21	0.40	1.70	0.60	0.65	0.22
BU3	0.22	0.21	72.1	0.10	0.18	0.33	3.07	1.05	0.61	0.21
BU4	0.24	0.21	34.6	0.05	0.28	0.53	0.70	0.24	1.11	0.45
BU5	0.37	0.31	72.1	0.10	0.18	0.33	3.07	1.05	0.61	0.21
BU6	0.34	0.31	37.5	0.15	0.25	0.42	0.61	0.20	2.67	0.83
BU7	0.43	0.31	30.6	0.15	0.38	0.55	0.49	0.17	2.60	0.81
BU8	0.81	2.59	72.1	0.10	0.18	0.33	3.07	1.05	0.61	0.21
BU9	0.73	2.59	37.5	0.15	0.25	0.42	0.61	0.20	2.67	0.83

Note: * V is reach-average velocity, Q is measured discharge at weir structure, and L is reach length.

The Spearman's correlation analysis results between the roughness height variables for all the reaches are shown in Figure 5. The obtained longitudinal IPR^L_{90}, and STD^L_z point-cloud data were inversely related to the cross-sectional profiles. Hydraulically, the cross-sectional roughness (IPR^x_{90}, STD^x_z) showed a strong correlation with the grain size D_{50} and a weak correlation with D_{84}.

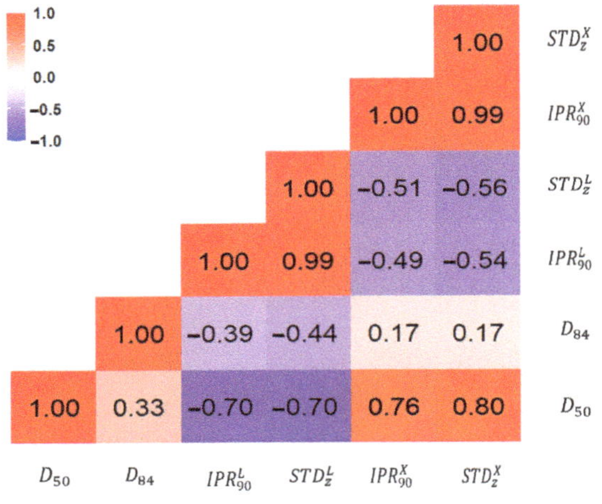

Figure 5. Correlation between the roughness height variables.

For mountain streams, the bed morphology is hydrologically characterized by the sediment supply limited conditions caused by large immobile boulders, which can form stable armored beds [2]. Thus, the grain size distribution may better reflect the lateral sediment inputs than hydraulic sorting processes. An armor layer may be present in mountain streams when fine bed materials are washed out during flooding events. Armored layers could be responsible for the small variation in the large grain size distribution observed, suggesting that the D_{80} values asymptotically approached a certain limit, regardless of the bed gradient.

3.2. Flow Discharge and Its Relation to Velocity

Different sets of flow measurements were performed during 2022–2023 using two mountain headwater streams. The reach-average discharge derived from the dilution of the dye tracer was compared to the point measurements. The measurements performed using dye tracing were generally in good agreement with the flow discharge measured at the weir of surface water monitoring stations. Rectangular broad-crest weirs were installed upstream of the GA and BU sites and the flow discharge was directly determined by measuring the hydraulic head difference. The above acquired data were used to examine the relationship between flow velocity and discharge.

The discharge measured with dye tracing ranged from a minimum of 1.01 m^3/s to a maximum of 2.45 m^3/s at the GA site, and from 0.91 m^3/s to 1.81 m^3/s at the BU site. Over the study period, the flow discharges measured concurrently with weirs varied from 0.94 m^3/s to 2.02 m^3/s at the GA site and 0.21 m^3/s to 2.59 m^3/s at the BU site.

Comparing the results obtained using the two methods, the average relative errors in the discharge were 67% and 38% at the GA and BU sites, respectively. As shown in Figure 6, the measurements using the tracer slightly underestimated the flow compared to the weir measurements during the study period. This was probably due to tracer losses between the injection point and the downstream measurement station, and due to incomplete mixing throughout the reach [28]. Figure 5 shows that the discrepancy between the measurements increased as the flow increased. Under flood conditions, it was impossible to measure the head difference of the weir with acceptable precision. Inevitably, some weir measurements at very high flows may be incorrect because of the difficulty in obtaining reliable measurements due to overflowing at the top of the weir instead of flowing through the weir section.

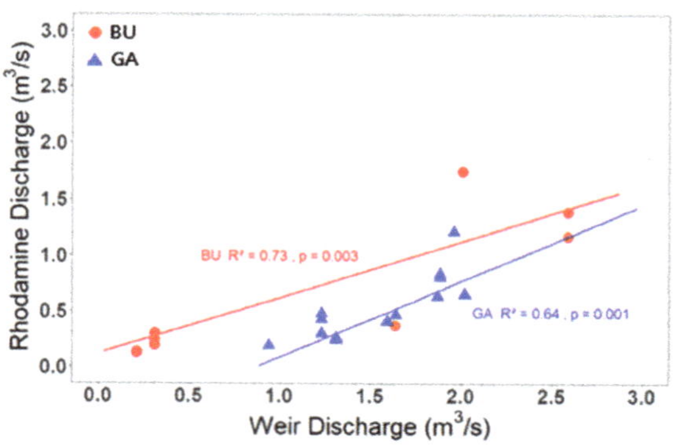

Figure 6. Relationship between weir discharge and rhodamine discharge.

An empirical relation between flow velocity and discharge was also analyzed. The measured reach-average flow velocities for each mountain stream exhibited curvilinear relationships with the flow discharges measured at the weirs, with $R^2 = 0.58$ ($p = 0.006$) for the GA site and $R^2 = 0.85$ ($p = 0.0002$) for the BU site. This result is in accordance with previous studies [5,27,29,30]. The flow velocities at the BU site were higher than those at the GA site at a certain discharge because of the steeper streambed gradient (S = 5–20%). A relatively wide variation in the velocity and discharge relationship was observed at the BU site compared with the respective variation at the GA site. This is probably due to the uncertainty of the reach-average velocity measurements associated with the free-surface distortion and hydraulic jumps that commonly occur in large-scale and undulated bed materials [10].

3.3. Relation between Flow Velocity and Hydraulic Roughness

Twenty-two reach-average hydraulic geometry data series were used, including a range of bed roughness values, grain sizes, and flow characteristics. Six roughness height models were calibrated using data from twenty-two field measurements, and the observed and predicted velocities were compared (Table 3). The dimensionless hydraulic geometry relationship that incorporated IPR^{L}_{90}, q and S, exhibited the largest R^2 between the observed and predicted data and the smallest RMSE, suggesting that this model is the best suited (Table 3). This result is consistent with those of Yang [30] and others [22]. Scatterplots of the flow velocities with the hydraulic roughness variables are shown in Figure 7.

Table 3. Reach-average velocity and discharge equations with roughness.

Roughness Variable	Hydraulic Geometry Relation	Measured–Estimated Velocity	
		R^2	RMSE (m/s)
IPR^{L}_{90}	$v_* = 0.53\, q_*{}^{0.46}$	0.68	0.07
STD^{L}_{z}	$v_* = 0.42\, q_*{}^{0.39}$	0.66	0.22
IPR^{X}_{90}	$v_* = 0.27\, q_*{}^{0.27}$	0.42	0.07
STD^{X}_{z}	$v_* = 0.27\, q_*{}^{0.27}$	0.42	0.53
D_{50}	$v_* = 0.31\, q_*{}^{0.25}$	0.31	0.91
D_{84}	$v_* = 0.30\, q_*{}^{0.29}$	0.28	0.16

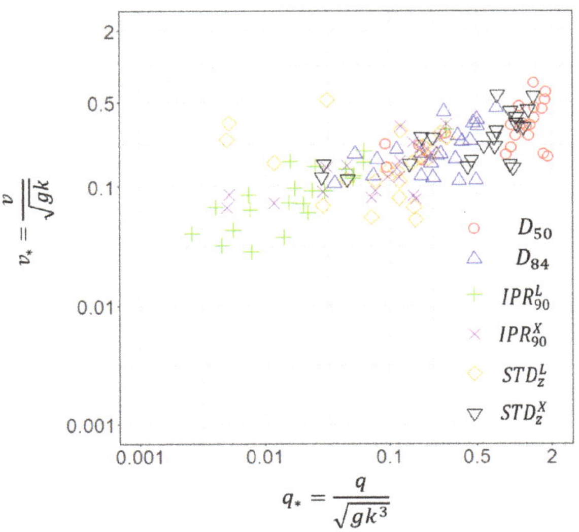

Figure 7. Dimensionless relationship between discharge and reach-average velocity with roughness.

Previous studies [2,12,13] demonstrated that the flow roughness in steep streams varied with the flow discharge or flow depth, as the large elements on the bed generally protrude above water surface at low flows, but to the contrary, submerge in the water during high flows. Therefore, the relations in Table 3 are valid for the measurement condition, where the discharge per unit width ranged from 0.04 m^2/s to 0.43 m^2/s. In many channels, the flow depth varies from location to location, and it may also change in time with rainfall. However, the variation in the flow rate was not considered in this study because the experiments were conducted over a limited distance in a relatively short amount of time.

The scattered patterns observed in Figure 7 can be attributed to the inability of the longitudinal IPR_{90} to fully capture the flow roughness of the streams. The importance of additional factors, such as the grain and spill resistances, has long been recognized in various contexts [13]. In the case of fluvial streams, the stream bedform is controlled by selective bedload transport during flood conditions. This significantly affects the grain size of transported materials and plays an important role in the determination of D_{84}.

Based on the results in Table 3, an equation for estimating the reach-average velocity was proposed:

$$v = 0.54g^{0.27}q^{0.46}S^{0.27}\left(IPR_{90}^{L}\right)^{-0.19} \quad (8)$$

Previous research has demonstrated that roughness heights derived from surface roughness, such as IPR_{90}^{L} and STD_{z}^{L}, are more suitable for representing the flow characteristics in mountain streams [1,5]. However, the difficulty in accurately measuring the streambed roughness of mountain streams can render their practical use unfeasible. In contrast, the grain size distribution can be easily measured and is commonly used to determine the roughness height in steep streams [15,17]. Recently, advanced photogrammetry and laser-scanning technologies have enabled researchers to obtain detailed topographical information and streambed roughness measurements [22]. The surface roughness derived from point-cloud data, especially IPR_{90}^{L}, can yield comparable results in flow estimations.

4. Discussion

4.1. Comparison of Point Velocity and Reach-Average Velocity

Accurate measurements of the flow velocity are necessary to understand the hydrologic, hydraulic, and sedimentary processes in mountain channels. The point-velocity method using a current meter has been widely used to quantify the flow velocity under both laboratory and field conditions. However, the presence of turbulence at very high flows can result in significant velocity measurement errors. In some special cases, this method underestimates the flow characteristics with errors greater than 50% [30]. Furthermore, the macroroughness of the streambed hinders the use of a current meter under low-flow conditions.

The conventional point-velocity method can measure the velocity at points of interest but is not suitable for measuring the reach-average velocity along streams. Thus, it is advisable to use alternative techniques for measuring the flow characteristics in mountain streams, which are characterized by coarse bed materials, steep slopes, and low depths. The dye trace method has shown favorable results in terms of accuracy when compared to current metering as a method for measuring flow in mountain channels, with its accuracy increasing in rivers with unsteady flow conditions and irregular streambed geometry. In any case, both methods can be complementary.

Each method has advantages and disadvantages, which limit their use. Given these considerations, it is necessary to select the appropriate type of measurement technique used to quantify water characteristics depending on the hydraulic and geomorphological conditions. Current meters are not suitable for extremely low or high flows in mountainous regions. The use of dye tracers in mountain streams is preferred, provided that tracer injection and detection are available.

4.2. Use of Dye Tracer

The use of artificial tracers for flow measurements is convenient due to their low cost, easy handling, low impact, and satisfactory results in mountain streams. The selection of the most suitable tracer depends on several factors, including the water quality, amount and type of suspended sediment, distance between the release and detection locations, and flow characteristics [31]. Before their use in the field experiments, water tracers, including a fluorescent (Rhodamine WT) and a chemical tracer (NaCl), were preliminarily compared to evaluate the differences in their behavior. Sodium chloride (NaCl) is the most frequently used tracer, providing the best results with a low environmental impact. However, its low detectability under high-flow conditions is a major drawback [4,30]. In contrast, even small quantities of Rhodamine WT are readily detectable, enabling good measurements to be acquired under very high-flow conditions. Given the chemical and spectral characteristics of the tracers, Rhodamine WT is highly suitable for use as a water tracer in steep, small mountain streams.

The field experiments also indicated that measurement errors can be mainly attributed to dye losses between the two measurement locations, resulting in poor measurement accuracy. Dye losses occur when excessively long reaches are involved or when finely suspended sediment exists, particularly clay or organic flocculant particles. Water-tracing dyes tend to adhere to the surfaces of landform materials such as stone, rock, and aquatic plants [32]. Previous studies have indicated that significant losses of Rhodamine WT may occur when diluted concentrations are exposed to direct sunlight for extended periods, owing to the photochemical decay of the dye [30]. This is a potential source of error that must be considered when performing dye-fluorometric studies. In this study, most exposure times for the dye experiments were short enough that the sorption and photochemical decay losses would be negligible.

4.3. Quantifying Hydraulic Roughness

Mountain streams have complex and rough bed morphology with immobile boulders, bedrock constrictions, or large woody debris. The streambed has long been regarded as the self-adjusted hydraulic system that produces to maximize flow resistance to maintain greater bed stability [13,16,21,22]. Bed stability is closely related to the grain sizes that can be entrained or deposited under various bed gradients and flow conditions. The grain size distribution (D_{50}, D_{84}) of the streambed is widely used to represent the flow resistance in natural streams due to its simplicity [1]. However, the field estimation of the grain size distribution in mountain headwater streams can result in large uncertainties due to operational bias, limited sample size, and spatially heterogeneous grain size distribution [22]. Alternatively, roughness height measures (IPR_{90}, STD_z) are preferable to quantify the flow resistance in steep mountain streams [19,30].

The scattering observations in Figure 6 suggested that IPR^L_{90} may not entirely capture the flow roughness in steep channels. It can be influenced by other factors, such as the grain and spill resistances [2,13]. Although the empirical Equation (8) is analogous to previous studies, it is helpful in cases of steep and rough channels in South Korea. Nonetheless, this study mainly contributes an alternative hydraulic relation that enhances our understanding of estimating the mean flow velocity in mountain streams. LiDAR is commonly used to collect high-resolution terrain data that capture the variation in geomorphologic characteristics, not only at a specific location but also along stream reach. A LiDAR survey provides new opportunities for accurately obtaining detailed topographic information and streambed roughness height measures, such as the interpercentile range IPR_{90} and the standard deviation STD_z. Moreover, there are time and costs saving advantages for LiDAR surveying applications. However, LiDAR imagery can often underestimate the depths of pockets between macroroughness elements due to shadow effects. Recently, an algorithm has been developed to reduce the shadow errors in mapping rough grain surface, but it is not considered in this study [33].

4.4. Limitations of the Study

In this study, the hydraulic roughness was quantified from field observations in mountain streams. However, this study has potential limitations. First, the hydraulic roughness is regarded as a constant parameter over time. However, every stream has a self-organized fluvial process to adjust the bed morphology to maximize the flow resistance [13,22]. Although the flow rate or water depth is assumed to be unchanged over the reach, the presence of protruding objects and large boulders can lead to a spatial variation in the bed roughness with the relative submergence of roughness elements (the ratio of flow depth to roughness height) in steep mountain channels [2,5,22]. This study does not deal with the hydraulic behavior of steep and rough mountain streams yet.

Field measurements were conducted in this study for low flows with a discharge per unit width of 0.04–0.43 m^2/s. It seems reasonable that within the range, the proposed hydraulic relationships might be valid in mountain headwater streams, such as the Gwanak and Baekun sites. Furthermore, the flow resistance may be influenced by other factors in addition to hydraulic roughness height, including the boulder arrangement and protrusion, energy dissipation, or erosional process. However, we have neglected these factors in this study.

5. Conclusions

This study focused on the influences of macroroughness elements on the flow in steep channels and their incorporation into hydraulic relations, and it examined a research framework for quantifying the hydraulic roughness of mountain streams based on limited field measurements. The hydraulic geometry relationships in mountain headwater streams were developed based on the reach-average velocity and associated hydraulic variables. A dye-tracing technique was used to measure the average velocity over the entire reach length. Various hydraulic roughness heights were expressed in terms of the grain (D_{50} and D_{84}) and form resistance (IPR_{90}, STD_z). Of the various measures for hydraulic roughness in the dimensionless hydraulic relations, IPR_{90}^L exhibited the largest R^2 (=0.68) and the smallest RMSE (=0.07 m/s) across 22 field measurements, implying that IPR_{90}^L was superior (in terms of the R^2 and RMSE) in explaining the variations in the reach-average velocity among mountain headwater streams. Consequently, the dimensionless hydraulic geometry relationship incorporating the discharge (Q), bed slope (S), Q, and longitudinal IPR_{90} was developed to accurately estimate the flow characteristics in steep and rough streams.

The flow rate is thought to have a significant influence on the flow resistance. The effect of hydraulic roughness on the flow increased as the relative submergence of the bed elements (roughness height/water depth) decreased, thus the proposed hydraulic relation between the reach-average velocity and hydraulic roughness might not be suitable for other regions. Nevertheless, the research framework for quantifying hydraulic roughness from local data can be applicable in mountain headwater streams.

In mountain streams with small flow depths, large protruding boulder elements and irregular bed morphology cause a high flow resistance, rendering many traditional approaches used for flatter streams and rivers inapplicable. The accuracy of the proposed hydraulic relation equation was evaluated using local data from field measurements; however, it was subject to considerable uncertainty with the varying flow rate. Furthermore, the utility of the proposed equations is constrained by the availability of data, while the flow characteristics in mountain streams may be influenced by factors other than the bed roughness. Further studies considering a wide spectrum of flow characteristics are needed to explore the flow behavior in mountain streams.

Author Contributions: Conceptualization, T.-H.K. and S.I.; methodology, T.-H.K. and S.I.; software, T.K.; validation, J.L. and T.K.; formal analysis, T.-H.K. and S.I.; investigation, T.K.; resources, J.L.; data curation, T.-H.K.; writing—original draft preparation, S.I.; writing—review and editing, J.L. and T.K.; visualization, T.-H.K.; supervision, H.T.C.; project administration, S.I.; funding acquisition, H.T.C. All authors have read and agreed to the published version of the manuscript.

Funding: This study was carried out with the support of the 'R&D Program for Forest Science Technology (Project No. 2021343C10-2323-CD01)' provided by the Korea Forest Service (Korea Forestry Promotion Institute).

Data Availability Statement: Data are contained within the article.

Conflicts of Interest: The authors declare no conflicts of interest.

References

1. Nitsche, M.; Rickenmann, D.; Kirchner, J.W.; Turowski, J.; Badoux, A. Macroroughness and variations in reach-averaged flow resistance in steep mountain streams. *Water Resour. Res.* **2012**, *48*, W12518. [CrossRef]
2. Rickenmann, D. *Methods for the Quantitative Assessment of Channel Processes in Torrents (Steep Streams)*; CRC Press: Boca Raton, FL, USA, 2016.
3. Holland, F.; Bragg, R. *Fluid Flow for Chemical Engineers*; Elsevier: Amsterdam, The Netherlands, 1995.
4. Richardson, M.E. Refinement of Tracer Dilution Methods for Discharge Measurements in Steep Mountain Streams. Master's Thesis, University of British Columbia, Vancouver, BC, Canada, 2015.
5. Aberle, J.; Smart, G. The influence of roughness structure on flow resistance on steep slopes. *J. Hydraul. Res.* **2003**, *41*, 259–269. [CrossRef]
6. Waldon, M.G. Estimation of average stream velocity. *J. Hydraul. Eng.* **2004**, *130*, 1119–1122. [CrossRef]
7. Wilson, J.; Cobb, E.; Kilpatrick, F. *Fluorometric Procedures for Dye Tracing*; Department of the Interior, US Geological Survey: Denver, CO, USA, 1986; Volume 31.
8. Moore, R. Slug injection using salt in solution. *Streamline Watershed Manag. Bull.* **2005**, *8*, 1–6.
9. Kilpatrick, F.A.; Wilson, J.F. *Measurement of Time of Travel in Streams by Dye Tracing*; US Government Printing Office: Washington, DC, USA, 1989; Volume 3.
10. Thome, C.R.; Zevenbergen, L.W. Estimating mean velocity in mountain rivers. *J. Hydraul. Eng.* **1985**, *111*, 612–624. [CrossRef]
11. de Lange, S.I.; Naqshband, S.; Hoitink, A. Quantifying hydraulic roughness from field data: Can dune morphology tell the whole story? *Water Resour. Res.* **2021**, *57*, e2021WR030329. [CrossRef]
12. Bathurst, J. Flow resistance estimation in mountain rivers. *J. Hydraul. Eng.* **1985**, *111*, 625–643. [CrossRef]
13. Zimmermann, A. Flow resistance in steep streams: An experimental study. *Water Resour. Res.* **2010**, *46*, W09536. [CrossRef]
14. Warmink, J.J.; Booij, M.J.; Van der Klis, H.; Hulscher, S.J. Quantification of uncertainty in design water levels due to uncertain bed form roughness in the Dutch river Waal. *Hydrol. Process.* **2013**, *27*, 1646–1663. [CrossRef]
15. Ferguson, R. Flow resistance equations for gravel- and boulder-bed streams. *Water Resour. Res.* **2007**, *43*, W05427. [CrossRef]
16. Rickenmann, D.; Recking, A. Evaluation of flow resistance in gravel-bed rivers through a large field data set. *Water Resour. Res.* **2011**, *47*, W07538. [CrossRef]
17. Lee, A.J.; Ferguson, R.I. Velocity and flow resistance in step-pool streams. *Geomorphology* **2002**, *46*, 59–71. [CrossRef]
18. National Institute of Forest Science. *Current Status of Nationwide Forest Watershed Test Sites*; National Institute of Forest Science: Seoul, Republic of Korea, 2022; pp. 60–101.
19. Yochum, S.E. *Flow Resistance Prediction in High-Gradient Streams*; Colorado State University: Fort Collins, CO, USA, 2010.
20. David, G.; Wohl, E.; Yochum, S. *Relationship of Bed and Bank Resistance to Total Flow Resistance in a High Gradient Stream, Fraser Experimental Forest, Colorado, USA*; AGU Fall Meeting Abstracts: San Francisco, CA, USA, 2010.
21. Buffington, J.M.; Montgomery, D.R. Effects of hydraulic roughness on surface textures of gravel-bed rivers. *Water Resour. Res.* **1999**, *35*, 3507–3521. [CrossRef]
22. Schneider, J.M.; Rickenmann, D.; Turowski, J.M.; Kirchner, J.W. Self-adjustment of stream bed roughness and flow velocity in a steep mountain channel. *Water Resour. Res.* **2015**, *51*, 7838–7859. [CrossRef]
23. Moran, G.W. Locally-Weighted-Regression Scatter-Plot Smoothing (LOWESS): A Graphical Exploratory Data Analysis Technique. Master's Thesis, Naval Postgraduate School, Monterey, CA, USA, September 1984.
24. Wolman, M.G. A method of sampling coarse river-bed material. *EOS Trans. Am. Geophys. Union* **1954**, *35*, 951–956.
25. Rice, S.; Church, M. Sampling surficial fluvial gravels; the precision of size distribution percentile sediments. *J. Sediment. Res.* **1996**, *66*, 654–665. [CrossRef]
26. Rickenmann, D. *Bedload Transport Capacity of Slurry Flows at Steep Slopes*; Swiss Federal Institute of Technology Zürich: Zürich, Switzerland, 1990.
27. Comiti, F.; Mao, L.; Wilcox, A.; Wohl, E.E.; Lenzi, M.A. Field-derived relationships for flow velocity and resistance in high-gradient streams. *J. Hydrol.* **2007**, *340*, 48–62. [CrossRef]
28. Sappa, G.; Ferranti, F.; Pecchia, G. Validation of salt dilution method for discharge measurements in the upper valley of aniene river (central italy). In Proceedings of the 13th International Conference on Environment, Ecosystem and Development, Athens, Greece, 29 June–2 July 2015.
29. Stewardson, M. Hydraulic geometry of stream reaches. *J. Hydrol.* **2005**, *306*, 97–111. [CrossRef]
30. Yang, H.; Lee, S.-J.; Im, S. Hydraulic Relation of Discharge and Velocity in Small, Steep Mountain Streams Using the Salt-dilution Method. *J. Korean Soc. For. Sci.* **2018**, *107*, 158–165.

31. Tazioli, A. Experimental methods for river discharge measurements: Comparison among tracers and current meter. *Hydrol. Sci. J.* **2011**, *56*, 1314–1324. [CrossRef]
32. Bencala, K.E.; Rathbun, R.E.; Jackman, A.P.; Kennedy, V.C.; Zellweger, G.W.; Avanzino, R.J. Rhodamine WT dye losses in a mountain stream environment. *J. Am. Water Resour. Assoc.* **1983**, *19*, 943–950. [CrossRef]
33. McDermott, M.; Rife, J. Mitigating shallows in LIDAR scan matching using spherical voxels. *IEEE Robot. Autom. Lett.* **2022**, *7*, 12363–12370. [CrossRef]

water

MDPI

Essay

Characteristics of Vegetation Resistance Variation in Muddy Water Flows

Xiaolei Zhang *, Yu Zhu, Haoran Wu, Zhengzheng Bi and Zhiheng Xu

Institute of Water Conservancy, North China University of Water Resources and Electric Power, Zhengzhou 450046, China; zy0900501@163.com (Y.Z.); wuhaorande@outlook.com (H.W.); bzz97@foxmail.com (Z.B.); 18237801369@163.com (Z.X.)

* Correspondence: zxl1334@163.com

Abstract: The shoal area of the lower Yellow River in China is not flooded with water during the dry season, so various plants can grow. When floods overflow the plains in the flood season, the complexity of water resistance is increased due to the resistance to water flow by vegetation, which directly affects flood discharge in the beach area. The drag force coefficient (C_D), Manning's roughness coefficient (n), and Darcy-Weisbach resistance coefficient (f) are commonly used to characterize vegetation drag force. Such studies are commonly conducted in clear water, but flood water in the lower Yellow River is generally muddy. In order to study the effect of the same sediment content and different sedimentation thicknesses on the resistance of muddy waters containing vegetation, this study conducted experiments in a flume (length = 28 m, width = 0.5 m, and height = 0.5 m) under different deposition thicknesses. The results showed that the vegetation drag force coefficient (C_D), vegetation roughness (n_b), and Darcy-Weisbach drag coefficient (f) all decreased logarithmically with increasing Reynolds number (Re) and Froude number (Fr). When Re > 30,000, under the conditions of different siltation thicknesses of vegetation, the vegetation roughness tended to stabilize near its minimum value. When the Reynolds number of the water flow is large (Re > 20,000), the variation of the Darcy-Weisbach drag coefficient f slows down with the Reynolds number Re. Logarithmic functions were established for the above resistance coefficients and flow coefficients, and the corresponding correlation coefficients were high, indicating that the conclusions were reliable.

Keywords: logarithmic fitting; muddy water containing vegetation; water resistance; water flume test

1. Introduction

Due to human activities, beaches are often interrupted by various kinds of vegetation and water-blocking structures, which increase surface roughness along beaches relative to the main trough [1–5], which directly affects the evolution of floodplains in beach areas. Floodplains are generally flooded only during the flood season, so when dry, various types of vegetation can grow in tidal areas. Beach flow resistance is closely related to the distribution and growth of beach vegetation and the distribution of various water-blocking structures [6–10]. The presence of vegetation has a particularly important effect on altering the movement characteristics of sand-carrying flow in beach areas as opposed to open channels, and the resistance of beach areas will be significantly increased by vegetation [11–15]. Similar processes occur in muddy areas, such as the Yellow River basin, so it is necessary to study the resistance characteristics of muddy substrates containing vegetation, which will provide a basis for understanding flood discharge and managing the ecological protection of muddy areas during flood seasons.

Vegetation can be divided into rigid vegetation and flexible vegetation. Vegetation with sufficient rigidity will not be deformed by water motion. Generally speaking, the structure of water flowing through a single plant can be regarded as the flow around a cylinder [16]. Huang et al. [17] used metal rods with different diameters to simulate the water flow resistance caused by planting trees in beach areas, and on this basis, they demonstrated that the resistance of the model and prototype were similar. Their work provided a guide for the design of water flow resistance experiments investigating the effects of vegetation. Stone and Shen [18] used rigid cylindrical rods to study the effect of vegetation diameter and density on resistance at different water depths and found that the combined action of the above three factors determined resistance. James et al. [19] showed that vegetation roughness changed linearly with water depth when vegetation was in a non-submerged state. To explore the change in vegetation roughness with water depth, Wu et al. [20] used a variable slope flume river model with horsehair instead of vegetation. They found that, in the case of submerged vegetation, the vegetative roughness coefficient is negatively correlated with the depth of the water. However, the roughness increased gradually with the depth of the water until the vegetation was completely submerged, at which point it began to decrease before eventually stabilizing and becoming constant. Järvelä [12] carried out flume experiments with willow branches, seagrass, and sedge. The resistance coefficients were calculated as the change in the water head measured by a pressure difference sensor, from which the influence rules of different plant types, densities, arrangements, and water depths on the resistance were obtained. Their results showed that the resistance coefficient of non-submerged willow branches increased linearly with water depth, and under the same arrangement, the resistance coefficients of branches with subbranches were 2~3 times those without. Lu et al. [21] studied the change rules of vegetation resistance coefficients under different vegetation densities, flow conditions, and water depths. They adopted the vegetation hydraulic radius as the characteristic length to represent vegetation and the Reynolds number to represent flow conditions when examining the relationship between the vegetation resistance coefficient and flow. Li and Shen [22] conducted an experiment on the influence of plants on water flow resistance and showed that the arrangement of vegetation had a significant effect. With the same vegetation density, a plum blossom crisscrossing vegetation arrangement had greater water flow resistance than the front and back parallel vegetation arrangements. Kouwen and Li [23] proposed that the factor affecting the boundary roughness was the bending degree of plants, and the greater the resistance coefficient, the greater the bending moment force exerted on plants. Therefore, both plant stiffness and density affect boundary roughness in water flow.

In summary, the research to date on resistance characteristics of water flow with vegetation has been relatively comprehensive, and the interaction between water flow and vegetation has been clearly established in terms of stiffness, flexibility, and vegetation. Several characteristic parameters, such as the drag coefficient, roughness coefficient, and Darcy-Weisbach resistance coefficient, have been utilized. However, although the above studies have been relatively comprehensive, they have all been based on clear water conditions. The flood of the lower Yellow River is generally muddy water with a relatively high sediment content, which further complicates the flow movement problem. On the one hand, the flow of sand can shape the boundary, which in turn affects the flow of water. Furthermore, the water flow carries sediment, and the presence of sediment in turn changes the physical properties and turbulent structure of the water flow, which in turn affects its energy loss, flow velocity distribution, and sand content distribution, and then affects the drag coefficient, roughness coefficient, and Darcy-Weisbach resistance coefficient. Moreover, the velocity of floodplain flood waters slows when flowing through intertidal areas, facilitating siltation and changing the topography of the intertidal area, which directly affects the evolution of floodplain floods in the intertidal area. The existing semi-empirical and semi-theoretical resistance formulas do not accurately describe the change laws of vegetation resistance in the beach areas of the lower Yellow River because models generally regard beach resistance as a fixed value in two-dimensional water and sediment that does not change with the water and sediment conditions. Furthermore, in the previous physical models, the beach morphology remained unchanged because they did not consider the deposition of sediment in the vegetation area. Therefore, in order to obtain accurate resistance characteristics for floodplain floods in the lower Yellow River, this investigation carried out a flume experiment with muddy water containing vegetation.

2. Materials and Methods

2.1. Design of the Experimental Installation

The basic theory of open channel flow was used in this study. Many tests on the resistance characteristics of water flow with vegetation have used rectangular flumes; therefore, to ensure test comparability and the reliability of the results, this test used a rectangular flume with a variable slope. The bottom slope could be adjusted between 0 and 1%. The structure of the test tank is shown in Figure 1. According to the purpose of the test, the following design points were determined:

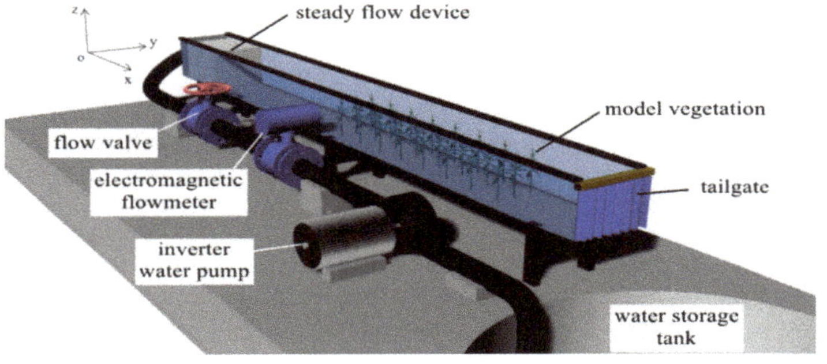

Figure 1. Experimental flume setup.

(1) The tank was 28 m in length, 0.5 m in width, and 0.5 m in height. Its side walls and bottom surfaces were made of rectangular sections of smooth tempered glass. In order to eliminate waves and stabilize water flow, the inlet of the water tank was the same size as the cross section of the tank. A 30-cm-long flow stabilizing device was used, composed of a number of PVC pipes placed inside the water tank. A 2.4-m-long, 0.5-m-wide, and 5-mm-thick PVC board was laid in the middle of the tank. Round holes are drilled into the board to fix the model plants.

(2) The model plants were typical maize vegetation found in the floodplain area, with a total vegetation height (H_v) of 21.5 cm, a lower main stem height of 5.5 cm, and a diameter of 0.15 cm. The upper branches and leaves had an average height of 12 cm. The length of the test control section was 1.4 m, and the width was the distance between the two inner walls of the water tank. The model plants were set in a rectangular arrangement (Figure 2), with a spacing of 10 cm between front and rear and left and right. The planting area per unit area was 9.091×10^{-3} m^2, and the plant density was 110 plants/m^2.

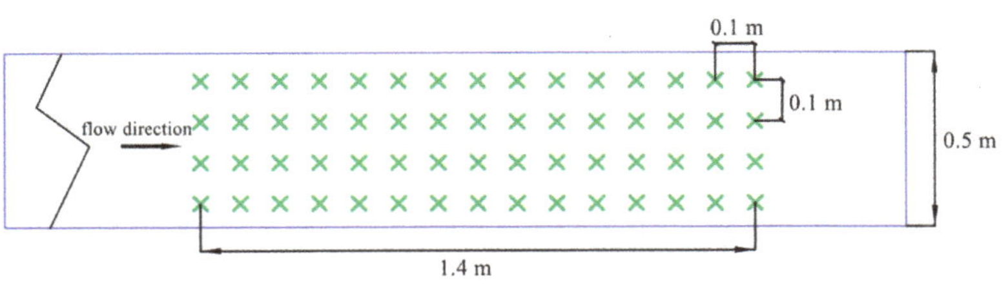

Figure 2. Schematic diagram of vegetation arrangement.

(3) The upper part of the tank had a movable bracket to facilitate the placement of measuring instruments. A tail door was set at the tail end of the tank to ensure that the water depth and flow were stable at a constant value. In order to ensure that the flow and water depth in the tank could be controlled based on operability, a centrifugal pump

was used to circulate the water in this flow tank. The water was pumped from a water storage mixing tank to the experimental tank through a pipeline. The maximum flow of the pump was 43.3 L/s, the head was 46 m, and a high-head, low-flow pump was used. The tank was equipped with a pump frequency conversion control cabinet and an E-MAG electromagnetic flowmeter to achieve accurate flow control. In this experiment, the flow range was 10~40 L/s and the water level range was 0.1~0.4 m. A water level gauge was placed on the outer wall of the glass tank. After the flow was stable, the water depth in the vegetation area was recorded, and then the flow and water level of each group were recorded.

(4) In the muddy water mixing tank, at the tail end of the tank, a mixing bar was constantly rotating during the test so that the sediment content in the test flow was maintained at a constant value of 16 kg/m^3. In addition, due to the muddy water flume test, three different silting thicknesses (i.e., 0, 6.5, and 11.5 cm) were used in the vegetation area to achieve the most realistic test results.

2.2. Hydraulic Parameter Calculation

The Reynolds and Froude numbers are two important hydraulic parameters. The Reynolds number is a dimensionless number that can be used to characterize fluid flow. The formula is $Re = VR/v$, where V is flow velocity (m/s); R is hydraulic radius (m); and v is the kinematic viscosity coefficient. The Froude number is used to represent the state of water flow. When $Fr = 1$, it means that the water flow is affected equally by the inertial force and the gravity action, and it is a critical flow state. When $Fr > 1$, the inertial force is greater than gravity (the inertial force plays a dominant role), and the water flow is a jet stream, also known as a high flow state. When $Fr < 1$, the inertial force action is less than the gravity action (gravity plays a dominant role), and the water flow is slow, also known as a low flow state. The formula is $Fr = V/\sqrt{gh}$, where V is flow velocity (m/s); g is gravitational acceleration (m/s^2); and h is test water depth (m). The calculated Fr and Re are shown in Table 1. It can be seen from Table 1 that the Froude numbers in the test were all less than 1, ranging from 0.11 to 0.19. This indicated that the test water flow was slow. The Reynolds number ranged from 9454 to 38062, which were all larger than the critical Reynolds number of an open channel. This indicated that the experimental water flow was turbulent. After the water pump was turned on and the flow meter reading stabilized, the water level scale was observed to determine when an approximately uniform flow had formed in the vegetation section. This was performed so that the relevant formula of constant uniform flow in the open channel could be used in the subsequent hydraulic calculation.

Table 1. Flow resistance test conditions.

Silting Thickness (cm)	Q (L/s)	H (m)	Fr	Re
0	7~43.06	0.118~0.31	0.1108~0.1594	9454~38,062
6.5	10~38.89	0.15~0.27	0.11~0.1771	12,376~37,023
11.5	10~35	0.15~0.24	0.11~0.1902	13,581~35,361

3. Experimental Results and Analyses

3.1. Drag Force Coefficient of Vegetation

When water flows through vegetation, vegetation has a drag effect on the water flow, resulting in a vegetation drag force. Usually, the vegetation drag force coefficient is expressed as a quantitative calculation of vegetation drag force. Water containing vegetation (Sections 1 and 2 in the flume) was taken as the control body to conduct the force analysis and obtain the expression of the vegetation drag force coefficient.

(1) Drag force of vegetation in a non-submerged state

Figure 3 shows the stress analysis of water in the control body section under the non-submerged vegetation state:

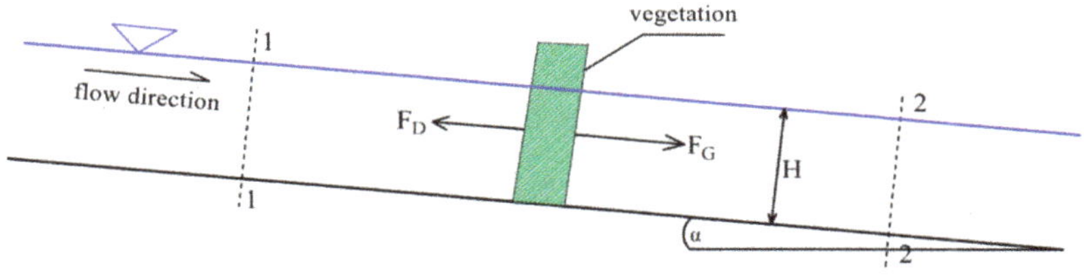

Figure 3. Vegetation stress in a non-submerged state.

Taking the water body in the control section (i.e., 1~2) as the research object, the force balance expression of the control body could be obtained as follows:

$$F_G = F_D + F_S \tag{1}$$

$$F_G = GSin\alpha = \rho_S g A J \tag{2}$$

$$F_D = C_D A_S N \frac{\rho_S U^2}{2} \tag{3}$$

where F_G is the component controlling weight force along the direction of water flow, N; F_D is the drag force of vegetation, N; F_S is the resistance generated by the bottom boundary; C_D is the drag force coefficient of vegetation, dimensionless and related to the shape of the object, angle of attack, and Reynolds number of incoming flow; N is the number of plants in the control body; A is the sectional area, $A = BH$, m^2; H is the water depth, m; A_S is the cross-sectional area of vegetation in the direction of vertical water flow, which is related to the shape of vegetation, m^2; U is the average velocity of the section, m/s; and ρ_S is the test flow density, kg/m^3. Note that, because the sink boundary was smooth glass and the bed surface was PVC board, the side wall resistance and bed surface resistance were very small relative to the vegetation drag force and could be ignored by comparison; therefore, the drag force was considered equal to the component force of gravity along the direction of water flow. Formulas (1)~(3) were combined to obtain the following:

$$C_D A_S N \frac{\rho_S U^2}{2} = \rho_S g A J \tag{4}$$

The expression of vegetation drag force can be simplified as follows:

$$C_D = \frac{2gJ}{NU^2} \frac{A}{A_S} \tag{5}$$

(2) Vegetation drag force in a submerged state

Figure 4 shows the stress analysis of water in the control section when vegetation is submerged.

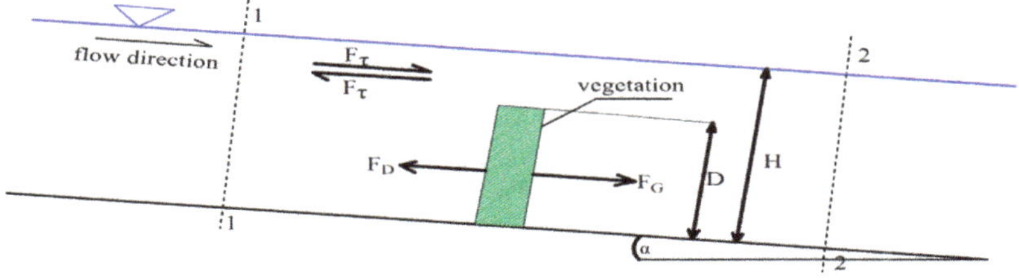

Figure 4. Stress of vegetation under submerged conditions.

When the water depth is greater than the vegetation height, the vegetation is considered submerged. In contrast to vegetation in a non-submerged state, shear stress occurs at the junction of vegetation and water flow when submerged, and there is no vegetation drag force in the water above the vegetation height.

$$F_{\tau} = \rho g B(H - D)J \tag{6}$$

The force analysis equation of the control body is:

$$F_G + F_{\tau} = F_D + F_S + F_{\tau} \tag{7}$$

Equations (2) and (3) can be combined to obtain the drag force coefficient of vegetation:

$$C_D = \frac{2gJ}{NU^2}\frac{BH}{A_S} \tag{8}$$

where J is the water surface slope. In two vegetation-submerged states, the calculation methods for the cross-sectional area (A_S) of the vegetation upflow surface are different. If the vegetation model is approximately cylindrical, the calculation method of A_S is relatively simple, i.e., $A_S = dH$ (d is the diameter of the cylinder; H is the water depth). Since the branches and leaves of the vegetation model in this test were mixed, A_S was more difficult to calculate. Therefore, the simplified vegetation model was used to determine the cross-section area in flow. The vegetation model was simplified as two cylinders of different sizes: the diameter d_1 = 8 cm and height h_1 = 12 cm of the upper branches, and the diameter d_2 = 0.2 cm and height h_2 = 5 cm of the lower branches, obtained through measurement. A_S is expressed as follows:

$$H \le h_2 \qquad A_S = d_2 H \tag{9}$$

$$h_2 < H \le (h_1 + h_2) \quad A_S = d_2 h_2 + \frac{(H - h_2)}{2}\left[2d_2 + \frac{d_1 - d_2}{h_1}(H - h_2)\right] \tag{10}$$

$$H > (h_1 + h_2) \qquad A_S = d_2 h_2 + \frac{1}{2}(d_1 + d_2)h_1 \tag{11}$$

Formula (9) was used to determine the cross-sectional area of vegetation (A_S). In order to obtain C_D, it is first necessary to obtain the number N of the plants in the unit control body, which can be determined by Formula (12):

$$N = \frac{LB}{\Delta x \Delta y} \tag{12}$$

where L and B are the length and width of the unit control body, respectively, m, and Δx and Δy are the spacing between the plants, m.

The drag force coefficient of vegetation under different siltation conditions can be calculated using the above formula. Figure 5a–f shows the relationship of the vegetation drag force coefficient with the water flow Reynolds number (Re) and Froude number (Fr) under different vegetation deposition thicknesses. The figure shows that under different deposition conditions, the vegetation drag force coefficient (C_D) exhibited similar distribution trends. In all cases, C_D decreased with increasing Re and Fr. This indicated that the greater the intensity of turbulence, the lower the drag force coefficients of vegetation. The amplitude tended to decrease, or become gentler, around Re = 15,000 and Fr = 0.13. However, if flow conditions are held constant, the C_D increases with increasing sediment thickness in the vegetation area. These results show that the vegetation drag coefficient C_D was related to both flow conditions (Re and Fr) and vegetation silting thickness. This was because, as the flow and water level increased, the branches and leaves of the plants were gradually submerged, which meant the turbulent intensity of the water flow in this area was relatively large and C_D was correspondingly reduced. Similarly, the C_D increased with increasing vegetation siltation thickness because, as the siltation of the vegetation area increased, the degree of inundation under the same water flow condition increased.

This resulted in more water flow through the vegetation's branch and leaf areas and led to a reduction in C_D. In addition, the water flow conditions (*Re* and *Fr*) had logarithmic relationships with C_D, and the fitting correlation coefficients were high in all conditions. The fitting formula and precision of the vegetation drag coefficient under various flow conditions are shown in Table 2.

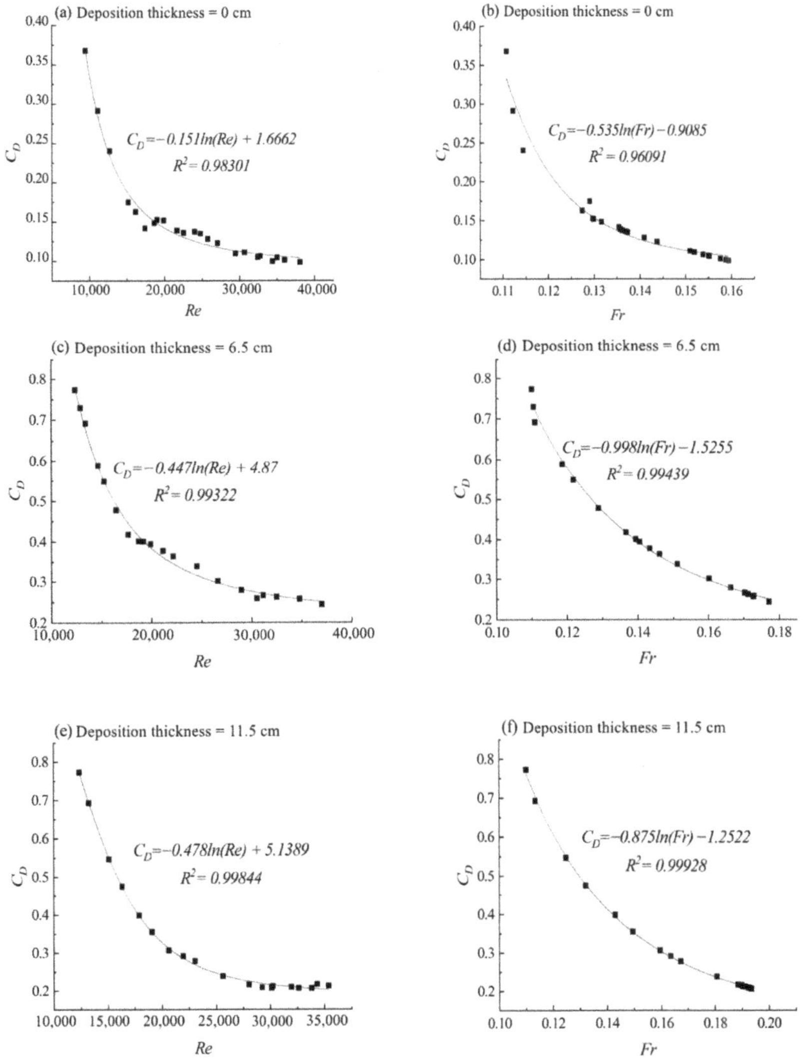

Fitting of C_D and *Re* Fitting of C_D and *Fr*

Figure 5. Relationship between the drag coefficient of vegetation and Reynolds number and Froude number at different sediment thicknesses.

Table 2. Fitting formula of the drag force coefficients of vegetation under different deposition thicknesses.

Silting Thickness	*Re*	R^2	*Fr*	R^2
0	$C_D = -0.151ln(Re) + 1.67$	0.98	$C_D = -0.535ln(Fr) - 0.91$	0.96
6.5	$C_D = -0.447ln(Re) + 4.87$	0.99	$C_D = -0.998ln(Fr) - 1.53$	0.99
11.5	$C_D = -0.478ln(Re) + 5.14$	0.99	$C_D = -0.875ln(Fr) - 1.25$	0.99

According to the fitted formula, when Re = 20,000 and Fr = 0.12, C_D changes with the change in deposition thickness, as shown in Figure 6.

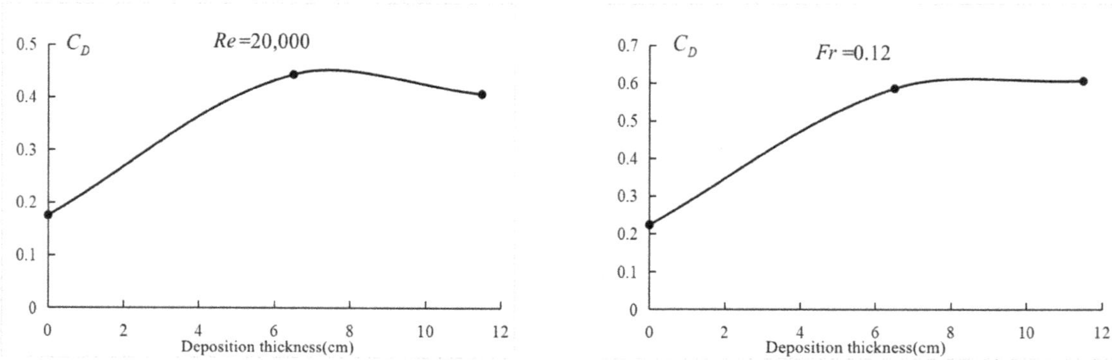

Re = 20,000 C_D changes with deposition thickness Fr = 0.12 C_D changes with deposition thickness

Figure 6. When Re = 20,000 and Fr = 0.12, C_D changes with deposition thickness.

3.2. Vegetation Roughness Coefficient

Manning's roughness coefficient is used to comprehensively reflect the resistance of pipes, channel walls, or rough water to water flow. It is also the most widely applied in resistance calculation, so this section analyzes how water conditions, sediment factors, and vegetation characteristics influence water flow using Manning's roughness coefficient. The analysis of muddy water containing vegetation was used to establish a fitting formula for vegetation roughness. Manning's roughness coefficient $n = \frac{1}{U}R^{2/3}J^{1/2}$, where U is the average velocity of the section, m/s; R is the hydraulic radius; and J is the water surface slope. Since the water flow in the vegetation section was approximately uniform, the water surface slope was equal to the change due to the flume slope.

The resistances in this paper included both riverbank and riverbed resistances, among which riverbed resistance included the resistance generated by the riverbed surface and the vegetation in the river. This can be expressed by the formula:

$$\tau = \tau_w + \tau_b \quad (13)$$

where τ is the total shear stress in the river, and τ_W and τ_b are riparian resistance and riverbed resistance, respectively. The resistance segmentation method was used. The expressions of riparian and riverbed resistances are as follows:

$$\tau_w = \gamma R_w J \quad (14)$$

$$\tau_b = \gamma R_b J \quad (15)$$

where R_w is the hydraulic radius relative to riverbank resistance; R_b is the hydraulic radius relative to riverbed resistance; and $R_w = A_w/2h$, $R_b = A_b/B$.

According to the flow continuity equation and Manning's formula:

$$AU = A_w U_w + A_b U_b \quad (16)$$

$$U_w = \frac{1}{n_w} R_w^{2/3} J^{1/2} \quad (17)$$

$$U_b = \frac{1}{n_b} R_b^{2/3} J^{1/2} \quad (18)$$

where n_w and n_b represent riverbank roughness and riverbed roughness, respectively. Since the bed surface was a smooth organic PVC board in this experiment, its roughness can be ignored relative to the vegetation's roughness. n_b is used to represent vegetation roughness, and it was assumed that the average velocity of each section in the river is equal, that is:

$$U = U_w = U_b \quad (19)$$

By combining Formulas (17)–(19), the following are obtained:

$$R_w = \left(\frac{n_w U}{J^{1/2}}\right)^{3/2} \quad (20)$$

$$R_b = h\left(1 - 2\frac{R_w}{B}\right) \quad (21)$$

Taking the edge wall roughness of the glass flume in the test as n_w = 0.01, then the vegetation roughness n_b can be calculated from Equations (17)–(21). The relationships between the vegetation roughness coefficient n_b, water flow rates *Re* and *Fr*, and vegetation sedimentation thickness were studied experimentally. Figure 7a–f shows how vegetation roughness and vegetation area deposition thickness affected the relationship of n_b with *Re* and *Fr*. The figure shows that, for all different vegetation siltation thicknesses, the vegetation roughness had negative correlations with *Re* and *Fr*, i.e., n_b decreased with increasing *Re* and *Fr*. In terms of the effect of vegetation deposition thickness, when *Re* and *Fr* are held constant, the larger the deposition thickness, the larger the roughness, and vice versa. However, when $Re \geq 30{,}000$, vegetation roughness tended to stabilize around its minimum value under all siltation thicknesses. The relationship between the vegetation roughness coefficient and *Re* and *Fr* was established as a logarithmic function. This was true across all experimental conditions, as can be seen by the high fitting correlation coefficients (R^2) across treatments. The fitting formula and precision of the vegetation roughness coefficient under various working conditions are shown in Table 3.

Table 3. Fitting formula for vegetation roughness coefficients under different deposition thicknesses.

Silting Thickness	*Re*	R^2	*Fr*	R^2
0	$n_b = -0.017ln(Re) + 0.26$	0.87	$n_b = -0.063ln(Fr) - 0.04$	0.98
6.5	$n_b = -0.034ln(Re) + 0.42$	0.98	$n_b = -0.075ln(Fr) - 0.06$	0.99
11.5	$n_b = -0.04ln(Re) + 0.48$	0.99	$n_b = -0.073ln(Fr) - 0.06$	0.99

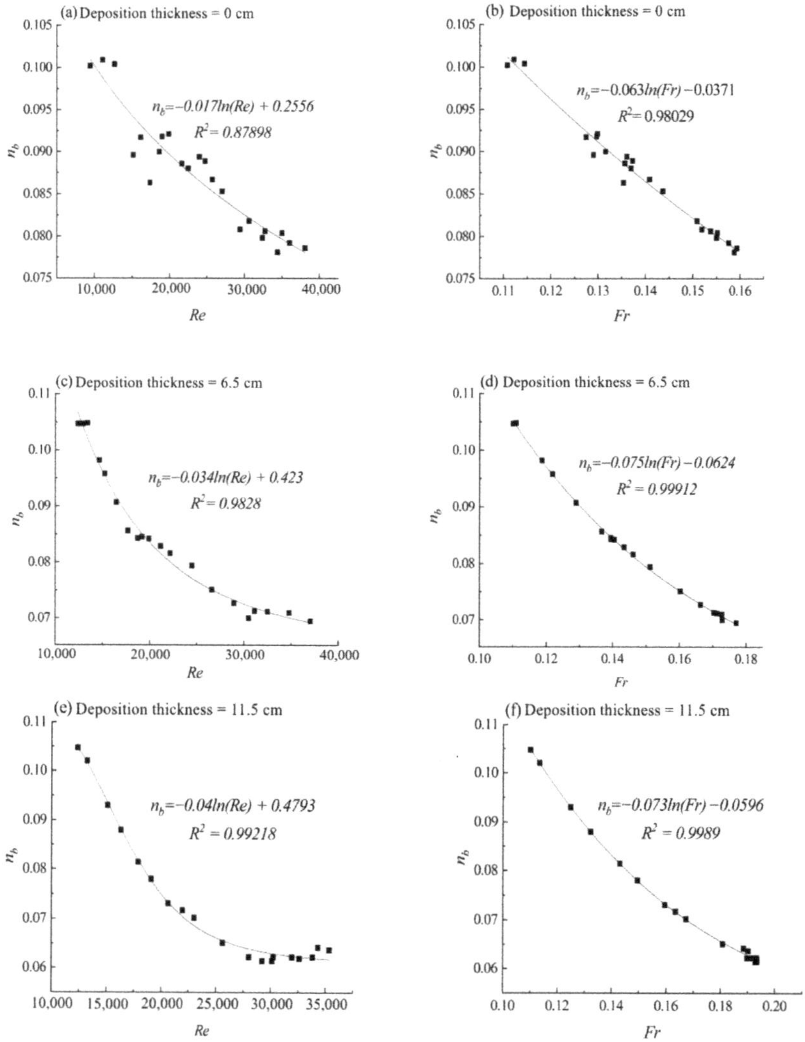

Fitting of n_b and *Re* Fitting of n_b and *Fr*

Figure 7. Relationship between the vegetation roughness coefficient and the Reynolds and Froude numbers with different sediment thicknesses.

In previous calculations of vegetation resistance, vegetation roughness was generally used to reflect the resistance of the vegetation area. Wu et al. [20] used horse hair as the model vegetation in flume tests and studied the change in vegetation resistance with water depth, comparing submerged and non-submerged conditions. Their results showed that the roughness of non-submerged vegetation n_b' decreased with increasing water depth to a certain vegetation height. When water depth was close to vegetation height, vegetation roughness n_b' increased with water depth. As the water depth continued to increase after submerging the vegetation, n_b' began to decrease and eventually became constant. The relationships between vegetation roughness and vegetation drag force coefficient were therefore established:

Unsubmerged condition:

$$n_b = \left(\frac{R_b^{2/3}}{\sqrt{2g}}\right)\sqrt{C'_D} \tag{22}$$

Submerged condition:

$$n_b = \left(\frac{D^{1/6}T^{1/2}}{\sqrt{2g}}\right)\sqrt{C'_D} \tag{23}$$

where R_b is the hydraulic radius corresponding to vegetation resistance, m; T is vegetation height, m; and C'_D is the drag force coefficient of vegetation. Cheng [24] studied the drag force of cylinders in water flow and proposed the concept of a generalized Reynolds number C'_D and drag force coefficient Re', wherein:

$$Re' = \frac{1+J}{1+80\lambda}Re \tag{24}$$

$$C'_D = 11Re'^{-0.75} + 0.9\left[1 - exp\left(-\frac{1000}{Re'}\right)\right] + 1.2\left[1 - exp\left(-\left(\frac{Re'}{4500}\right)^{0.7}\right)\right] \tag{25}$$

The expression of the vegetation resistance coefficient n_b can be obtained by transforming the above-mentioned resistance formula:

$$n_b = \left(\frac{R_b{}^{2/3}}{\sqrt{2g}}\right)\sqrt{C'_D} \tag{26}$$

where R_b is the hydraulic radius corresponding to vegetation resistance, m, and λ is the coefficient related to vegetation, λ = 0.03.

3.3. *Darcy-Weisbach Resistance Coefficient*

The experimental data were used to analyze the vegetation resistance in the beach area. According to the Darcy-Weisbach formula, the Darcy-Weisbach resistance coefficient can be derived, i.e., $f = 2gRJ/U^2$ where R is hydraulic radius, J is hydraulic slope, and U is the average velocity of section. The relationship between Darcy-Weisbach resistance coefficient f and water flow rates Re and Fr under different vegetation deposition conditions was established as shown in Figure 8a–f. The figure shows that the relationship between the Darcy-Weisbach resistance coefficient and flow Re and Fr in the vegetation area conformed to the Nikolaze curve under all different vegetation silting thickness conditions. With increasing vegetation area silting thickness, the resistance coefficient was adjusted to some extent due to the change in hydraulic conditions. When the Re of the experimental flow was small (Re < 20,000), the drag coefficient varied greatly with Re. When the Re continued to increase (Re > 20,000), the drag coefficient decreased with increasing Re. As the Reynolds number continued to increase, the resistance coefficient did not increase and eventually stabilized. Interestingly, the relationship between Fr and the Darcy-Weisbach resistance coefficient in the vegetation area was slightly different from the above. Under all vegetation deposition conditions, with increasing Fr, the Darcy-Weisbach coefficient f exhibited a nonlinear decrease. If flow and water level were held constant, the Darcy-Weisbach resistance coefficient would not change due to changes in vegetation siltation thickness. The logarithmic function correlation coefficient between water flow conditions and vegetation resistance coefficient was high, indicating reliable fitting accuracy. The fitting formula and precision of the vegetation roughness coefficient under various working conditions are shown in Table 4.

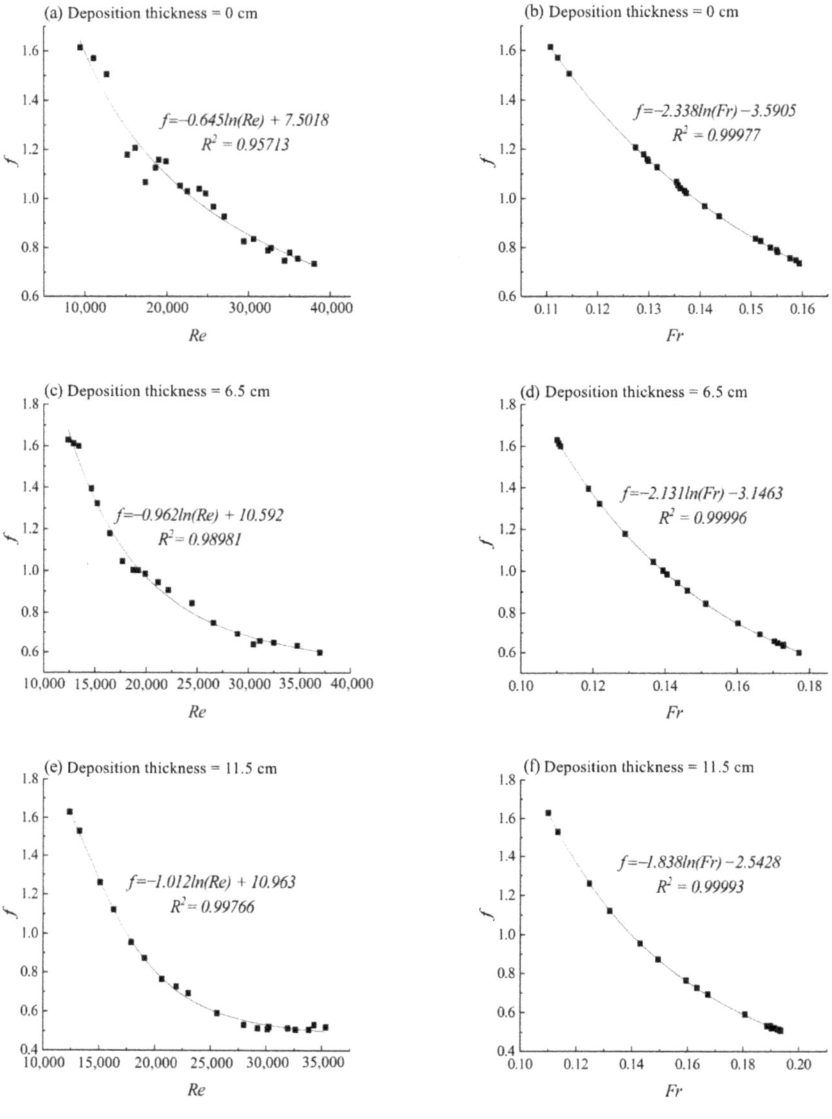

Figure 8. Relationship between the Darcy-Weisbach coefficient and Reynolds number and Froude number with different sediment thicknesses.

Table 4. Fitting formula for Darcy-Weisbach drag coefficient under different deposition thicknesses.

Silting Thickness	*Re*	R^2	*Fr*	R^2
0	$f = -0.645ln(Re) + 7.50$	0.95	$f = -2.338ln(Fr) - 3.59$	0.99
6.5	$f = -0.962ln(Re) + 10.59$	0.98	$f = -2.131ln(Fr) - 3.14$	0.99
11.5	$f = -1.012ln(Re) + 10.96$	0.99	$f = -1.838ln(Fr) - 2.54$	0.99

4. Conclusions

Taking maize-typical vegetation in the Lower Yellow River as the model vegetation, the influence of different sediment thicknesses on vegetation resistance characteristics in the muddy water area was simulated and studied by carrying out muddy water trough experiments with the same sand content and different silt thicknesses. The changes in drag force coefficient, roughness coefficient, and Darcy-Weisbach resistance coefficient with changes in the depth of sediment deposition were studied. This information has enriched our understanding of vegetation resistance in lower Yellow River beach areas and provided a scientific basis and reference for future studies of flood evolution in the lower Yellow River beach area. The main conclusions were as follows:

(1) For all vegetation deposition thicknesses (i.e., 0, 6.5, and 11.5 cm), the drag force coefficient of the vegetation decreased with increasing Reynolds and Froude numbers. In other words, with increased flow and water level, the drag force coefficient of vegetation decreased, and the decreased drag force variability around $Re = 15{,}000$ and $Fr = 0.13$ was indicative of gentler conditions. However, when the water flow conditions are the same, the vegetation drag coefficient increases with the increase in deposition thickness in the vegetation area.

(2) To analyze the resistance in river and beach areas, resistance is decomposed into riparian resistance and riverbed resistance, and the final calculation of vegetation roughness is obtained using the resistance segmentation method. With the same deposition thickness in the same vegetation area, the vegetation roughness coefficient was negatively correlated with Reynolds and Froude numbers. When $Re > 30{,}000$, under all deposition thicknesses, the vegetation roughness tended to stabilize near its minimum value. Logarithmic functions were established between the vegetation roughness coefficients and the Reynolds and Froude numbers of the different experimental water flow conditions, and all had high correlation coefficients.

(3) Based on the experimental data and the Darcy-Weisbach resistance formula, the expression of the Darcy-Weisbach resistance coefficient was deduced and analyzed. The results showed that when the Reynolds number was small ($Re < 20{,}000$), the Darcy-Weisbach resistance coefficient f varied greatly with Re. When the Re number was large ($Re > 20{,}000$), the Darcy-Weisbach resistance coefficient f decreased with Re. With further increases in flow, or Reynolds number, the resistance coefficient tended to stabilize after reaching a certain value. Under the conditions of constant flow and water level, there was no significant correlation between the Darcy-Weisbach resistance coefficient and the thickness of vegetation deposition.

Author Contributions: Conceptualization, X.Z. and Y.Z.; methodology, H.W.; software, Z.B.; validation, Y.Z., H.W. and Z.X.; formal analysis, X.Z.; investigation, Y.Z.; resources, H.W.; data curation, X.Z.; writing—original draft preparation, Y.Z.; writing—review and editing, X.Z.; visualization, X.Z.; supervision, X.Z.; project administration, X.Z.; funding acquisition, X.Z. All authors have read and agreed to the published version of the manuscript.

Funding: This research was supported by the National Natural Science Foundation of China under contract No. 41930643, "Study on carbon and nitrogen process and its effect in the Lower Yellow River"; the National Natural Science Foundation of China, U2243220, "Efficient sand transport mechanism and critical regulation of sand content in the Lower Yellow River under changing circumstances"; and the Henan Province's key research, development, and promotion projects in 2022 (scientific and technological breakthroughs) under contract No. 222102320195. The authors are thankful to the anonymous reviewers for their valuable comments and suggestions to improve the quality of the paper.

Data Availability Statement: Data available in a publicly accessible repository that does not issue DOIs. Publicly available datasets were analyzed in this study.

Conflicts of Interest: The authors declare no conflict of interest.

References

1. Wilson, C.A.M.E. Flow resistance models for flexible submerged vegetation. *J. Hydrol.* **2007**, *342*, 213–222. [CrossRef]
2. Rhee, D.S.; Woo, H.; Kwon, B.A.; Ahn, H.K. Hydraulic resistance of some selected vegetation in open channel flows. *River Res. Appl.* **2008**, *24*, 673–687. [CrossRef]
3. Li, G.; Wang, X.; Zhao, X.; Huang, E.; Liu, X.; Cao, S. Flexible and rigid vegetation in overland flow resistance. *Trans. ASABE* **2013**, *56*, 919–926.
4. Box, W.; Järvelä, J.; Västilä, K. Flow resistance of floodplain vegetation mixtures for modelling river flows. *J. Hydrol.* **2021**, *601*, 126593. [CrossRef]
5. Okhravi, S.; Schügerl, R.; Velísková, Y. Flow Resistance in Lowland Rivers Impacted by Distributed Aquatic Vegetation. *Water Resour. Manag.* **2022**, *36*, 2257–2273. [CrossRef]
6. Stephan, U.; Gutknecht, D. Hydraulic resistance of submerged flexible vegetation. *J. Hydrol.* **2002**, *269*, 27–43. [CrossRef]
7. Musleh, F.A.; Cruise, J.F. Functional relationships of resistance in wide flood plains with rigid unsubmerged vegetation. *J. Hydraul. Eng.* **2006**, *132*, 163–171. [CrossRef]
8. Jordanova, A.; James, C.; Birkhead, A. Practical estimation of flow resistance through emergent vegetation. *Water Manag.* **2006**, *159*, 173–181. [CrossRef]
9. Li, Y.H.; Xie, L.; Su, T.C. Resistance of open-channel flow under the effect of bending deformation of submerged flexible vegetation. *J. Hydraul. Eng.* **2018**, *144*, 04017072. [CrossRef]
10. Chiaradia, E.A.; Gandolfi, C.; Bischetti, G.B. Flow resistance of partially flexible vegetation: A full-scale study with natural plants. *J. Agric. Eng.* **2019**, *50*, 55–65. [CrossRef]
11. Kouwen, N. Effect of riparian vegetation on flow resistance and flood potential. *J. Hydraul. Eng.* **2000**, *126*, 954. [CrossRef]
12. Järvelä, J. Flow resistance of flexible and stiff vegetation: A flume study with natural plants. *J. Hydrol.* **2002**, *269*, 44–54. [CrossRef]
13. Järvelä, J. Effect of submerged flexible vegetation on flow structure and resistance. *J. Hydrol.* **2005**, *307*, 233–241. [CrossRef]
14. Kothyari, U.C.; Hayashi, K.; Hashimoto, H. Drag coefficient of unsubmerged rigid vegetation stems in open channel flows. *J. Hydraul. Res.* **2009**, *47*, 691–699. [CrossRef]
15. Zhao, C.; Gao, J.E.; Huang, Y.; Wang, G.; Zhang, M. Effects of vegetation stems on hydraulics of overland flow under varying water discharges. *Land Degrad. Dev.* **2016**, *27*, 748–757. [CrossRef]
16. Bulling, J.; John, V.; Knobloch, P. Isogeometric analysis for flows around a cylinder. *Appl. Math. Lett.* **2017**, *63*, 65–70. [CrossRef]
17. Huang, B.S.; Lai, G.W.; Qiu, J.; Wan, P. Experimental Study on the Effect of Planting Trees in the River Beach on Flood Flow. *Chin. J. Hydrodyn.* **1999**, *14*, 468–474.
18. Stone, B.M.; Shen, H.T. Hydraulic resistance of flow in channels with cylindrical roughness. *J. Hydraul. Eng.* **2002**, *128*, 500–506. [CrossRef]
19. James, C.S.; Birkhead, A.L.; Jordanova, A.A.; O'sullivan, J.J. Flow resistance of emergent vegetation. *J. Hydraul. Res.* **2004**, *42*, 390–398. [CrossRef]
20. Wu, F.C.; Shen, H.W.; Chou, Y.J. Variation of roughness coefficients for unsubmerged and submerged vegetation. *J. Hydraul. Eng.* **1999**, *125*, 934–942. [CrossRef]
21. Lu, Y.; Wang, W.X.; Lu, Y.J.; Song, Y.T.; Yan, D.L. Theoretical and experimental study on resistance characteristics in shrub vegetation distribution area. *Adv. Water Sci.* **2020**, *31*, 556–564.
22. Li, R.M.; Shen, H.W. Effect of tall vegetations on flow and sediment. *J. Hydraul. Div.* **1973**, *99*, 793–814. [CrossRef]
23. Kouwen, N.; Li, R.M. Biomechanics of vegetative channel linings. *J. Hydraul. Div.* **1980**, *106*, 1085–1103. [CrossRef]
24. Cheng, N.S. Calculation of drag coefficient for arrays of emergent circular cylinders with pseudofluid model. *J. Hydraul. Eng.* **2013**, *139*, 602–611. [CrossRef]

Journal of
*Marine Science
and Engineering*

MDPI

Article

Revised Friction Groups for Evaluating Hydraulic Parameters: Pressure Drop, Flow, and Diameter Estimation

Dejan Brkić [1,2]

[1] Faculty of Electronic Engineering, University of Niš, 18000 Niš, Serbia; dejan.brkic@elfak.ni.ac.rs
[2] IT4Innovations, VSB—Technical University of Ostrava, 708 00 Ostrava, Czech Republic; dejan.brkic@vsb.cz

Abstract: Suitable friction groups are provided for solving three typical hydraulic problems. While the friction group based on viscous forces is used for calculating the pressure drop or head loss in pipes and open channels, commonly referred to as the Type 1 problem in hydraulic engineering, additional friction groups with similar behaviors are introduced for calculating steady flow discharge as the Type 2 problem and, for estimating hydraulic diameter as the Type 3 problem. Contrary to the viscous friction group, the traditional Darcy–Weisbach friction factor demonstrates a negative correlation with the Reynolds number. This results in curves that slope downward from small to large Reynolds numbers on the well-known Moody chart. In contrast, the friction group used here, based on viscous forces, establishes a more appropriate relationship. In this case, the friction and Reynolds number are positively correlated, meaning that both increase or decrease simultaneously. Here, rearranged diagrams for all three mentioned problems show similar behaviors. This paper compares the Moody diagram with the diagram for the viscous force friction group. The turbulent parts of both diagrams are based on the Colebrook equation, with the newly reformulated version using the viscous force friction group. As the Colebrook equation is implicit with respect to friction, requiring an iterative solution, an explicit solution using the Lambert W-function for the reformulated version is offered. Examples are provided for both pipes and open channel flow.

Keywords: Colebrook function; Moody chart; pipe flow; hydraulic resistance

1. Introduction

This paper presents suitable friction groups for addressing three common problems in hydraulic engineering [1], i.e., it offers procedures for the following three types of problems:

1. Pressure drop, Δp, i.e., of head loss Δh—Type 1;
2. Steady flow discharge Q—Type 2;
3. Diameter of pipe D [2], i.e., hydraulic diameter in open channel flow D_H—Type 3.

The procedures shown in this paper are given as an alternative to the classical approach given, e.g., in [1].

The Darcy–Weisbach flow friction factor f is defined based on inertial forces [3,4] leading to its negative correlation with the Reynolds number, Re. In the widely recognized Moody chart [5–8], the turbulent Darcy–Weisbach friction factor f generates curves that slope downward from small to large Reynolds numbers Re. However, a positive correlation with the Reynolds number Re shows similar behavior as electrical resistance with electrical current. This paper presents diagrams constructed using suitable friction groups that demonstrate this positive correlation for all three mentioned types of hydraulic problems.

Among others, this paper gives a comparison between the traditional Moody diagram and the new diagram based on the rearranged viscous friction group λ (viscous friction group λ is used here for solving the Type 1 problem). The turbulent portion of both diagrams relies on the empirical Colebrook function [9,10]. The Colebrook equation is implicit with respect to friction, and it requires an iterative solution in both [11]; its standard version is based on the Darcy–Weisbach factor f and its transformed version is based on the rearranged viscous friction group, λ. This paper presents an explicit reformulation of the transformed version of the Colebrook function using the Lambert W-function [12]. An accurate and computationally efficient approximation is also given [13].

Viscous friction group λ is replaced with the friction group $f \cdot Re^2$ for solving the Type 2 problem, and with f/Re for solving the Type 3 problem. All three diagrams Re→λ (for Type 1), Re→ $f \cdot Re^2$ (for Type 2), and Re→ f/Re (for Type 3), have a similar shape. This paper presents an alternative approach to the traditional methods for solving three common types of hydraulic problems, yielding identical final results.

Following the introduction in Section 1, the Type 1 problem is examined in Section 2, Type 2 in Section 3, and Type 3 in Section 4. Concluding remarks are presented in Section 4, followed by notes about open channel flow in Appendix A and with some computational examples for both pipes and open channels in Appendix B. Finally, a list of symbols from this paper is given after Appendices.

In the context of marine engineering, solutions presented in this paper can find applications in offshore pipelines, etc.

2. Pressure Drop (Head Loss)—Type 1

The flow friction factor based on inertial forces, which is commonly used in engineering practice, is known as the Darcy–Weisbach, also referred to as the Moody or Darcy friction factor f. It is given in Equation (1):

$$f = \frac{\Delta h}{\frac{L}{D}\frac{\upsilon^2}{2g}} \tag{1}$$

The pressure loss, Δp, is connected to head loss, Δh, as $\Delta p = \rho g \Delta h$, while flow discharge Q with the velocity $\upsilon = \frac{4 \cdot Q}{D^2 \cdot \pi}$.

The dimensionless Fanning friction factor, f_f, shares the same physical meaning as the Darcy–Weisbach friction factor f but is equal to one-fourth of it; $f_f = f/4$. The less commonly used dimensionless Stanton and Pannel friction factor, Φ, also has the same physical interpretation, but is half of the Fanning factor ($\Phi = f_f/2$) and one-eighth of the Darcy–Weisbach factor ($\Phi = f/8$). These differences arise from definitions based on varying but related quantities, such as pipe diameter versus pipe radius. All of them are defined through shear stress on the surface [14,15]; τ, $\frac{\tau}{\rho \upsilon^2} = \Phi = \frac{f_f}{2} = \frac{f}{8}$.

The Darcy–Weisbach friction, f, the Fanning friction, f_f, and the Stanton and Pannel friction, Φ, are negatively correlated with the Reynolds number Re. On the other hand, suitable rearrangements given in Equation (2) leads to positive correlation of friction with the Reynolds number Re. A friction group, λ, based on viscous forces [7] is given in Equation (2):

$$\lambda = \frac{\rho g \Delta h}{32 \eta \upsilon}\frac{D^2}{L} = \frac{\Delta p \pi}{128 \eta Q}\frac{D^4}{L} \tag{2}$$

To avoid mistakes, it is crucial to clearly define the friction factor when using it to calculate pressure drops, Δp, and related head losses, Δh.

Both Equations (1) and (2) can be used equally for solving the Type 1 problem, but the approach through Equation (1) is more common [1].

2.1. *Calculation through the Standard Darcy–Weisbach Flow Friction Factor*

2.1.1. Standard Colebrook Function

The Darcy–Weisbach flow friction factor, f, for laminar flow is given as $f = 64/Re$, while for turbulent flow most often it is given through the empirical Colebrook function, as shown in Equation (3):

$$\frac{1}{\sqrt{f}} = -2\log_{10}\left(\frac{2.51}{Re}\frac{1}{\sqrt{f}} + \frac{\varepsilon/D}{3.7}\right) \tag{3}$$

The equation is developed by Colebrook solely [9], but it is based on his joint experiment with White [10].

The Reynolds number is defined as Re = $\upsilon D/\mu$, where $\mu = \eta/\rho$ and $Q = \upsilon D^2\pi/4$. The absolute value of the inner pipe roughness, ε [m], depends on the type and condition of the used material of which pipes are manufactured [1].

The Colebrook equation is implicit with respect to the unknown friction factor, f, and can be solved only iteratively (few iterations are usually sufficient) [16–18]. Also, the Colebrook equation can be expressed explicitly through the Lambert W-Function [12,13].

2.1.2. Moody Chart

The Moody chart is typically used for the graphical interpretation of the Darcy–Weisbach flow friction factor f (its version is given in Figure 1). In the Moody diagram, as the value of the Reynolds number, Re, increases the Darcy–Weisbach flow friction factor, f, decreases, and vice versa.

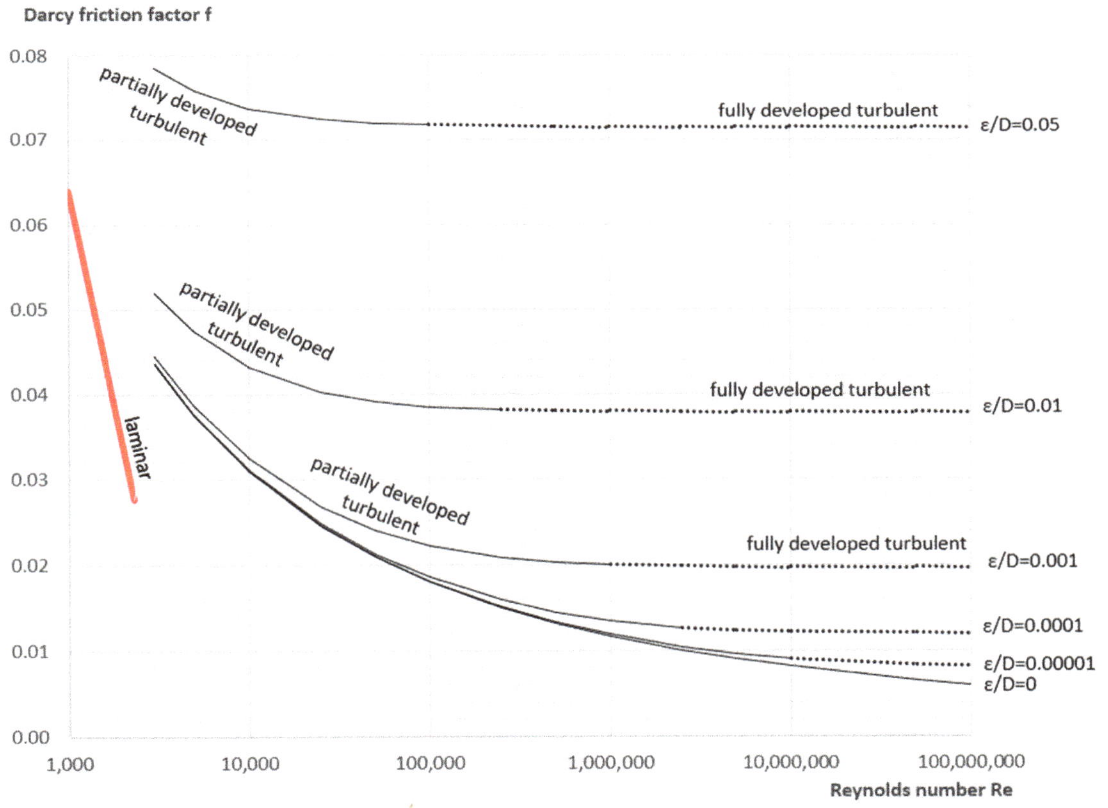

Figure 1. Moody chart for the Darcy–Weisbach flow friction factor, f, as a function of the Reynolds number, Re, and the relative roughness of inner pipe surface, ε/D. The red line represents laminar flow, full black lines represent partially developed turbulent flow, while point-lines represent fully turbulent flow.

The Moody chart is based on the Reynolds number Re, which is governed by the velocity of the moving fluid υ, i.e., by flow $Q = \upsilon D^2\pi/4$. This means that the Reynolds number Re, together with the relative roughness ε/D [19–24], is in the function of diameter,

D. It implies that if the diameter D and the flow Q are known, the pressure drop Δp (or head loss Δh) can be easily computed using Equation (1).

2.2. Calculation through Viscous Friction Group

The Colebrook function can be rewritten [7] in terms of the viscous friction group λ. Therefore, it can be graphically presented in a chart in Figure 2, following the same principles as in the widely recognized Moody chart.

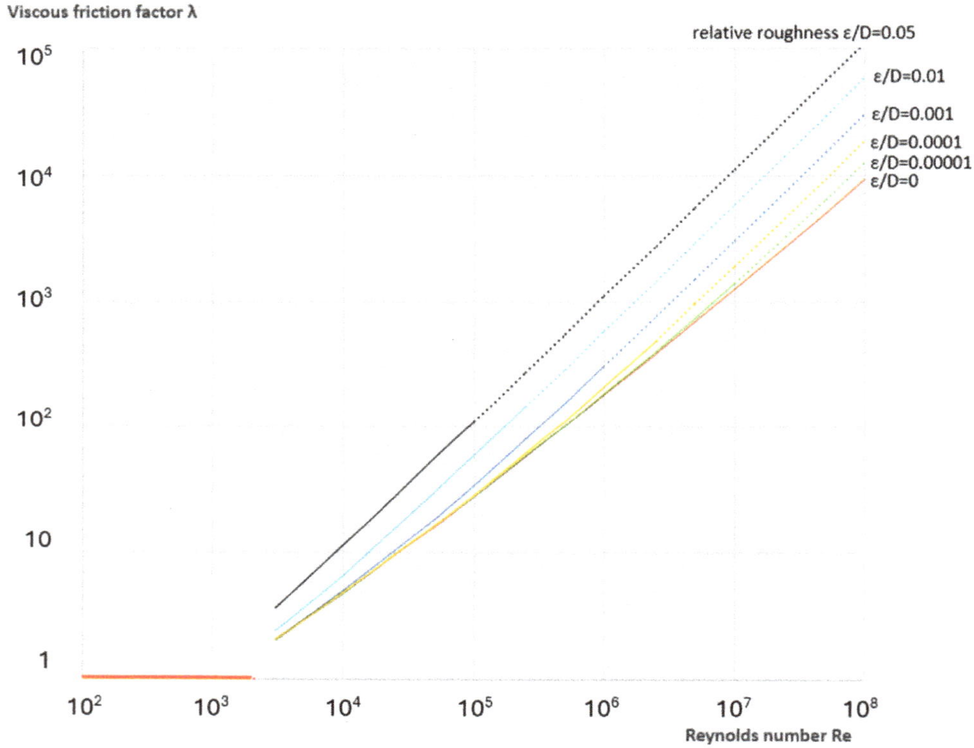

Figure 2. Transformed Moody chart based on viscous flow friction factor λ: Its turbulent part is based on the transformed Colebrook equation—The red line represents laminar flow, full lines represent partially developed turbulent flow, while point-lines represent fully turbulent flow.

2.2.1. Transformed Colebrook Function

The Colebrook equation transformed using viscous friction group λ [7] is given in Equation (4):

$$\frac{\sqrt{Re}}{\sqrt{\lambda}} = -16\log_{10}\left(\frac{0.314}{Re}\frac{\sqrt{Re}}{\sqrt{\lambda}} + \frac{\varepsilon/D}{3.7}\right) \tag{4}$$

2.2.2. Modified Moody Diagram through the Viscous Friction Group

The transformed Moody chart based on the viscous friction group λ is given in Figure 2. In the laminar region of the diagram, λ = 1, while in the turbulent region, represented by the transformed Colebrook equation from Equation (4), the curves follow a continuous upward trend more accurately representing the physical behavior of flow friction phenomena.

In the transformed Moody chart, it becomes challenging to pinpoint the boundary where the partially turbulent regime transitions into the fully turbulent regime as the

Reynolds number Re increases. In the original Moody chart, curves become straight horizontal lines for the fully developed turbulent regime. For example, this boundary for $\varepsilon/D = 0.05$ is around $Re = 1 \times 10^5$, while for $\varepsilon/D = 0.001$, it is around $Re = 1 \times 10^6$.

Equally, as the original Moody chart, the diagram in Figure 2 is suitable for the calculation of pressure drop Δp (and subsequently of head loss Δh). It can be completed through Equation (2).

2.2.3. Explicit Expression of the Transformed Colebrook Equation

Similar to the standard Colebrook function, which is based on the Darcy–Weisbach friction factor f, the transformed Colebrook function expressed in terms of the viscous friction group λ is also implicitly defined with respect to the unknown variable, and it can only be solved iteratively. Fortunately, the iterative procedure in this case converges fast and requires only a few iterations. However, the Colebrook equation can be transformed in an explicit form using the Lambert W-function, which can be further solved using established mathematical procedures [25–27]. The Lambert W-function is suitable for the transformation of implicit function in an equivalent explicit form [28].

The graph of the Lambert W-function is given in Figure 3 [25,27].

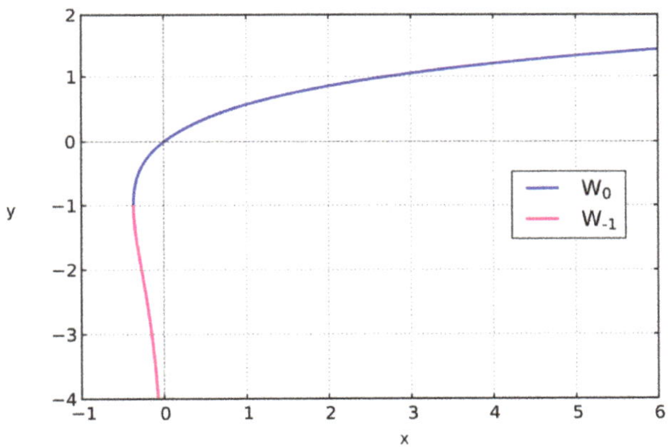

Figure 3. Branches of the Lambert W-function.

The transformed Colebrook function based on the viscous friction group λ in terms of the Lambert W-function is given in Equation (5) [12,13]:

$$\left.\begin{array}{c} \frac{\sqrt{Re}}{\sqrt{\lambda}} = \frac{Z}{0.314}(B + y) \\ Z = \frac{16 \cdot 0.314}{\ln(10)} \approx 2.1819 \\ A = \frac{Re}{Z}\frac{\varepsilon/D}{3.7} \\ B = \ln(Re) - \ln(Z) \\ x = A + B \\ y = W(e^x) - x \end{array}\right\} \quad (5)$$

However, in the closed-form solution from Equation (5), e^x grows so rapidly that calculations on computers can fail for certain combinations of the Reynolds number Re and the relative roughness of the pipe's inner surface ε/D, which is commonly used in engineering practice. This issue arises due to overflow errors caused by the limited size of typical computer registers [29,30]. The limitations are presented in Figure 4.

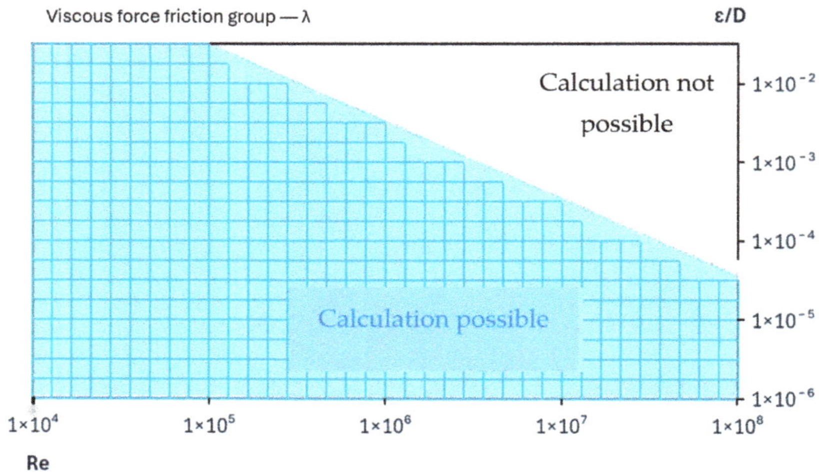

Figure 4. Limitations for verification of $W(e^x)$ in computer registers.

However, limitations for the verification of $W(e^x)$ can be overwhelmed by using the shifted Lambert function, i.e., the Wright ω-function $W(e^x) - x = \omega(x) - x$, which is automatically completed in some software tools such as Matlab. It can be approximated as $W(e^x) - x \approx \ln(x)\left(\frac{1}{x} - x\right)$, based on the upper branch of the Lambert W-function, denoted as W_0 in Figure 3 by the blue curve.

Similar approaches are available in [31–34].

2.2.4. An Approximation of the Transformed Colebrook Equation

To develop efficiently an approximation of the transformed Colebrook equation through the viscous friction group λ, the upper branch W_0 of the Lambert W-function should be approximated only in the range typically used in engineering practice. Based on Equation (5), this range is only between x_{min} = 6.97 and x_{max} = 619,363.78 for $W(e^x) - x$. This gives an approximation, which is given in Equation (6):

$$\left.\begin{array}{c} \frac{\sqrt{Re}}{\sqrt{\lambda}} \approx 6.94871 \cdot \left(B - C + \frac{C}{x - 0.5588 \cdot C + 1.2079}\right) \\ A \approx \frac{Re \cdot \varepsilon / D}{8.0897} \\ B \approx \ln(Re) - 0.779626 \\ C = \ln(x) \\ x = A + B \end{array}\right\} \quad (6)$$

In any case, restriction shown in Figure 4 cannot be associated with Equation (6).

The relative error introduced by Equation (6) compared with Equation (5) is negligible up to 0.2% [12,13].

3. Flow Discharge—Type 2

Using the Darcy–Weisbach flow friction factor f, a direct noniterative closed-form solution for the Type 2 problem is available [1,35].

It is also possible to rearrange the diagram from Figure 2 for the calculation of flow discharge Q by replacing the viscous friction group λ with the friction group $f \cdot Re^2$. The Darcy–Weisbach flow friction factor f in Equation (7) should be calculated using Equation (3), with the main purpose of this transformation being used to produce a friction group that is independent of the velocity of the moving fluid υ, i.e., on the unknown flow discharge Q.

$$f\cdot Re^2 = \frac{\frac{\Delta p}{\rho\cdot g}}{\frac{L}{D}\frac{\upsilon^2}{2g}}\cdot\left(\frac{\rho\cdot\upsilon\cdot D}{\eta}\right)^2 = \frac{2\cdot D^3\cdot\Delta p\cdot\rho}{L\cdot\eta^2}, \tag{7}$$

The new chart suitable for the calculation of flow discharge Q is given in Figure 5.

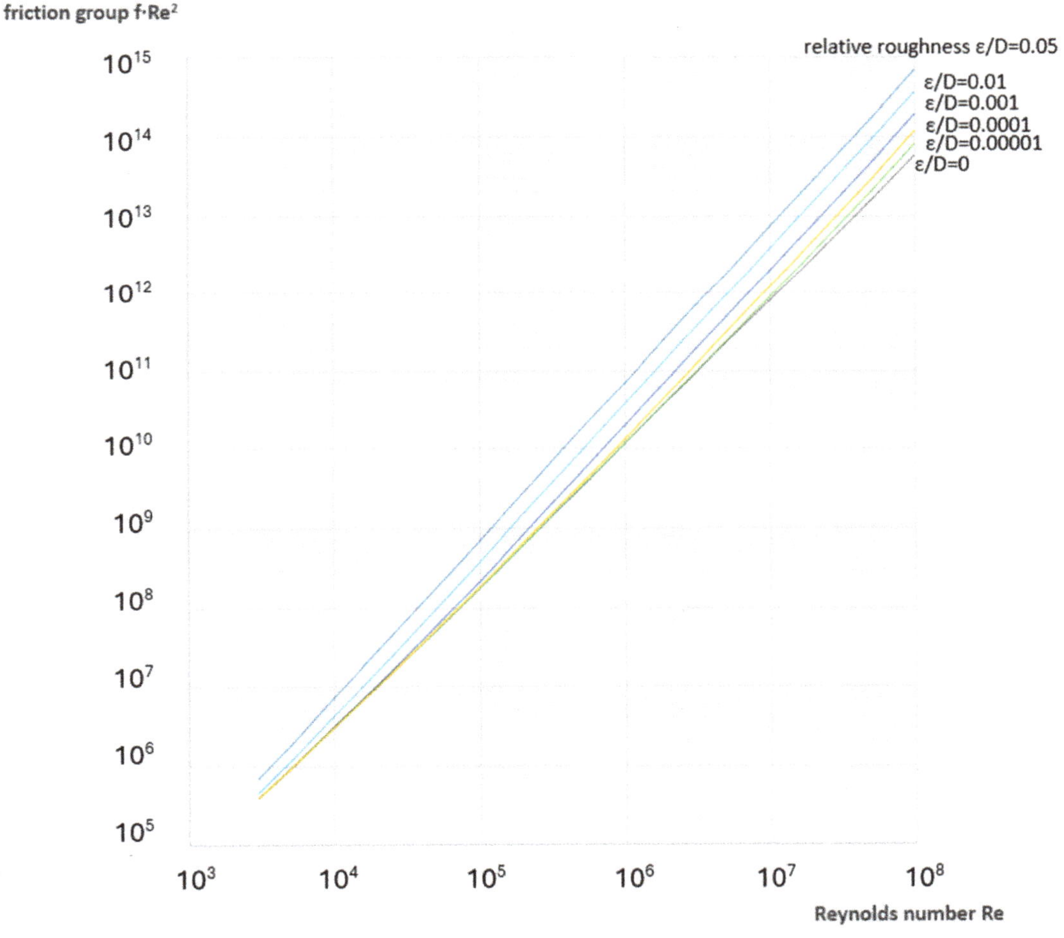

Figure 5. Chart suitable for determination of flow discharge.

The calculation should be completed through the following next steps:

1. Calculate the value of the friction group $f\cdot Re^2$ using Equation (7); $\frac{2\cdot D^3\cdot\Delta p\cdot\rho}{L\cdot\eta^2}$;
2. Find the corresponding intersection with the curve for the appropriate relative roughness ε/D;
3. Read the value of the Reynolds number Re from the abscissa;
4. Finally, calculate the flow discharge Q using Equation (8)

$$\upsilon = \frac{Re \cdot \eta}{\rho \cdot D} \rightarrow Q = \upsilon \frac{D^2 \cdot \pi}{4} \tag{8}$$

4. Diameter Estimation—Type 3

For the estimation of diameter D, the viscous friction group λ should be replaced with the new friction group f/Re as given in Figure 6. This approach works if the velocity of the moving fluid υ is known or can be estimated.

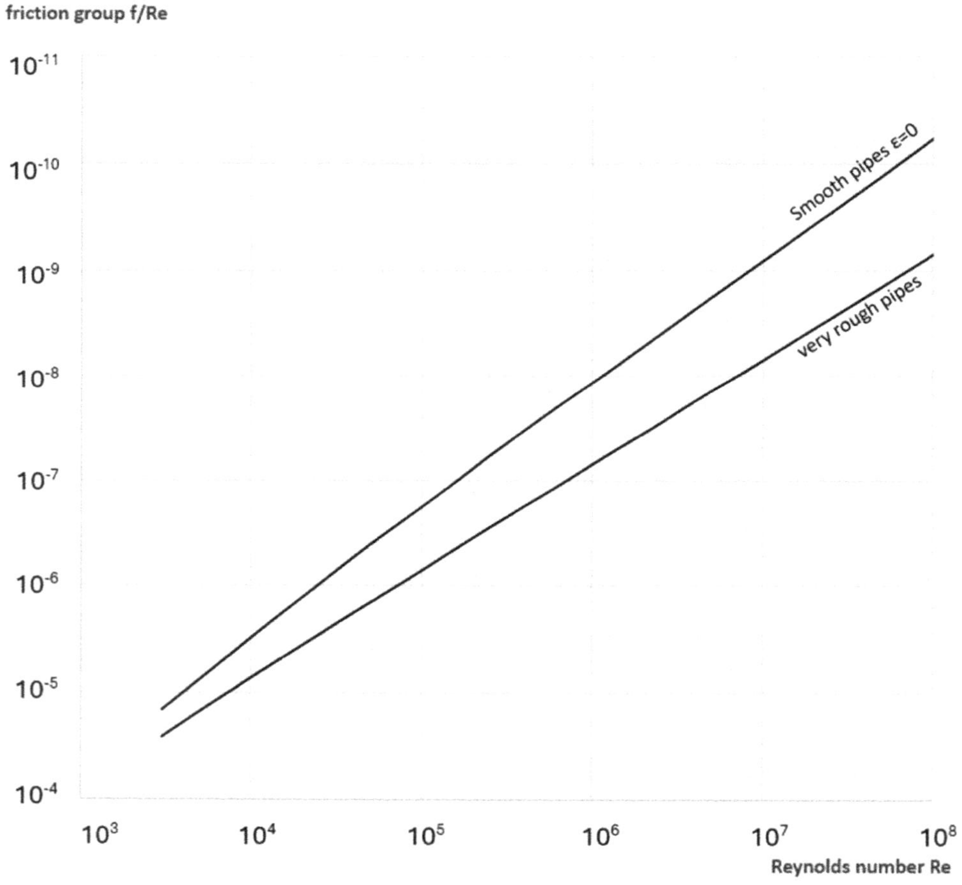

Figure 6. Chart suitable for estimation of pipe diameters.

To use the chart from Figure 6, the friction group f/Re, which is in its final form free of diameter D, should be calculated using Equation (9):

$$\frac{f}{Re} = \frac{\frac{\frac{\Delta p}{\rho \cdot g}}{\frac{L}{D}\frac{\upsilon^2}{2g}}}{\frac{\rho \cdot \upsilon \cdot D}{\eta}} = \frac{2 \cdot \Delta p \cdot \eta}{L \cdot \rho^2 \cdot \upsilon^3} \tag{9}$$

The relative roughness ε/D is not independent of the unknown diameter D, which can only be estimated within a range for smooth and very rough pipes.

Figure 6 should be used with the following steps:

1. Calculate the value of the friction group f/Re using Equation (9); $\frac{2\cdot\Delta p\cdot\eta}{L\cdot\rho^2\cdot\upsilon^3}$;
2. Find the corresponding intersection with the curves for two extremums for smooth and very rough pipes;
3. Read the value of the Reynolds number Re from the abscissa;
4. Finally, calculate diameter D using Equation (10)

$$D = \frac{Re\cdot\eta}{\rho\cdot\upsilon} \tag{10}$$

For real engineering applications, the pipes are manufactured with standard sizes, so the nearest size based on Equation (10) should be selected (lower or upper as requested by design and in accordance with expertise about the condition of the inner pipe surface).

An estimation of diameter D, in the case of the known flow discharge Q instead of the velocity of the moving fluid υ, with the appropriate friction group is given in Equation (11):

$$f\cdot Re^5 = \frac{\Delta h}{\frac{8\cdot Q^2\cdot L}{D^5\cdot g\cdot\pi^2}}\cdot\left(\frac{4\cdot\rho\cdot Q}{\eta\cdot D\cdot\pi}\right)^5 = \frac{128\cdot\Delta h\cdot g}{Q^2\cdot L\cdot\pi^3}\cdot\left(\frac{\rho\cdot Q}{\eta}\right)^5 \tag{11}$$

5. Conclusions

In this paper, among others, a new method to calculate the friction factor is proposed, in which the relationship between the friction and the Reynolds number Re are positively correlated. A new version of Lambert-W function for calculating the viscous friction group λ is represented, which is an explicit expression of the Colebrook function.

This paper uses an alternative for the traditional strategies for solving three common types of hydraulic problems, where the final results are identical for both approaches.

The viscous friction group λ can be used for pipe and open channel sizing in a manner similar to the established Darcy–Weisbach friction factor f. The Moody diagram, traditionally used for the Darcy–Weisbach friction factor f, has been reformulated in terms of the viscous friction group λ. This rearrangement presents the turbulent friction factor as increasing with the Reynolds number Re, providing a correlation similar to that between electrical resistance and current.

Diagrams are provided for the three common hydraulics problems: the calculation of Δp or Δh—Type 1, calculation of Q—Type 2, and calculation of D—Type 3. The solution for Type 1 is not limited to graphical methods. However, graphical solutions for Types 2 and 3 are based on the same principles as those used for the viscous friction group λ, using the appropriate friction groups for Type 1: $f\cdot Re^2$ and for Type 3: f/Re for the known flow υ velocity (or $f\cdot Re^5$ for the known flow discharge Q).

The accuracy of reading from charts remains a main challenge associated with graphical solutions [36,37].

Although the new version of the calculation method of the friction works well, some possible disadvantages can be envisaged, which are especially related to the newly proposed solution of the Type 1 problem:

1. The existing Darcy–Weisbach friction factor f and its calculation methods have been widely and successfully applied in so many fields, so it can be costly to replace the existing calculation system with the new version of the friction factor calculation method and it may be consequently impracticable;
2. Although a positive correlation relationship between the friction factors and Reynolds number may be good, it may bring some numerical calculation problems in hydraulic iteration calculations.

Funding: Financial support was received from the Ministry of Science, Technological Development and Innovation of the Republic of Serbia through institutional financing, from the European Union under the REFRESH—Research Excellence For REgion Sustainability and High-tech Industries, project number CZ.10.03.01/00/22_003/0000048 via the Operational Programme Just Transition, and from the Ministry of Education, Youth and Sports of the Czech Republic (ID: MC2104) and e-INFRA CZ (ID:90254).

Institutional Review Board Statement: Not applicable.

Informed Consent Statement: Not applicable.

Data Availability Statement: All data for repeat computations are given in the text.

Conflicts of Interest: The author declares no conflicts of interest.

Notations

Δp	pressure loss in [Pa]
Δh	the head loss in [m]
Q	flow discharge in [m^3/s]
D	pipe diameter in [m]
D_H	hydraulic diameter in [m]
f	the Darcy–Weisbach, Moody, or Darcy flow friction factor [dimensionless]
Re	the Reynolds number [dimensionless]
λ	viscous friction group [dimensionless]
W	the Lambert function
L	length of pipe in [m]
υ	velocity of the moving fluid in [m/s]
g	gravitational acceleration $\approx 9.81\ m/s^2$
ρ	fluid density in [kg/m^3]
f_f	the Fanning flow friction factor [dimensionless]
Φ	the Stanton and Pannel flow friction factor [dimensionless]
τ	shear stress on surface in [N/m^2]
η	dynamic fluid viscosity in [Pas]
ε/D	relative roughness of the inner pipe surface [dimensionless]
μ	kinematic fluid viscosity [m^2/s]
π	the Ludolph's number, $\pi \approx 3.1415\ldots$ [dimensionless]
Z, B, y, A, and x	auxiliary parameters defined in Equation (5)
ω	the Wright function (shifted Lambert function)
C	auxiliary parameters defined in Equation (6)
S	cross-section area of the fluid flow given in [m^2]
l_w	the wetted perimeter in [m]
α and β	angles in [radians]
b	length of the channel base in [m]
h	level of the water in channel in [m]
e	exponential function
$\log_{10}$	logarithmic function of based 10
ln	natural logarithm

Appendix A —Notes on Open Channel Flow

All principles from this paper can be used also for open channel flow after following adaptations (although the issue can be much more complex [38–45]).

The general fluid flow situation, with respect to open channels, is given in Figure A1.

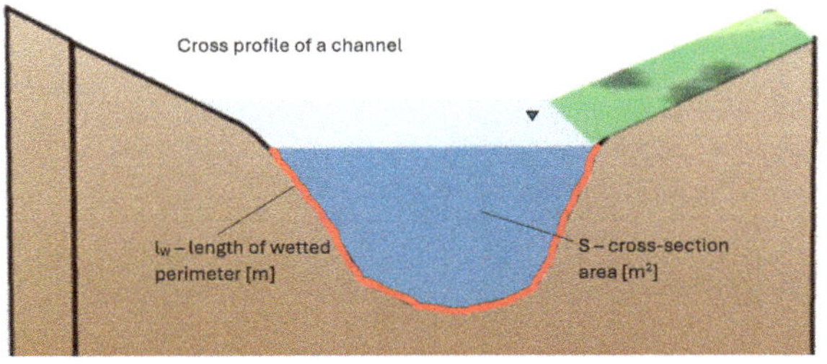

Figure A1. Elements for defining the hydraulic diameter.

Both the Darcy–Weisbach and the viscous friction group can be used not only for the simulation of flow in circular pipes, but also for open channel flow. For the open channel flow, as referred to Figure A1, the diameter of the pipe should be replaced with a hydraulic diameter, as given in Equation (A1):

$$D_H = \frac{4 \cdot S}{l_w}, \tag{A1}$$

In pipe flow, the hydraulic diameter is equal to pipe diameter; D_H = D. All equations developed for pipe flow can be equally used for open channel flow, where hydraulic diameter D_H as defined in Equation (A1) should be used instead of diameter D.

Specific examples for the calculation of hydraulic diameters are given for flow in the following:

- Partially filled circular pipes in Figure A2a and Equation (A2);
- Trapezoidal channels in Figure A2b and Equation (A3).

$$\left.\begin{array}{r} S = (\alpha - \sin\alpha)\frac{D^2}{8} \\ l_w = \frac{\alpha \cdot D}{2} \\ D_H = D \cdot \left(1 - \frac{\sin\alpha}{\alpha}\right) \end{array}\right\}, \tag{A2}$$

The angle α outside of trigonometric function is by definition always in radians, even if the calculation of trigonometric functions is settled to degrees.

$$\left.\begin{array}{r} S = h(h \cdot \cot\beta + b) \\ l_w = b + \frac{2h}{\sin\beta} \\ D_H = \frac{4h(h\cot\beta + b)}{b + \frac{2h}{\sin\beta}} \end{array}\right\}, \tag{A3}$$

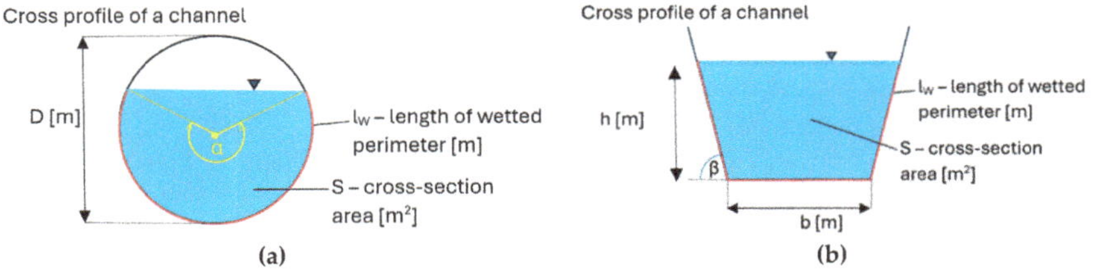

Figure A2. Hydraulic diameter in specific cases: (**a**) partially filled circular pipes—Equation (8) and (**b**) trapezoidal channels—Equation (9).

Appendix B —Computational Examples

Type 1: Here are given two examples for the calculation of head loss Δh, i.e., pressure drop Δp, through (a) the Darcy–Weisbach friction and (b) the viscous friction, for the following:

1. Circular pipe: $Re = 2 \times 10^6$ and $\varepsilon/D = 0.03$.

The pipe length L is 100 m, velocity of the moving fluid $\upsilon = 10$ m/s, while kinematic viscosity of transported water is $\mu = 1.002 \times 10^{-6}$ m^2/s with relative density $\rho = 1000$ kg/m^3 (this example is based on the last computational example of this in Appendix B—See Type 2 example).

From the definition of $Re = \upsilon D/\mu \rightarrow D = 0.2004$ m

(a) Darcy–Weisbach flow friction:
Equation (3): $f = 0.057189361$;
Equation (1): $\Delta h = 145.4516081$ m, $\Delta p = \rho g \Delta h = 14.3$ MPa.

(b) Viscous flow friction:
Equation (4): $\lambda = 1787.167928$;
Equation (5): $Z = 2.180158299$, $A = 7438.09118$, $B = 13.72926025$, $x = 7451.82044$, $y \approx \ln(x)\left(\frac{1}{x} - x\right) \approx -8.915017121$ giving $\lambda \approx 1787.167957$;
Equation (6): $A = 7416.838696$, $B = 13.72903174$, $C = 8.913357545$, $x = 7430.567728$ giving $\lambda \approx 1785.216882$;
Equation (2): $\Delta h = 145.4516392$, $\Delta p = \rho g \Delta h = 1.43$ MPa.

In the case of open channel flow in a partially filled circular pipe with $D_H = 0.2004$ m and if the angle of wetting perimeter from Figure A2a is $\alpha = 120^\circ = 2\pi/3$, the real diameter of the pipe should be $D = 0.341686$ m—from Equation (8).

2. Trapezoidal open channel: $Re = 5 \times 10^5$ and $\varepsilon/D = 0.0015$.

The pipe length L, velocity of the moving fluid υ, kinematic viscosity μ, and density ρ of transported water are the same as from the previous example.

From definition of $Re = \upsilon D_H/\mu \rightarrow D_H = 0.08542151$ m

(a) Darcy–Weisbach flow friction:
Equation (3): $f = 0.022178284$;
Equation (1): $\Delta h = 225.6270747$ m, $\Delta p = \rho g \Delta h = 2.21$ MPa.

(b) Viscous flow friction:
Equation (4): $\lambda = 173.2705591$;
Equation (5): $Z = 2.180158299$, $A = 92.9761397$, $B = 12.34296589$, $x = 105.319105$, $y \approx \ln(x)\left(\frac{1}{x} - x\right) \approx -4.612776894$ giving $\lambda \approx 173.29306$;
Equation (2): $\Delta h = 225.63061$, $\Delta p = \rho g \Delta h = 22.1.3$ MPa.

In the case of open channel flow in the trapezoidal channel with $D_H = 0.2004$ m, if the angle from Figure A2b is $\beta = 60^\circ$ and $b = 0$, the $h = D_H = 0.2004$ m—from Equation (9).

Type 2: Flow discharge Q should be estimated for the diameter of pipe $D = 0.2004$ m, head loss $\Delta h = 145.4516081$ m, relative roughness of inner pipe surface $\varepsilon/D = 0.03$, kinematic viscosity of fluid $\mu = 1.002 \times 10^{-6}$ m^2/s and its density $\rho = 1000$ kg/m^3, pipe length $L = 100$ m, and gravitational acceleration $g = 9.81$ m/s^2 (this example is based on the first computational example of this in Appendix B—See Type 1).

Equation (7): $f \cdot Re^2 = 2.29258 \times 10^{11}$.

From Figure 5, from the intersection with curve $\varepsilon/D = 0.03$, Re is estimated to 2×10^6.

Equation (8): $\upsilon = 10$ m/s, $Q = 0.315417159$ m^3/s.

Type 3: Using the data from the previous example, the pipe diameter for smooth and the very rough pipes should be estimated: Head loss $\Delta h = 145.45$ m, velocity of the moving fluid $\upsilon = 10$ m/s, pipe length $L = 100$ m, kinematic viscosity of fluid $\mu = 1.002 \times 10^{-6}$ m^2/s and its density $\rho = 1000$ kg/m^3, and gravitational acceleration $g = 9.81$ m/s^2.

Equation (9): $f/Re = 2.859 \times 10^{-8}$.

Figure 6 for smooth pipes: $Re = 5 \times 10^5$ gives, using Equation (10), $D = 0.0501$ m.

Figure 6 for very rough pipes: Re = 2.5×10^6 gives, using Equation (10), D = 0.2505 m.

References

1. Brkić, D.; Stajić, Z.; Živković, M. Sizing pipes without iterative calculus: Solutions for head loss, flow discharge and diameter. In Proceedings of the 24th International Carpathian Control Conference (ICCC), Miskolc-Szilvásvárad, Hungary, 12–14 June 2023; IEEE: New York, NY, USA, 2023; pp. 71–76. [CrossRef]
2. Brkić, D.; Praks, P.; Praksová, R.; Kozubek, T. Symbolic regression approaches for the direct calculation of pipe diameter. *Axioms* **2023**, *12*, 850. [CrossRef]
3. Brown, G.O. The history of the Darcy-Weisbach equation for pipe flow resistance. In Proceedings of the Environmental and Water Resources History, Washington, DC, USA, 3–7 November 2002; pp. 34–43. [CrossRef]
4. Fiorillo, F.; Esposito, L.; Ginolfi, M.; Leone, G. New insights into turbulent and laminar flow relationships using Darcy–Weisbach and Poiseuille laws. *Water* **2024**, *16*, 1452. [CrossRef]
5. Moody, L.F. Friction factors for pipe flow. *Trans. Am. Soc. Mech. Eng.* **1944**, *66*, 671–678. [CrossRef]
6. LaViolette, M. On the history, science, and technology included in the Moody diagram. *J. Fluids Eng.* **2017**, *139*, 030801. [CrossRef]
7. de Souza Mendes, P.R. A note on the Moody diagram. *Fluids* **2024**, *9*, 98. [CrossRef]
8. Mishra, R.; Ojha, C.S.P. Application of AI-based techniques on Moody's diagram for predicting friction factor in pipe flow. *J* **2023**, *6*, 544–563. [CrossRef]
9. Colebrook, C.F.; White, C.M. Experiments with fluid friction in roughened pipes. *Proc. R. Soc. Ser. A Math. Phys. Sci.* **1937**, *161*, 367–381. [CrossRef]
10. Colebrook, C.F. Turbulent flow in pipes, with particular reference to the transition region between the smooth and rough pipe laws. *J. Inst. Civ. Eng.* **1939**, *11*, 133–156. [CrossRef]
11. Praks, P.; Brkić, D. Choosing the optimal multi-point iterative method for the Colebrook flow friction equation. *Processes* **2018**, *6*, 130. [CrossRef]
12. Brkić, D.; Praks, P. Accurate and efficient explicit approximations of the Colebrook flow friction equation based on the Wright ω-function. *Mathematics* **2019**, *7*, 34. [CrossRef]
13. Praks, P.; Brkić, D. Review of new flow friction equations: Constructing Colebrook's explicit correlations accurately. *Rev. Int. Métodos Numéricos Cálculo Diseño Ing.* **2020**, *36*, 41. [CrossRef]
14. Coulson, J.M.; Richardson, J.F.; Backhurst, J.R.; Marker, J.H. In Chemical Engineering, Volume 1, Sixth edition, Fluid Flow, Heat Transfer and Mass Transfer: Flow of Liquids in Pipes and Open Channels, 1999. Butterworth-Heinemann: A division of Reed Educational and Professional Publishing Ltd. Reed Elsevier plc group. Available online: https://shop.elsevier.com/books/chemical-engineering-volume-1/backhurst/978-0-08-050101-7 (accessed on 31 August 2024).
15. Stanton, T.E.; Pannell, J.R. Similarity of motion in relation to the surface friction of fluids. *Philosophical Transactions of the Royal Society of London A, Math. Phys. Eng. Sci.* **1914**, *214*, 199–224. [CrossRef]
16. Vatankhah, A. Approximate analytical solution for the Colebrook equation. *J. Hydraul. Eng.* **2018**, *144*, 06018007. [CrossRef]
17. Lamri, A.; Easa, S.M. Computationally efficient and accurate solution for Colebrook equation based on Lagrange theorem. *J. Fluids Eng.* **2022**, *144*, 014504. [CrossRef]
18. Ferreri, G.B. A new approach for explicit approximation of the Colebrook–White formula for pipe flows. *J. Hydroinformatics* **2024**, *26*, 1558–1571. [CrossRef]
19. Santos-Ruiz, I.; López-Estrada, F.-R.; Puig, V.; Valencia-Palomo, G. Simultaneous optimal estimation of roughness and minor loss coefficients in a pipeline. *Math. Comput. Appl.* **2020**, *25*, 56. [CrossRef]
20. Carvajal, J.; Zambrano, W.; Gómez, N.; Saldarriaga, J. Turbulent flow in PVC pipes in water distribution systems. *Urban Water J.* **2020**, *17*, 503–511. [CrossRef]
21. Flack, K.A.; Schultz, M.P. Hydraulic characterization of sandpaper roughness. *Exp. Fluids* **2023**, *64*, 3. [CrossRef]
22. Madeira, A.A. Major and minor head losses in a hydraulic flow circuit: Experimental measurements and a Moody's diagram application. *Eclética Química* **2020**, *45*, 47–56. [CrossRef]
23. Muzzo, L.E.; Matoba, G.K.; Ribeiro, L.F. Uncertainty of pipe flow friction factor equations. *Mech. Res. Commun.* **2021**, *116*, 103764. [CrossRef]
24. Muzzo, L.E.; Pinho, D.; Lima, L.E.; Ribeiro, L.F. Accuracy/Speed analysis of pipe friction factor correlations. In *INCREaSE 2019, Proceedings of the 2nd International Congress on Engineering and Sustainability in the XXI Century 2020, Faro, Portugal, 9–11 October 2019*; Springer International Publishing: Cham, Switzerland, 2019; pp. 664–679. [CrossRef]
25. Corless, R.M.; Gonnet, G.H.; Hare, D.E.; Jeffrey, D.J.; Knuth, D.E. On the Lambert W function. *Adv. Comput. Math.* **1996**, *5*, 329–359. [CrossRef]
26. Wu, B.; Zhou, Y.; Lim, C.W.; Zhong, H. Analytical approximations to the Lambert W function. *Appl. Math. Model.* **2022**, *104*, 114–121. [CrossRef]
27. Barry, D.A.; Parlange, J.Y.; Li, L.; Prommer, H.; Cunningham, C.J.; Stagnitti, F. Analytical approximations for real values of the Lambert W-function. *Math. Comput. Simul.* **2000**, *53*, 95–103. [CrossRef]
28. Hayes, B. Why W? *Am. Sci.* **2005**, *93*, 104–108. Available online: https://www.americanscientist.org/article/why-w (accessed on 1 September 2024).

29. Sonnad, J.R.; Goudar, C.T. Constraints for using Lambert W function-based explicit Colebrook–White equation. *J. Hydraul. Eng.* **2004**, *130*, 929–931. [CrossRef]
30. Brkić, D.; Stajić, Z. Excel VBA-Based User Defined Functions for highly precise Colebrook's pipe flow friction approximations: A comparative overview. *Facta Univ. Ser. Mech. Eng.* **2021**, *19*, 253–269. [CrossRef]
31. Sonnad, J.R.; Goudar, C.T. Turbulent flow friction factor calculation using a mathematically exact alternative to the Colebrook–White equation. *J. Hydraul. Eng.* **2006**, *132*, 863–867. [CrossRef]
32. Biberg, D. Fast and accurate approximations for the Colebrook equation. *J. Fluids Eng.* **2017**, *139*, 031401. [CrossRef]
33. Clamond, D. Efficient resolution of the Colebrook equation. *Ind. Eng. Chem. Res.* **2009**, *48*, 3665–3671. [CrossRef]
34. Rollmann, P.; Spindler, K. Explicit representation of the implicit Colebrook–White equation. *Case Stud. Therm. Eng.* **2015**, *5*, 41–47. [CrossRef]
35. Swamee, P.K.; Jain, A.K. Explicit equations for pipe-flow problems. *J. Hydraul. Div.* **1976**, *102*, 657–664. [CrossRef]
36. Huang, S. Reading the Moody chart with a linear interpolation method. *Sci. Rep.* **2022**, *12*, 6587. [CrossRef]
37. Yıldırım, G. Computer-based analysis of explicit approximations to the implicit Colebrook–White equation in turbulent flow friction factor calculation. *Adv. Eng. Softw.* **2009**, *40*, 1183–1190. [CrossRef]
38. Sidiropoulos, E.; Vantas, K.; Hrissanthou, V.; Papalaskaris, T. Extending the applicability of the Meyer–Peter and Müller bed load transport formula. *Water* **2021**, *13*, 2817. [CrossRef]
39. Kim, T.-H.; Lee, J.; Kim, T.; Choi, H.T.; Im, S. A Framework for quantifying reach-scale hydraulic roughness in mountain headwater streams. *Water* **2024**, *16*, 647. [CrossRef]
40. Zhang, X.; Zhu, Y.; Wu, H.; Bi, Z.; Xu, Z. Characteristics of vegetation resistance variation in muddy water flows. *Water* **2023**, *15*, 2238. [CrossRef]
41. Yen, B.C. Open channel flow resistance. *J. Hydraul. Eng.* **2002**, *128*, 20–39. [CrossRef]
42. McGahey, C.; Knight, D.W.; Samuels, P.G. Advice, methods and tools for estimating channel roughness. *Proc. Inst. Civ. Eng.-Water Manag.* **2009**, *162*, 353–362. [CrossRef]
43. Chen, Y.; Cao, F.; Cheng, W.; Liu, B. Enhancing the Accuracy of Water-Level Forecasting with a New Parameter-Inversion Model for Estimating Bed Roughness in Hydrodynamic Models. *Appl. Sci.* **2023**, *13*, 4551. [CrossRef]
44. Chen, Z.; Zhou, J.; Chen, Q. Research and application of the calculation method of river roughness coefficient with vegetation. *Water* **2023**, *15*, 2638. [CrossRef]
45. Coscarella, F.; Penna, N.; Ferrante, A.P.; Gualtieri, P.; Gaudio, R. Turbulent flow through random vegetation on a rough bed. *Water* **2021**, *13*, 2564. [CrossRef]

water

MDPI

Article

Effect of Median Soil–Particle Size Ratio on Water Storage Capacity of Capillary Barrier

Honghua Liu [1,2,3], Jie Dong [1,2,3], Qiang Liu [4,*], Lin Geng [1,2,3], Zhongsheng Wang [1,2,3] and Chong Sun [5]

1 Key Laboratory of Geological Safety of Coastal Urban Underground Space, Ministry of Natural Resources, Qingdao 266100, China; liuhonghua567@163.com (H.L.); 13853286966@163.com (J.D.); 13963981027@163.com (L.G.); 18253279528@163.com (Z.W.)

2 Qingdao Geo-Engineering Surveying Institute (Qingdao Geological Exploration Development Bureau), Qingdao 266101, China

3 Qingdao Key Laboratory of Groundwater Resources Protection and Rehabilitation, Qingdao 266100, China

4 College of Earth Science and Engineering, Shandong University of Science and Technology, Qingdao 266590, China

5 College of Environmental Science and Engineering, Ocean University of China, Qingdao 266100, China; 11181022035@ouc.edu.cn

* Correspondence: sunnyseasea@sdust.edu.cn

Abstract: Capillary barriers are widely used as a cover system to enhance the upper-soil-layer water storage capacity and reduce water infiltrate into the lower soil layer. In this paper, the effects of the median soil–particle size ratio on the water storage capacity of capillary barriers were studied using a series of indoor one-dimensional soil column infiltration tests. The results show that the water storage capacity rises with an increase in the median soil–particle size ratio until it exceeds 10. The variation in the total water storage capacity is related to not only the median soil–particle size ratio but also the particle size of coarse-grained soil or fine-grained soil. When the fine-grained soil-layer particle size is constant, the total water storage first increases, then decreases, and finally remains constant after increasing the median soil–particle size ratio. In contrast, when the coarse-grained soil layer particle size is constant, the relationship between the capillary barrier's total water storage and median soil–particle size ratio can be defined as a power function. Using the capillary barrier can increase coarse-grained sand by 90% in water storage capacity and can only increase fine-grained sand by 7% in water storage capacity. The breakthrough time increases with the increase in the median soil–particle size ratio. The presence of the coarse and fine-grained soil layer interface in the capillary barrier can affect the fine-grained soil layer infiltration rate.

Keywords: capillary barrier; quartz sand; median soil–particle size ratio; breakthrough time; water storage

1. Introduction

The capillary barrier is a soil system composed of a fine-grained soil layer covered by a coarse-grained soil layer. Because of the difference in the permeability between the coarse-grained and fine-grained layers under the same matric suction condition, the capillary barrier not only cuts off the water moving downward into the coarse-grained soil layer at the top surface of the coarse-grained soil layers but also the capillary water moving upward into the fine-grained layer at the bottom surface of the coarse-grained soil layers. Because the capillary barrier has these functions, it has been used in several different fields, including as the cover in landfill and nuclear waste treatment plants instead of the traditional compacted clay gap [1–5], as the enhanced agricultural layer in an arid area to improve the root-zone water content [6–8], as a protection layer from rainfall in slope engineering [9,10], and as an obstruction layer to stop salted water from flowing upward into cultivated soil [11,12].

Several factors, such as the shape of the interface [13], the thickness of fine- and coarse-grained soil layers [14–19], the particle size of each soil layers [20], the slope [21–23], the rainfall intensity [24], the gradation [25], the temperature difference [26], the compaction [27], the organic [28], and the structure [29–32], affect the water storage capacity, lateral diversion, and breakthrough time, which will impact the performance of capillary barriers in practice. Among these factors, the particle sizes of two layers are important indicators for the preliminary evaluation of the capillary barrier. Smesrud et al. [33] used numerical simulations and found that 90% of seepage water laterally divert when the median soil–particle size ratio exceeds 5. Apambilla et al. [34] found that when the median soil–particle size ratio exceeds 8, the capillary barrier can effectively prevent the groundwater from rising. The mechanism of the influence of the mean ratio of soil–particle size on the prevention of infiltration is not clear. Moreover, the effect of the mean ratio of soil-particle size on the water storage capacity needs to be discussed further.

In order to obtain the effect of the median soil–particle size ratio on the water storage capacity and the breakthrough time, a series of indoor 1D ponding infiltration tests were conducted. Quartz sands of different grain sizes were used as the materials for the fine-grained layer and the coarse-grained layer. The results show that the water storage capacity increases with the decrease in the grain size of the fine-grained layer. However, when the grain size of the fine-grained layer is constant, the water storage capacity no longer increases with the median soil–particle size ratio, which exceeds by 10 times. The breakthrough time is mainly determined by the size of the fine-grained layer, when the diameter of the particles of the fine-grained layer is smaller than 0.075 mm, the breakthrough time linearly increases with the median soil–particle size ratio; otherwise, the breakthrough time remains almost same.

2. Materials and Methods

2.1. Materials and Equipment

The physical properties of quartz sand used in the test are shown in Table 1. The maximum compaction density was determined using a proctor test [35] [ASTM D7263-2009 (R2018)], the specific gravity of different soil separates was determined using the pycnometer method [36] [ASTM D854-2014], and the saturated permeability coefficient was determined using the constant/variable head permeability coefficient method.

Table 1. Basic physical properties of quartz sand.

Sand Diameter (mm)	Specific Gravity	Maximum Compaction Density (g/cm^3)	Saturated Hydraulic Conductivity (cm/s)	Porosity
<0.075	2.80	1.66	2.11×10^{-4}	0.408
0.075–0.1	2.77	1.56	3.16×10^{-3}	0.439
0.1–0.25	2.88	1.57	4.39×10^{-3}	0.454
0.25–0.5	2.68	1.57	4.11×10^{-2}	0.416
0.5–1	2.64	1.57	1.69×10^{-1}	0.408
1–2	2.63	1.56	3.61×10^{-1}	0.406
2–5	2.64	1.65	1.43	0.378

As shown in Figure 1, the soil column of the infiltration experiments is made of acrylic transparent tubes with a 1 cm thick wall and 14.4 cm internal diameter. The cylinders are 50 cm high and easy to combine and install. Five holes with a 40 cm diameter were located at 5 cm, 15 cm, 25 cm, 35 cm, and 45 cm heights of the acrylic cylinder wall. SM150T moisture meter (Delta-T Devices Ltd., Cambridge, UK) was adopted, with the measurement range of 0–0.7 $m^3 \cdot m^{-3}$ and accuracy of $\pm$ 0.03 $m^3 \cdot m^{-3}$. The datalogger (GP2, Delta-T Devices Ltd., Cambridge, UK) was used to automatically collect and record the readings of the moisture meter. The Mariotte bottle was used as the water source to provide a constant 5 cm water head at the top of the soil column, and a weighing device was set under the Mariotte bottle to record the infiltration water quantity. The soil column was connected to a percolation collector at the bottom using a water outlet, and the percolation collector was set on an electronic balance to record the mass of seepage water.

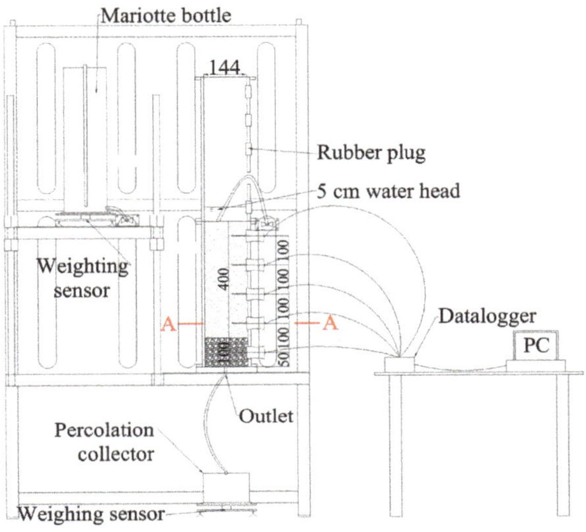

Figure 1. Setup of the 1D ponding infiltration test.

To ensure a uniform sand distribution, a consistent compaction effort was given at each layer with a thickness of 5 cm, and a measured mass of quartz sand was densified with a rubber hammer to achieve maximum compaction density. The maximum compaction density is listed in Table 1.

The capillary barrier was tested using two soil layers. The coarse-grained soil layer had a thickness of 10 cm, while the fine-grained soil layer was 40 cm thicker than the coarse-grained soil layer. A non-woven geotextile was placed between the fine- and coarse-grained soil layers to prevent the fine-grained soil layer particles from entering the coarse-grained soil layer.

2.2. Experiment

In this paper, the median particle size (D_{50}) is an index that evaluates the difference in particle sizes between the coarse- and fine-grained soil layers; the screened quartz sand is used in the experiment to prevent the influence of the gradation of natural sand. The quartz sand is screened from well-continuous graded quartz sand; therefore, the median particle size can be approximated to the logarithmic average of the minimum and maximum particle sizes of screened quartz sand. Table 2 show the 21 infiltration experiments and specific configurations.

The experiment XLC1–XLC5′s coarse-grained soil-layer particle size is 2–5 mm, and XLC6–XLC9 is 1–2 mm, the fine-grained soil-layer quartz-sand particle size decrease led to the increase in the median soil–particle size ratio. The fine-grained soil-layer particle size of CLC1–CLC5 and CLC6–CLC10 are 0.1–0.25 mm and 0.075–0.1 mm, respectively. The coarse-grained soil-layer quartz-sand particle size increase led to the increase in the median soil–particle size ratio. JZT1–JZT6 are homogeneous soil columns.

In order to verify the influence of non-woven geotextile on the hydraulic properties of the soil column, infiltration experiments of 20 cm homogeneous quartz sand were carried out. The quartz-sand particle size is 0.1–0.25 mm, non-woven geotextile is set at 10 cm of the soil column in the experiment group, and no non-woven geotextile is in the control group. The 5 cm water head was applied on the soil column. Then, we monitored the outflow time from the bottom of the column and recorded the water storage of the soil column. The results show that the infiltration rate and final water storage capacity of the soil column with or without non-woven geotextile are basically the same as those shown in Figure 2. Therefore, using non-woven geotextile in the partition layer does not affect the test results.

Table 2. Scheme of ponding infiltration experiments.

Number	Coarse Quartz Sand Grade/mm	Fine Quartz Sand Grade/mm	Median Soil-Particle Size Ratio
XLC1	2–5	1–2	2.24
XLC2		0.5–1	4.45
XLC3		0.25–0.5	9.03
XLC4		0.1–0.25	19.75
XLC5		0.075–0.1	36.7
XLC6	1–2	0.5–1	2
XLC7		0.25–0.5	3.99
XLC8		0.1–0.25	8.95
XLC9		0.075–0.1	16.25
CLC1	0.25–0.5	0.1–0.25	2.19
CLC2	0.5–1		4.44
CLC3	1–2		8.81
CLC4	2–5		19.75
CLC5	5–10		44.75
CLC6	0.1–0.25	0.075–0.1	1.78
CLC7	0.25–0.5		4.02
CLC8	0.5–1		8.16
CLC9	1–2		16.20
CLC10	2–5		36.70
JZT1	2–5	2–5	1
JZT2	1–2	1–2	
JZT3	0.5–1	0.5–1	
JZT4	0.25–0.5	0.25–0.5	
JZT5	0.1–0.25	0.1–0.25	
JZT6	0.075–0.1	0.075–0.1	

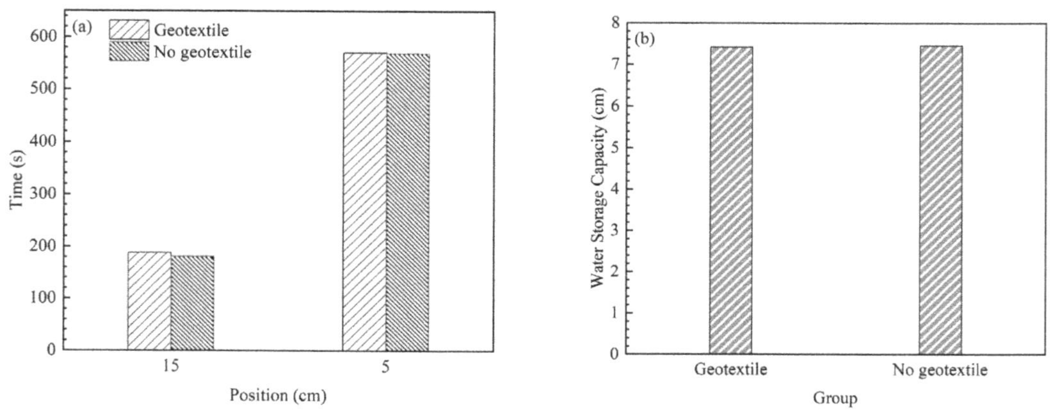

Figure 2. Effect of non-woven geotextile on hydraulic properties of soil column. (**a**) Detection time of water at different positions. (**b**) water storage capacity with and without geotexile.

Stormont and Morris [37] defined the water held in the fine-grained soil layer before it flows into the coarse-grained soil layer as the water storage capacity of the capillary barrier. In this paper, the water storage capacity and breakthrough time are the two main factors used to judge the capillary barrier's effectiveness. We use the following methods to determine the water storage capacity and the breakthrough time. The total water storage capacity was defined as the total volume per unit area of water that is stored in both the fine-grained soil layer and the coarse-grained soil layer. It was calculated by the difference between the record of the two electronic balances after testing, of which one was used to weigh the variation in the Mariotte bottle, representing the water quantity of the infiltration, and the other one was used to weigh the mass of the outflow. The water storage of the coarse-grained soil layer was measured by drying the coarse-grained soil after the test by disassembling the columns. The water storage capacity (of the fine-grained soil layer) was calculated by the difference between the total water storage capacity and the water storage of the coarse-grained soil layer.

In the experiment, the moisture meter of the coarse-grained soil layer is arranged in the middle of coarse layer, and the thickness of the coarse-grained soil layer is only 10 cm. The coarse-grained soil layer used in the experiment is coarser than 0.25–0.5 mm, the saturated hydraulic conductivity is large, as shown in Table 1, and the sensing range of SM150T is 35 mm. When the wetting front reaches 15 mm below the interface [38], the SM150T can record the water content. The maximum lag is less than 40 s at 0.25–0.5 mm quartz sand as the coarse-grained soil layer. Therefore, the time of the moisture content monitored using the moisture meter in the coarse-grained soil layer is regarded as the approximate breakthrough time.

3. Results and Discussion

3.1. Influence of Median Soil–Particle Size Ratio on Water Storage Capacity of Capillary Barrier

The relationships of the water storage capacity and total water storage capacity with the median soil–particle size ratio are shown in Figure 3. In Figure 3a,b, the coarse-grained soil layer is 2–5 mm and 1–2 mm, respectively. The finer the fine-grained soil-layer particle size, the larger the median soil–particle size ratio. In Figure 3c,d, the fine-grained soil layer is 0.1–0.25 mm and 0.075–0.1 mm, respectively. The coarser the coarse-grained layer particle size, the larger the median soil–particle size ratio. The circle hollow dot is the total water storage of the capillary barrier, the square dot is the water storage of the capillary barrier, the triangle dot is soil saturation at section A-A shown in Figure 1, and section A-A is 5 cm above the fine- and coarse-grained layer interface, which is the moisture meter monitoring section closest to the interface of the capillary barrier.

It can be seen from Figure 3a,b, when the coarse-grained soil-layer particle size is constant, the total water storage capacity and water storage increase with the increased in the median soil–particle size ratio of about 10–20 times. The coarse-grained soil layer is made of 2–5 mm quartz sand, the total water storage capacity and water storage reach the maximum value at a median soil–particle size ratio of 20. When the coarse-grained soil layer is 1–2 mm, the maximum total water storage capacity and water storage were reached at a median soil–particle size ratio of 10. The rule of saturation at section A-A (Figure 1) is the same as the water storage capacity law: when the median soil–particle size ratio reaches 20 times and 10 times, respectively, the volumetric water content of quartz sand at A-A section is close to saturation.

In Figure 3c,d, with the increase in the median soil–particle size ratio, the total water storage capacity first increases, then decreases, and finally stabilizes, while the water storage capacity first increases before eventually stabilizing. When the fine-grained soil layer is 0.1–0.25 mm quartz sand, the water storage capacity and total water storage capacity reach a stable state at the median soil–particle size ratio of 10 times; when the fine-grained soil layer is 0.075–0.1 mm quartz sand, the water storage capacity and total water storage capacity of the capillary barrier reach a stable state at the median soil–particle size ratio of 20 times. The rule of saturation at profile A-A is the same as the water storage capacity law: when the median soil–particle size ratio reaches 10 and 20 times, respectively, the A-A profile is close to saturation.

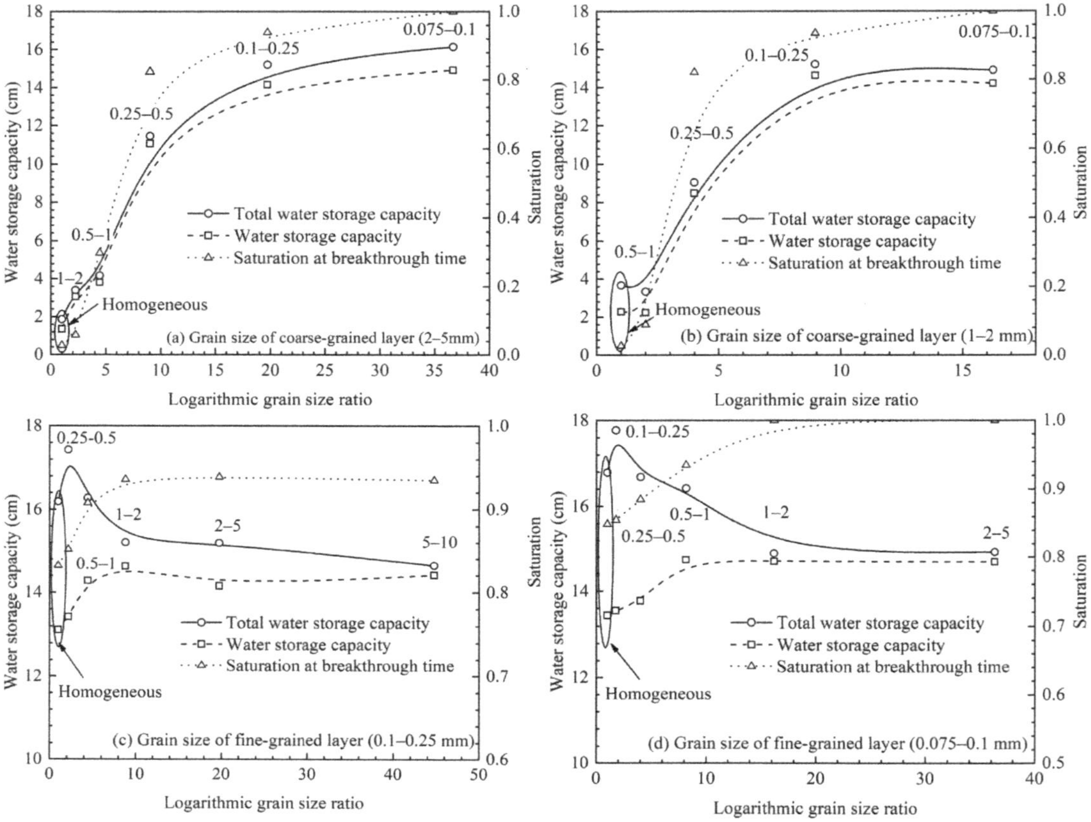

Figure 3. Variation curve of median soil–particle size ratio and water storage capacity of capillary barrier.

In conclusion, with the increase in the median soil–particle size ratio, the capillary barrier water storage capacity increases. When the median soil–particle size ratio increases to 10–20 times, the water storage capacity reaches its maximum capacity. Because the A-A section has been fully saturated, it means the bottom of the fine-grained soil layer has been saturated. When the fine-grained soil-layer particle size is finer and the coarse-grained soil-layer particle size remains constant, the total water storage of the capillary barrier increases with the increase in the median soil–particle size ratio. When the fine-grained soil-layer particle size remains constant and the coarse-grained soil-layer particle size is coarser, with the increase in the median soil–particle size ratio, the total water storage of capillary barrier first increases, then decreases, and finally stabilizes.

The water storage capacity of capillary barriers with coarse-grained particle sizes of 2–5 mm and 1–2 mm quartz sand compared to the corresponding fine-grained soil-layer particle size of homogeneous sand is shown in Figure 4.

In general, regardless of the fine-grained soil-layer particle size, the water storage capacity of the capillary barrier that used 2–5 mm quartz sand as the coarse-grained soil layer is higher than the 1–2 mm quartz sand used as the coarse-grained soil layer. When fine-grained soil layer is made of quartz sand coarser than 0.25–0.5 mm, the capillary barrier that used 2–5 mm quartz sand as the coarse-grained soil layer increases the water storage capacity by more than 20–90% compared with the 1–2 mm quartz sand used as the coarse-grained soil layer. However, when the fine-grained soil layer is made of quartz sand finer than 0.25–0.5 mm, the coarse-grained layer that used 2–5 mm quartz sand only increases the water storage capacity by 4% than the 1–2 mm quartz sand used as the coarse-grained soil layer.

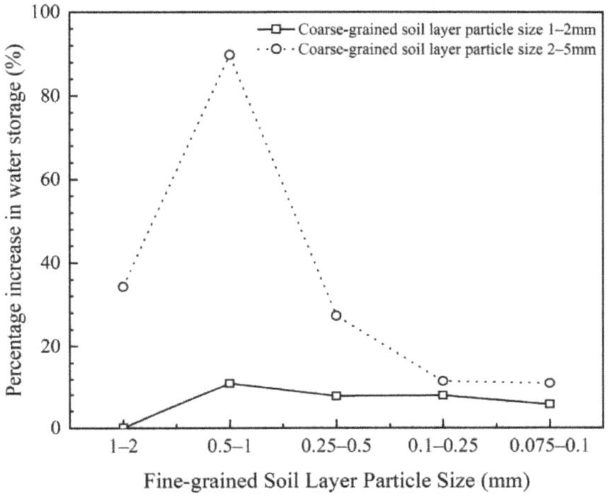

Figure 4. The effect of capillary barrier on increased water storage capacity compared with homogeneous soil.

This is because when the particle size is coarser than 0.25–0.5 mm, the water content of the fine-grained soil layer is sensitive to the matrix potential. As shown in Figure 5, the water entry value of 2–5 mm quartz sand as the coarse-grained soil layer is −1 kPa, only 1 kPa higher than the 1–2 mm quartz sand used as the coarse-grained soil layer. When the fine-grained soil-layer soil particle size is coarser than 0.25–0.5 mm, the water storage capacity increases by more than 20% in the 1 kPa matrix suction. When the fine-grained soil layer particle size is finer than 0.25–0.5 mm, the fine-grained soil layer has been almost saturated at 1–2 mm quartz sand coarse-grained soil layer's water entry value; the water content of the fine-grained soil layer does not substantially increase when the 2–5 mm quartz sand is used as coarse-grained soil layer. According to Figure 3a,b, when the coarse-grained soil layer remains constant, the fine-grained soil-layer particle size is coarser than 0.25–0.5 mm, regardless of whether the coarse-grained soil layer uses 2–5 mm or 1–2 mm quartz sand, the median soil–particle size ratio is lower than 10; the water storage capacity of capillary barrier that uses 2–5 mm quartz sand as the coarse-grained soil layer is far better than the 1–2 mm quartz sand used as the coarse-grained soil layer. When the fine-grained soil-layer particle size is finer than 0.25–0.5 mm, the median soil–particle size ratio is greater than 10; the water storage capacity of capillary barrier that uses the 2–5 mm quartz sand as the coarse-grained layer is slightly higher than the 1–2 mm quartz sand used as the coarse-grained soil layer. Similarly, it can be seen from Figure 3c,d that the fine-grained soil layer remains constant when the median soil–particle size ratio is lower than 10, and the water storage of capillary barrier increased with the increase in the median soil–particle size ratio; when the median soil–particle size ratio is higher than 10, the water storage of capillary barrier does not increase with the increase in the median soil-particle size ratio. Therefore, increasing the coarse-grained soil-layer particle size cannot increase the water storage capacity of the capillary barrier infinitely. When the median soil–particle size ratio is higher than 10, the water storage of the capillary barrier only slightly increases.

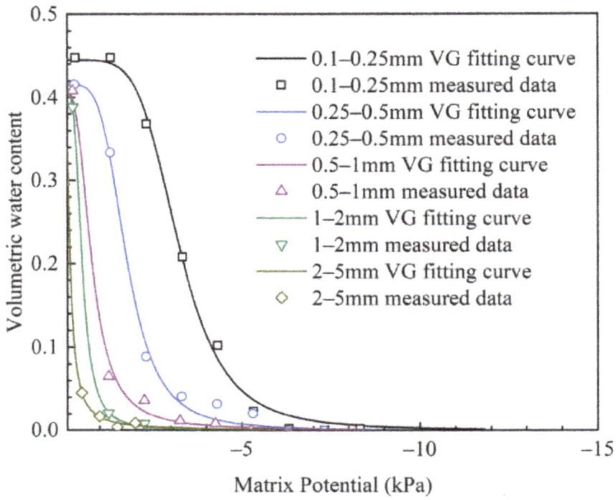

Figure 5. Wetting soil–water characteristic curve.

3.2. *Effect of Median Soil-Particle Size Ratio on Breakthrough Time of Capillary Barrier*

It can be seen from Figure 6a that when the coarse-grained soil-layer particle size is constant and the fine-grained soil-layer particle size decreases, the breakthrough time increases with the increase in the median soil–particle size ratio. The coarse-grained soil-layer particle size remains constant, and the water entry value is certain. With the decrease in the fine-grained soil-layer particle size, the saturated hydraulic conductivity of the fine-grained soil layer decreases, and a higher water content is required to achieve the water entry value, which leads to breakthrough time increasing with the increase in the median soil–particle size ratio. Figure 6b shows that when the fine-grained soil layer is made of 0.075–0.1 mm quartz sand, the breakthrough time increases with the increase in the median soil–particle size ratio. But, when the fine-grained soil layer used 0.1–0.25 mm quartz sand, the breakthrough time almost does not increase with the increase in the median soil–particle size ratio. As shown in Figure 6, when the coarse-grained soil-layer particle size is coarser, ranging from 0.25–0.5 mm to 2–5 mm, the water entry value range is −3.2 kPa–−0.7 kPa; 0.1–0.25 mm quartz sand is close to saturation at −4 kPa; and the water content is insensitive to the increased water entry value. So, the water entry value increase only results in a slight water storage increase when 0.1–0.25 mm quartz sand is used in the fine-grained soil layer, and the breakthrough time almost does not increase with the increase in the median soil–particle size ratio. When the fine-grained soil layer is made of 0.075–0.1 mm quartz sand, the saturated hydraulic conductivity is low. The water content at −3.2–0.7 kPa (coarse-grained soil layer's water entry value) was not saturated, the coarse-grained soil layer used coarser quartz sand, and the fine-grained soil layer needed more water to obtain a higher water entry value. So, when the fine-grained soil layer is made of 0.075–0.1 mm quartz sand, it takes a long time to reach the corresponding breakthrough water content with the given coarser coarse-grained soil-layer particle size.

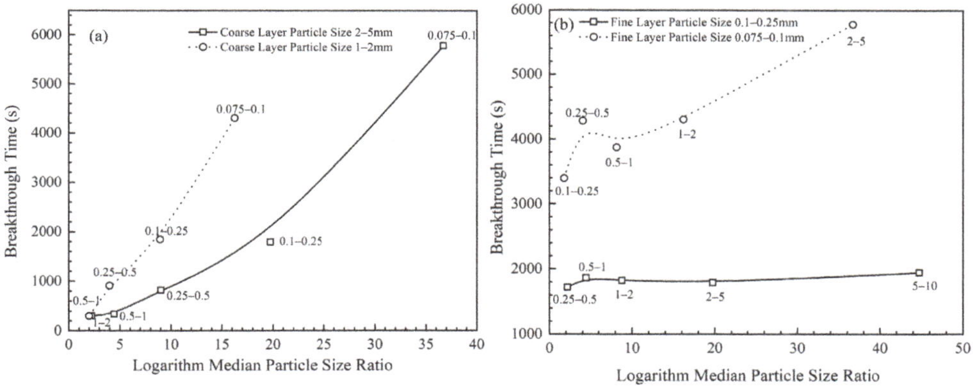

Figure 6. The effect of median soil–particle size ratio on breakthrough time of capillary barrier. (**a**) different partical size of coarse-grained soil layer. (**b**) different partical size of fine-grained soil layer.

3.3. Effect of Capillary Barrier on Infiltration Rate of Fine-Grained Soil Layer

The moisture meter can record a moving wetting front in the fine-grained soil layer with time. As shown in Figure 7a, capillary barrier has a weak effect on the infiltration rate below 15 cm in fine-grained soil layer when the fine-grained soil layer is made of 0.1–0.25 mm quartz sand. When the coarse-grained soil-layer particle size is coarser, the infiltration rate decreases slightly, and the time for the wetting front to reach the coarse- and fine-grained soil-layer interface increases slightly. It can be seen in Figure 7b that the capillary barrier has a significant effect on the infiltration rate l below 25 cm in the fine-grained soil layer, and the infiltration rate decreases significantly when the coarse-grained soil-layer particle size is coarser and the fine-grained soil layer is made of 0.075–0.1 mm quartz sand. The results show that the capillary barrier can influence the infiltration rate of the fine-grained soil layer, and the coarse-grained soil-layer particle size and fine-grained soil-layer particle size are major factors. The coarse-grained soil-layer particle size controls the infiltration rate of the fine-grained soil layer, and the fine-grained soil-layer particle size determined the range of the fine-grained soil layer affected by the capillary barrier; that is, the coarser the coarse-grained soil-layer particle size, the slower the infiltration rate of the fine-grained soil layer, and the finer the fine-grained soil-layer particle size, the greater the range of the infiltration rate of the fine-grained soil layer affected by the capillary barrier.

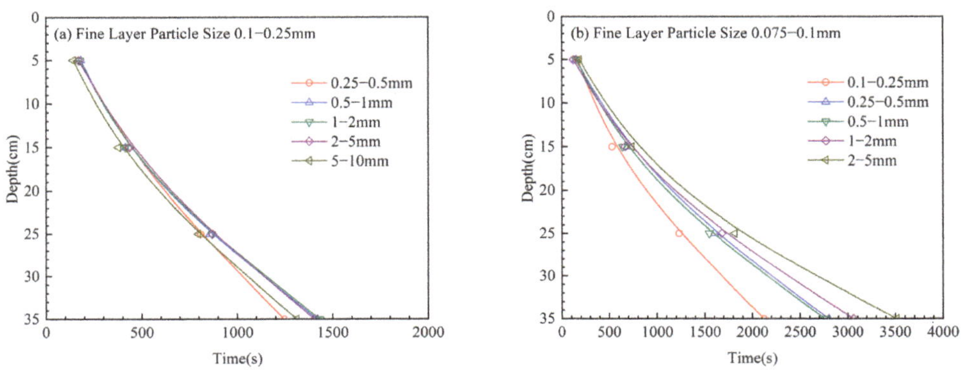

Figure 7. Relationship between wetting front and infiltration time.

4. Conclusions

The ponding infiltration tests conducted on the quartz sand capillary barrier with different grain sizes show that the water storage capacity is influenced by the grain size of the fine-grained layer and the median soil–particle size ratio. When the grain size of the coarse-grained layer is constant, the water storage capacity increases with the decrease in the grain size of the fine-grained layer and is more affected by it. However, when the grain size of the fine-grained layer is constant, the water storage capacity no longer increases with the median soil–particle size ratio that exceeds 10 times.

The total water storage is related to the median soil–particle size ratio change mode. When the fine-grained soil-layer particle size remains constant and the coarse-grained soil-layer particle size is coarser, with the increase in the median soil-particle size ratio, the total water storage first increases, then decreases, and finally stabilizes. When the fine-grained soil-layer particle size is finer and the coarse-grained soil-layer particle size remains constant, the relationship between the total water storage of the capillary barrier total and the median soil–particle size ratio can be defined a power function. The capillary barrier can increase coarse-grained sand by 20–90% in field capacity compared to the corresponding homogeneous sand, but it can only increase fine-grained sand by 7% in field capacity because the field capacity of homogenous fine-grained sand is close to saturation.

The breakthrough time increases with the increase in the median soil–particle size ratio. Therefore, enlarging the median soil–particle size ratio can improve the effective time of the capillary barrier. The coarser the coarse-grained soil-layer particle, the slower the infiltration rate of the fine-grained soil layer. The finer the fine-grained soil-layer particle size, the greater the range of the infiltration rate of the fine-grained soil layer affected by the capillary barrier.

In conclusion, when using capillary barriers in engineering, the median soil–particle size ratio should greater than 10 to obtain the maximum water storage capacity and a long effective time.

Author Contributions: Conceptualization, Q.L. and C.S.; Methodology, Q.L. and C.S.; Validation, C.S.; Formal analysis, C.S.; Writing—original draft, C.S.; Supervision, J.D. and Q.L.; Project administration, H.L., Q.L., L.G. and Z.W.; Funding acquisition, Q.L. All authors have read and agreed to the published version of the manuscript.

Funding: The financial support is provided by the Shandong Provincial Natural Science Foundation (ZR2021MD021) and the Key Laboratory of Geological Safety of Coastal Urban Underground Space, Ministry of Natural Resources (BHKF2021Z05).

Data Availability Statement: The data presented in this study are available on request from the corresponding author.

Conflicts of Interest: The authors declare no conflicts of interest.

References

1. Benson, C.H. Final covers for waste containment systems: A North American perspective. In Proceedings of the XVII Conference of Geotechnics of Torino, Control and Management of Subsoil Pollutants, Italian Geotechnical Society, Torino, Italy, 23–25 November 1999.
2. Dwyer, S.F. Water Balance Measurements and Computer Simulations of Landfill Covers. Ph.D. Dissertation, The University of New Mexico, Albuquerque, NM, USA, 2003.
3. Simon, F.G.; Müller, W.W. Standard and alternative landfill capping design in Germany. *Environ. Sci. Policy* **2004**, *7*, 277–290. [CrossRef]
4. Forman, A.D.; Anderson, J.E. Design and performance of four evapotranspiration caps. *Pract. Period. Hazard. Toxic Radioact. Waste Manag.* **2005**, *9*, 263–272. [CrossRef]
5. Hey, C.; Simms, P. Preliminary assessment of biosolids in covers with capillary barrier effects. *Eng. Geol.* **2021**, *280*, 105973. [CrossRef]
6. Al-Mayahi, A.; Al-Ismaily, S.; Al-Maktoumi, A.; Al-Busaidi, H.; Imnek, J. A smart capillary barrier-wick irrigation system for home gardens in arid zones. *Irrig. Sci.* **2020**, *38*, 235–250. [CrossRef]

7. Menezes-Blackburn, D.; Al-Ismaily, S.; Al-Mayahi, A.; Al-Siyabi, B.; Al-Kalbani, A.; Al-Busaid, H.; Al-Naabi, I.; Al-Mazroui, M.; Al-Yahyai, R. Impact of a nature-inspired engineered soil structure on microbial diversity and community composition in the bulk soil and rhizosphere of tomato grown under saline irrigation water. *J. Soil Sci. Plant Nutr.* **2021**, *21*, 173–186. [CrossRef]
8. Noguchi, K.; Saito, H.; Saefuddin, R.; Šimůnek, J. Evaluation of Subsurface Drip Irrigation Designs in a Soil Profile with a Capillary Barrier. *Water* **2021**, *13*, 1300. [CrossRef]
9. Rahardjo, H. Application of unsaturated soil mechanics in capillary barrier system. In Proceedings of the 3rd Asian Conference on Unsaturated Soils, Nanjing, China, 21–23 April 2007.
10. Satyanaga, A.; Zhai, Q.; Rahardjo, H.; Gitirana, G.d.F.N., Jr.; Moon, S.-W.; Kim, J. Performance of capillary barrier as a sustainable slope protection. In Proceedings of the 3rd Pan-American Conference on Unsaturated Soils, Rio de Janeiro, Brazil, 26–28 July 2021.
11. Guo, G.; Araya, K.; Jia, H.; Zhang, Z.; Ohomiya, K.; Matsuda, J. Improvement of Salt-Affected Soils, Part 1: Interception of Capillarity. *Biosyst. Eng.* **2006**, *94*, 139–150. [CrossRef]
12. Chavez-Garcia, E.; Christina, S. Rehabilitation of a highly saline-sodic soil using a rubble barrier and organic amendments. *Soil Tillage Res.* **2019**, *189*, 176–188. [CrossRef]
13. Selker, J. Design of interface shape for protective capillary barriers. *Water Resour. Res.* **1997**, *33*, 259–260. [CrossRef]
14. Stormont, J.C.; Anderson, C.E. Capillary barrier effect from underlying coarser soil layer. *J. Geotech. Geoenvironmental Eng.* **1999**, *125*, 641–648. [CrossRef]
15. Khire, M.V.; Benson, C.H.; Bosscher, P.J. Capillary barriers: Design variables and water balance. *J. Geotech. Geoenvironmental Eng.* **2000**, *126*, 695–708. [CrossRef]
16. Fala, O.; Molson, J.; Aubertin, M.; Bussiere, B. Numerical modelling of flow and capillary barrier effects in unsaturated waste rock piles. *Mine Water Environ.* **2005**, *24*, 172–185. [CrossRef]
17. Parent, S.E.; Cabral, A. Design of Inclined Covers with Capillary Barrier Effect. *Geotech. Geol. Eng.* **2006**, *24*, 689–710. [CrossRef]
18. Smagin, A.V.; Khakimova, G.M.; Khineeva, D.A.; Sadovnikova, N.B. Gravity factor of the formation of the field and capillary water capacities in soils and artificial layered soil-like bodies. *Eurasian Soil Sci.* **2008**, *41*, 1189–1197. [CrossRef]
19. Qian, T.; Huo, L.; Zhao, D. Laboratory investigation into factors affecting performance of capillary barrier system in unsaturated soil. *Water Air Soil Pollut.* **2010**, *206*, 295–306. [CrossRef]
20. Li, J.H.; Du, L.; Chen, R.; Zhang, L.M. Numerical investigation of the performance of covers with capillary barrier effects in south china. *Comput. Geotech.* **2013**, *48*, 304–315. [CrossRef]
21. Walter, M.T.; Kim, J.S.; Steenhuis, T.S.; Parlange, J.Y.; Heilig, A.; Braddock, R.D. Funneled flow mechanisms in a sloping layered soil: Laboratory investigation. *Water Resour. Res.* **2000**, *36*, 841–849. [CrossRef]
22. Tami, D.; Rahardjo, H.; Leong, E.C.; Fredlund, D.G. Design and laboratory verification of a physical model of sloping capillary barrier. *Can. Geotech. J.* **2004**, *41*, 814–830. [CrossRef]
23. Chen, R.; Liu, J.; Ng, C.; Chen, Z.K. Influence of slope angle on water flow in a three-layer capillary barrier soil cover under heavy rainfall. *Soil Sci. Soc. Am. J.* **2019**, *83*, 1637–1647. [CrossRef]
24. Aubertin, M.; Cifuentes, E.; Martin, V.; Apithy, S.; Maqsoud, A. An investigation of factors that influence the water diversion capacity of inclined covers with capillary barrier effects. In *Unsaturated Soils 2006*; American Society of Civil Engineers: Reston, VA, USA, 2006.
25. Tojo, Y.; Tanaka, N.; Matsuto, T.; Matsuo, T. An application of capillary barrier as a leachate drainage system at bottom of landfill. numerical analysis. *Proc. Jpn. Soc. Civ. Eng.* **2010**, *692*, 53–62.
26. Gong, Z.; Huang, Y.; Zhou, C.; Wang, L.; Tan, C. Influence of temperature change on the diversion of CCBE layer. *J. Water Resour. Archit. Eng.* **2019**, *17*, 125–130+142.
27. Lipiec, J.; Wójciga, A.; Horn, R. Hydraulic properties of soil aggregates as influenced by compaction. *Soil Tillage Res.* **2009**, *103*, 170–177. [CrossRef]
28. Scott, N.A.; Cole, C.V.; Elliott, E.T.; Huffman, S.A. Soil Textural Control on Decomposition and Soil Organic Matter Dynamics. *Soil Sci. Soc. Am. J.* **1996**, *60*, 1102–1109. [CrossRef]
29. Baker, R.S.; Hillel, D. Laboratory tests of a theory of fingering during infiltration into layered soils. *Soil Sci. Soc. Am. J.* **1990**, *54*, 20–30. [CrossRef]
30. Morris, C.E.; Stormont, J.C. Parametric study of unsaturated drainage layers in a capillary barrier. *J. Geotech. Geoenvironmental Eng.* **1999**, *125*, 1057–1065. [CrossRef]
31. Zhan, L.T.; He, L.; Jia, G.W.; Chen, Y.M.; Fredlund, D.G. Physical and numerical study of lateral diversion by three-layer inclined capillary barrier covers under humid climatic conditions 1. *Can. Geotech. J.* **2014**, *51*, 1438–1448. [CrossRef]
32. Ng, C.; Coo, J.L.; Chen, Z.K.; Chen, R. Water infiltration into a new three-layer landfill cover system. *J. Environ. Eng.* **2016**, *142*, 04016007. [CrossRef]
33. Smesrud, J.K.; Selker, J.S. Effect of soil-particle size contrast on capillary barrier performance. *J. Geotech. Geoenvironmental Eng.* **2001**, *127*, 885–888. [CrossRef]
34. Apambilla, A.J.; Pax, C.L.; Abdul, A.M. Capillary cut design for soil-groundwater salinity control. *J. Fac. Environ. Sci. Technol. Okayama Univ.* **2009**, *14*, 17–22.
35. *ASTM D7263-2009(R2018)*; Standard Test Methods for Laboratory Determination of Density (Unit Weight) of Soil Specimens. ASTM International: West Conshohocken, PA, USA, 2009.

36. *ASTM D854-2014*; Standard Test Methods for Specific Gravity of Soil Solids by Water Pycnometer. ASTM International: West Conshohocken, PA, USA, 2014.
37. Stormont, J.C.; Morris, C.E. Method to estimate water storage capacity of capillary barriers. *J. Geotech. Geoenvironmental Eng.* **1998**, *124*, 297–302. [CrossRef]
38. Liu, Q.; Xi, P.; Miao, J.; Li, X.; Wang, K. Applicability of wetting front advancing method in the sand to silty clay soils. *Soils Found.* **2020**, *60*, 1215–1225. [CrossRef]

fluids

MDPI

Article

An Experimental Study on the Effect of Distance and Sheltering Area of a Group of Linearly Arranged Sacrificial Piles on Reducing Local Scour around a Circular Bridge Pier under Clear-Water Conditions

Subodh Guragain and Norio Tanaka *

Graduate School of Science and Engineering, Saitama University, 255 Shimo-Okubo, Sakura Ku, Saitama Shi 338-8570, Japan; subodhguragain2015@gmail.com

* Correspondence: tanaka01@mail.saitama-u.ac.jp

Abstract: One of the major problems associated with bridge piers is ensuring their safety against local scouring caused by the erosive action of flow. Numerous countermeasures have been developed and tested to solve this problem, among which sacrificial piles are highly recognized due to their high performance, economy, durability, and ease of construction. Several factors affect the performance of sacrificial piles, such as their number, size, degree of submergence, and geometric arrangement parameters. In this study, the performance of a group of linearly arranged cylindrical sacrificial piles in reducing local scour around a circular bridge pier was investigated by varying the number of piles (or sheltering area) and distance between piles and the pier under clear-water conditions. Three values of distance between piles and the pier and three values of sheltering area (or number of piles) were tested. The efficiencies of sacrificial piles in different configurations were presented in terms of the percentage reduction in maximum scour depth at an unprotected pier under the same hydraulic conditions. The results of this experiment show that when linearly arranged sacrificial piles are placed close to the pier (at distance D; D is the pier diameter), an increase in the number of piles (or sheltered area) results in an increased scour depth, and when placed far from the pier ($2D$ and $3D$), an increase in the number of piles results in a decrease in scour depth around the pier. In addition, for 40% and 60% sheltering conditions, scour depth increased with an increase in the spacing between piles and the pier, while for 80% sheltering conditions, optimum protection was observed at a distance of $2D$. Overall, two piles placed at distance D provided optimum protection with a scour depth reduction of 41.6%, while minimum protection was recorded when the same were placed at a spacing of $3D$ from the pier (25.5%).

Keywords: bridge pier; sacrificial piles; linear arrangement; sheltered area; scour depth; contour; horseshoe vortex; wake region; boundary layer

1. Introduction

Bridge piers located within the waterway of a river are often subjected to the erosive action of the flowing water. Local scour, in its simplest form, can be defined as the process of the removal of bed material from the vicinity of the pier by the local flow field components. Removal of bed materials from around the bridge pier can cause buckling failure of the pier [1], thereby leading to huge damage to properties and the loss of human lives. Historical evidence shows that more than half of the bridge failures around the world are due to scours [2]. Hence, local scour is recognized as one of the major problems in hydraulic engineering. Local scour at bridge piers is a result of three-dimensional boundary-layer separation leading to the formation of a scour hole at the base of a pier. The main components of the local flow field include the downflow, horseshoe vortex, flow separation, wake vortices, and surface rollers (bow wave) [3]. Whenever a blunt body such as a pier is placed in a river, the velocity of approach flow becomes zero at the front face of the pier, known as the stagnation point, due to which the fluid pressure increases at the pier's front face. Since the velocity of flow decreases from the surface to the bottom along the flow depth, the pressure also decreases in the same fashion, thereby developing a vertical pressure gradient, which is responsible for the downflow. The downflow, acting as a vertical jet of water, hits the bed and rolls upwards, which, upon meeting the downward flow, leads to a rolling flow that exerts shear stress on the soil grains. This shear stress, upon exceeding the shear resistance of the bed material, causes the removal of the bed soil, and consequently, a horseshoe vortex is formed at the front bed, which is the most dominant feature in the process of the formation of a scour hole. The boundary-layer separation on either side of the pier leads to the formation of wake vortices on either side behind the pier, and these wake vortices are responsible for the removal of bed sediment from the back of the pier by suction action. The periodical shedding of the wake eddies leads to the transportation of sediment particles, which are deposited at some distance behind the pier. These three vortices acting together are responsible for the development and expansion of a scour hole around the pier [3]. Past findings have shown that the local scour is a complex phenomenon affected by several parameters related to flow characteristics, bed sediment characteristics, pier geometry characteristics, and time [4].

The control of local scouring around bridge piers has been the field of interest of many researchers, and several countermeasures have been developed and tested for this purpose. These countermeasures involve strengthening the bed soil against the erosive action of the flow (bed armoring devices/passive countermeasures) and reducing the erosive force of the flow (flow altering devices/active countermeasures). One of the popular countermeasures is the use of sacrificial piles upstream of a bridge pier. These piles of small diameter placed upstream of a bridge pier are found to reduce the local scouring around the pier by deflecting the high-velocity flow and creating a low-velocity wake region behind them. The effectiveness of these sacrificial piles in reducing the local scour depends on the number of piles, the size of the piles relative to the pier, the protrusion of the piles, and the geometric arrangement of the piles [5].

Laboratory studies on mitigating local scour around bridge piers by using sacrificial piles have been reported by several researchers in the past by adopting different shapes, sizes, flow parameters, and geometric arrangements of sacrificial piles. Ref. [6] conducted a laboratory experiment to study the effect of gap width and pile diameter on the efficiency of a single sacrificial pile placed in front of a bridge pier. Their results showed that when a single sacrificial pile was placed close to the pier (at $1D$; D = pier diameter), the highest protection was obtained, while the result of increasing the pile diameter at the optimum distance of $1D$ was a reduction in the protection provided by the sacrificial pile. Ref. [7] conducted experiments using single sacrificial piles of different diameters placed in front of piers of diameters (D) 5 cm and 10 cm at the distance of $2D$ and $3D$. According to their results, when the distance between piles and pier was increased, the countermeasure effect of sacrificial piles decreased for both pier sizes. In addition, the scour depth reduction effect of sacrificial piles increased with the diameter of the piles (d) up to $d/D = 0.40$, and a slightly decreasing trend was observed for d/D between 0.4 and 0.8. Ref. [8] studied the effect of blockage ratio (or sheltering area) on the efficiency of a group of sacrificial piles linearly arranged at a gap width of two times the pier diameter (D) in front of a circular cylinder. Their results showed that at a 60% blockage ratio using three piles, the scour-reduction efficiency was 32.2%, while at a 100% blockage ratio using five piles, the efficiency was reduced to 5.5%. Other values of blockage ratio and the effect of distance between sacrificial piles and the pier were not investigated. Ref. [9], from their experimental results, found that in the case of complex piers, three collinear piles in the transverse direction of a stream can provide better protection than other numbers and arrangements of piles, and the optimum distance for placing the piles was double the outer width of the complex piers. Ref. [5] conducted an extensive study using a triangular arrangement of sacrificial piles and found that a triangular configuration pointing upstream, with a distance of the apex from the upstream face of the pier equal to two times the diameter of the pier, a wedge angle equal to 30°, spacing between piles equal to 0.67 times the pier diameter, and a pile diameter equal to one-sixth of the pier diameter, resulted in a 56% reduction in the maximum scour depth around the pier for an aligned flow under clear-water conditions. Ref. [10] performed numerical simulations to evaluate the effect of the size and spacing of sacrificial piles from the pier on reducing local scour. They reported that maximum reduction was observed at distance $X = 3.5D$ (D is the pier diameter) and pile diameter (d) = $(1/3)D$. Ref. [11] conducted an experiment on reducing local scour using permeable sacrificial piles and found that the scour depth around sacrificial piles itself can be significantly minimized by using permeable sacrificial piles, while the reduction in local scour around the pier was same as that of a solid pile. The optimum protection with a permeable sacrificial pile was observed when the pile was placed at a distance of three times the pier diameter. Ref. [12] undertook numerical and experimental research to determine how well sacrificial piles can work to reduce local scour around pile groups. Their results showed that sacrificial piles performed better when applied to protect a group of service piles than when applied to a single pier. Ref. [13] studied the effect of debris accumulation at sacrificial piles on the bridge pier scour and reported an increase in the local scour depth due to debris trapping.

In past studies, the relationship between the sheltering area and the distance between linearly arranged sacrificial piles was not fully established in the case of an isolated circular pier. The efficiency of sacrificial piles, linearly arranged in front of a pier, is likely to depend on the number of piles as well as on the gap width between piles and the pier. To obtain optimum protection from sacrificial piles, a proper understanding of these parameters associated with their functioning is essential. Although some studies can be found on the linear arrangement of sacrificial piles, they lack sufficient information to relate the effect of sheltering area and gap width to the scour-reduction efficiency. The present study aims to elucidate the relationship between sheltering area and distance (gap width) on the efficiency of a linearly arranged group of sacrificial piles in protecting a circular bridge pier against local scouring by changing both the sheltered area provided by the piles to the pier (i.e., pile numbers) and the distance of the pile group from the pier under clear-water conditions. A single-pier scour experiment was used as a control experiment, and sacrificial piles in emergent conditions were used. The percentage reduction in maximum scour depth using sacrificial piles under a variety of configurations was determined, and the effects of distance and sheltered area were analyzed and are herein presented. The results of this experiment are expected to help in better understanding the performance of linearly arranged sacrificial piles.

2. Materials and Methods

2.1. Experimental Facilities, Materials Selection, and Installation

Each experiment in this research was performed on a horizontal bed flume of 5 m length having a rectangular cross-section of 0.7 m × 0.5 m (channel width (B) × channel height (H)) and located in the hydraulic engineering laboratory at Saitama University, Japan. The sides of the flume were made of glass panels to facilitate observation during the experiment, and a 5.5 kW centrifugal pump (manufactured by Tokyo Ebara Corporation, Tokyo, Japan) recirculated water between a head tank and a sump tank attached to either end of the flume. A manually operated gate valve was used to control the flow into the channel, on which graduations were made to mark the desired openings. On the upstream and downstream parts of the channel, wooden false beds were placed to create a 2.355 m long and 18.4 cm deep recess in the middle part of the flume length for placing the sediment, as shown in Figure 1. A tailgate was mounted at the end of the downstream wooden bed to maintain the desired flow depth. Honeycomb grids were placed at the outlet of the head tank to minimize inflow turbulence, thereby ensuring smooth flow into the channel. An electromagnetic velocity meter (EVM) (manufactured by KENEK Co., Ltd., Tokyo, Japan) was used to measure the inflow velocity of the water in both streamwise and cross-stream directions up to an accuracy of +/− 2%. The device had a main unit VM2000 and detector unit VMT2-200-04P with an L-type probe. A laser displacement gauge (LK-500, manufactured by KEYENCE International, Osaka, Japan) was used to digitize the sand-bed level before and after each test. Scour depths at key points were confirmed using a point gauge. A mercury thermometer was used to measure the temperature of the water during the experiment, and the average water temperature was 24 °C.

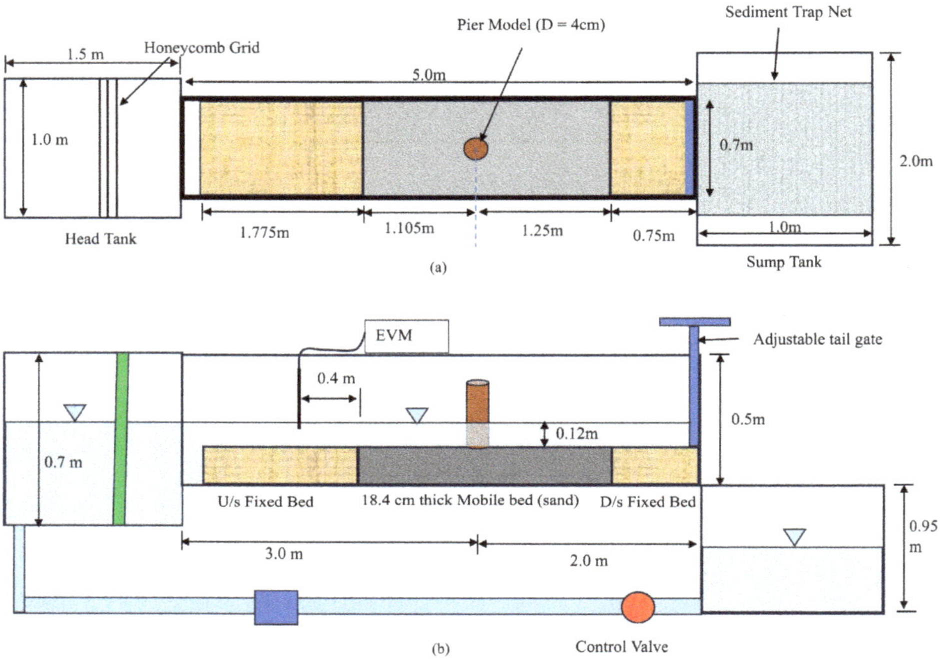

Figure 1. Schematic of the experimental setup: (**a**) plan view; (**b**) side view.

In order to obtain maximum scour depth under clear-water conditions, experiments should be performed using non-ripple-forming sediments with median grain size $d_{50} > 0.7$ mm [14]. This is because sand particles smaller than that are found to form bedforms such as ripples, which reduce the scour depth. In addition, uniform sediments with geometric standard deviation of grain size $(\sigma_g) < 1.3$ are recommended by [15] to avoid the armoring effect caused by sediment non-uniformity. Since the maximum scour depth for uniform sediment occurs near threshold conditions, the experiments should be performed with the mean flow velocity (V) slightly less than the critical velocity (Vc) corresponding to incipient motion of bed materials ($0.9 < V/Vc < 1$). In order to neglect the effect of flow deflections from the side walls of the flume, $B/D \geq 10$ is recommended [16]. According to [4], scour depth reduces when $D/d_{50} < 8$ because, in that case, the bed sediments are significantly coarser compared to the pier size, while [17] found a reduction in scour depth for prototype-sized piers when $D/d_{50} > 50$. Hence, a value of $D/d_{50} = 25$ to 50 is recommended to neglect the influence of sediment coarseness on scour depth. For smaller values of y/D (y = flow depth, D = pier diameter), the surface roller weakens the downflow, and a value of $y/D \geq 2.5$ is needed to neglect the flow shallowness effect on the pier scour experiments [18]. All these past results were considered while selecting experimental parameters for the present experiment.

Hence, the recess was filled with silica sand no. 4 (manufactured by Mikawa Silica Co., Ltd., Aichi, Japan) having a mean grain size (d_{50}) of 0.87 mm and a geometric standard deviation (s.d.) of grain size distribution ($\sigma_g = \left(\frac{d_{84}}{d_{16}}\right)^{0.5}$; d_{84} and d_{16} = grain size such that 84% and 16% particles are finer than that size, respectively) equal to 1.27. A constant flow depth (y) of 12 cm was selected for all experimental cases, and a wooden cylinder of diameter (D) = 4 cm was selected for the pier model considering the criteria for side wall effect, flow shallowness, and sediment coarseness parameters mentioned above. The selection of various parameters and dimensions is summarized in Table 1.

Table 1. Summary of selection of experimental parameters.

Selection	Parameters	Selection Criteria	Suggested By	Purpose	Selected Value	Remarks
Sand	Mean size (d_{50})	$d_{50} > 0.7$ mm	[14]	To avoid ripple formation	$d_{50} = 0.87$ mm	Silica sand no. 4
	Geometric s.d. (σ_g)	$\sigma_g < 1.3$	[15]	To avoid the armoring effect	$\sigma_g = 1.27$	
Pier	Diameter (D)	$B/D \geq 10$	[16]	To avoid side wall effect	$D = 4$ cm	$B = 70$ cm
		$D/d_{50} = 25$–50	[4]	To avoid the effect of sediment size	$D = 4$ cm	
Flow	Depth (y)	$y/D \geq 2.5$	[18]	To avoid the effect of flow shallowness	$y = 12$ cm	
	Velocity (V)	$0.9 < V/Vc < 1$	[8]	Maximum scour under clear-water conditions	0.92	

2.2. Flow Conditions

All the experiments in this study were performed under steady, uniform, subcritical, and turbulent flow conditions with a constant flow depth of 12 cm. Since the maximum scour depth for uniform sediments occurs at near-threshold conditions, first, the critical velocity causing the threshold of sediment motion was determined. The critical shear velocity (u^*_c) corresponding to the incipient motion condition was first approximated theoretically using Shield's diagram. A useful approximation to Shield's diagram for quartz sediments in water was given by [19] as follows:

$$u^*_c = 0.0115 + 0.0125\, {d_{50}}^{1.4}\ (0.1\text{ mm} < d_{50} < 1\text{ mm}) \quad (1)$$

where u^*_c is in m/s, and d_{50} is in mm. The corresponding mean velocity (V_C) was determined using logarithmic velocity distribution suggested by [15]:

$$\frac{Vc}{u_c^*} = 5 \cdot 75log\left(5 \cdot 53\frac{y}{d_{50}}\right) \quad (2)$$

The theoretical calculation showed that the incipient motion condition was likely to occur at a flow velocity of 36.1 cm/s. For validating the theoretically obtained value of critical velocity experimentally, the 18.4 cm deep recess was filled by silica sand no. 4 without placing the pier model, and the sand bed was properly leveled. At the beginning of the experiment, the sand bed was fully saturated by applying a very small discharge without disturbing the sand grains. The tailgate was fully closed, and the flow depth of 12 cm was established without disturbing the mobile bed. The discharge into the channel was then gradually increased while maintaining the flow depth of 12 cm by adjusting the tailgate opening. The incipient condition of the sand grains was confirmed by visual observation. The average flow velocity at this stage was measured using an EVM. The EVM was always placed at a fixed distance of 40 cm from the beginning of the mobile bed, and the measurements were always made in the middle of channel section at mid-flow depth to avoid any errors due to the measurement location. Since the EVM readings fluctuated over time, the readings were recorded for 1 min duration and then averaged, with the average value calculated being 30.1 cm/s. This average velocity at which incipient motion was observed was lower than the theoretical approximation. This was probably because the theoretical calculation used the mean grain size for the approximation; however, the grains smaller than the average size started moving at a lower velocity. The value obtained from practical observation was selected as the critical velocity, as it was more accurate than the theoretical approximation. After that, the valve opening was slightly reduced, maintaining the flow depth so that the incipient motion condition of the sand grains was stopped. The valve opening was then marked for the rest of the experiments. At this valve opening, the average flow velocity was measured, and the average value was equal to 27.6 cm/s such that the flow intensity value (V/Vc), when calculated, was found to be 0.92. For the rest of the experimental cases, the valve opening was made exactly at that marking such that the average flow velocity was constant in all the experimental cases. By adopting a scale of 1:25, the flow conditions are summarized in Table 2 below. On the same scale, the pier diameter of 4 cm and the sacrificial piles of 8 mm diameter represented the prototype pier diameter of 1 m and the prototype piles of 20 cm diameter, respectively.

Table 2. Flow conditions adopted.

Parameters	Model	Prototype	Remarks
Flow depth (y)	12 cm	3 m	Lr (Length ratio)
Critical flow velocity (Vc)	30.1 cm/s	1.50 m/s	$Vr = Lr^{0.5}$
Mean flow velocity (V)	27.6 cm/s	1.38 m/s	$Vr = Lr^{0.5}$
Design discharge (Q)	23.2 lps	72.4 m^3/s	$Qr = Lr^{5/2}$
Froud no., Fr = $V/(gy)^{0.5}$	0.254	0.254	Froude similarity
Pier Reynold's no., Re = VD/v	11,040	1.38×10^{-6}	

2.3. Single Pier Scour Experiments

The local scour experiment for a single pier without placing sacrificial piles was performed at a flow intensity of $V/Vc = 0.92$. The results of this experiment served as the control experiment for scour experiments with sacrificial piles. For this experiment, the pier model was placed in the middle of the mobile bed, and the sand bed was properly leveled. Very thin masking tapes were attached to the four faces of the pier model before placing it, to facilitate the scour depth readings during and after the experiment. The sand bed was digitized using a laser displacement gauge before the experimental run. The same procedure of increasing the flow discharge while maintaining the flow depth (as explained in incipient motion test above) was followed until the valve opening reached the previously marked position corresponding to the desired average velocity (27.6 cm/s). The flow velocity was confirmed from the EVM readings each time. The first test was performed for a total duration of 3.5 h, and the instantaneous maximum scour depth readings were recorded at 0.5, 1, 2, 3, 4, 5, 6, 7, 8, 9, 10, 11, 12, 15, 30, 45, 60, 90, 120, 150, 180, and 210 min. An inspection camera capable of working under water was used to record and read the instantaneous scour depth values from the graduated tape attached to the pier model. The device had a small camera fitted at the tip of a long cable, which was connected to the main component where visuals could be seen on a screen. The camera also facilitated a closer look at the scour process. The temporal scour evolution data showed that after 3 h, the increase in scour depth was very small, and hence, an experimental duration of 3 h was adopted for all the remaining experiments in this study. For all experimental cases, after 3 h, the tailgate opening was gradually reduced, and the valve was closed accordingly, maintaining the flow depth such that when the valve was fully closed, the tailgate was also fully closed. At the end, the tailgate was slowly opened to drain the water without disturbing the scour and deposition geometry. A perforated drainage pipe placed at the bottom of the sand bed was used to drain the water from the sand. Finally, the scour hole and deposition zone geometry were scanned using the laser scanner at 2 cm intervals along the length, and the scour hole profile and cross-section data as well as scour depths at key points were also taken using a point gauge for cross-verification. The graduated tape readings were also recorded.

To understand the effect of flow intensity on scour depth, more tests using flow intensity values of 0.88, 0.78, and 0.68 were also conducted, each for a duration of 3 h. Hence, the experimental cases with a single pier are summarized in Table 3 below.

Table 3. Single pier experimental cases.

Experimental Case Name	Pier Diameter (*D*) [cm]	Flow Depth (*y*) [cm]	Flow Intensity (*V*/*Vc*)
S-0.92	4.0	12.0	0.92
S-0.88	4.0	12.0	0.88
S-0.78	4.0	12.0	0.78
S-0.68	4.0	12.0	0.68

In the naming of the experimental cases, "S" refers to the single-pier scour experiment, and the numeric value after it refers to the flow intensity at which the experiment was conducted. For example, "S-0.92" refers to the scour experiment with a single pier at flow intensity value of 0.92.

2.4. Scour Experiments with Linear Sacrificial Piles

Experiments were conducted by placing a linearly arranged group of sacrificial piles in front of the pier. Two main parameters associated with the sacrificial piles were varied: (a) the number of piles, i.e., N (or sheltered area, As), and (b) the distance of the piles from the face of the pier (X). In total, nine different cases consisting of a pier model protected by sacrificial piles were studied. All the experiments were performed at flow depth y = 12 cm and flow intensity V/Vc = 0.92, similar to the control experiment. The sacrificial piles were modeled by using 8 mm diameter wooden cylinders in emergent condition in all the experimental cases. The same experimental procedure that was followed in the case of a single-pier scour experiment was adopted. The efficiency of a sacrificial pile arrangement was determined in terms of % decrease in the value of the maximum depth of local scour near the pier with sacrificial piles in comparison with the maximum scour depth in the case of an unprotected pier in the control experiment (V/Vc = 0.92).

$$\% \text{ efficiency } (E) = 100 \times (ds_1 - ds_2)/ds_1 \quad (3)$$

where

ds_1 = maximum scour depth in the case of a single pier in the absence of protection;
ds_2 = maximum scour depth at the pier with sacrificial pile configuration.

Figure 2 shows the general layout of the sacrificial pile arrangement.

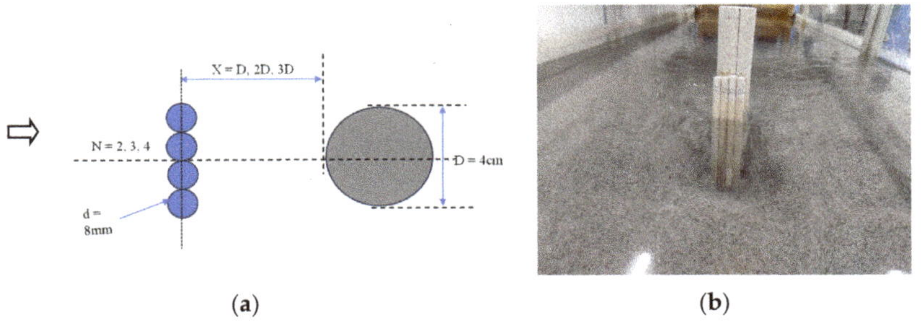

Figure 2. Scour experiment with a linear arrangement of sacrificial piles: (**a**) general layout; (**b**) a picture of the experimental run.

The experimental cases that were performed with linear arrangements of sacrificial piles are presented in Table 4. Case 1 to case 9 were performed with sacrificial piles placed in front of the pier. Cases 10, 11, and 12 were performed with sacrificial piles only (without placing the pier) to have a better understanding of the scour and deposition characteristics of the linear arrangement of sacrificial piles.

Table 4. Experimental cases with linear arrangements of sacrificial piles.

Experimental Case Name	No of Piles (N)	Sheltered Area (As) [%]	Diameter. of Pile (d) [mm]	Distance of Piles from the Face of the Pier (X)	Remarks
L-P2-D	2	40	8	D	
L-P2-2D	2	40	8	2D	
L-P2-3D	2	40	8	3D	
L-P3-D	3	60	8	D	
L-P3-2D	3	60	8	2D	
L-P3-3D	3	60	8	3D	
L-P4-D	4	80	8	D	
L-P4-2D	4	80	8	2D	
L-P4-3D	4	80	8	3D	
L-P2	2		8		Piles only
L-P3	3		8		Piles only
L-P4	4		8		Piles only

The sheltering area (*As*) provided by the sacrificial piles is calculated as follows:

$$As = 100 \times (N \times d)/D \quad (4)$$

In the naming of the experimental cases, "L" refers to the linear arrangement of sacrificial piles; "P2", "P3", and "P4" refer to the number of piles used in the configuration; and "D", "2D", and "3D" refer to the distance of the piles from the upstream face of the pier. For example, the configuration name "L-P2-D" refers to a linear arrangement of sacrificial piles consisting of two piles placed at a distance *D* from the upstream face of the pier.

3. Results and Discussion

3.1. Single-Pier Scour Experiments

Figure 3 illustrates the outcome of the temporal scour evolution test for a single pier. From the plot, it can be seen that there was a rapid development of a scour hole during the initial period, and about 80% of the maximum scour depth was attained in the first 30 min. Results also show that after 3 h from the beginning of the experiment, the increase in scour depth was very small. During the experiment, the first sign of movement of bed material was seen behind the pier due to the lee-wake vortices. However, the formation of the scour hole began once the horseshoe vortex was formed. The horseshoe vortex stretched from the sides to the front of the pier, and there was a rapid growth in scour depth at the front. A rapid entrainment of the bed materials due to the formation of a horseshoe vortex was observed, and the scour hole increased in all three dimensions. With the increase in scour depth, the strength of the horseshoe vortex gradually decreased, due to which there was a fall in the rate of expansion of scour hole as time elapsed. Finally, the horseshoe vortex was no longer capable of removing a significant volume of bed materials, and equilibrium was assumed to be reached, although the scouring never stopped. The lee-wake vortices, formed behind the cylinder as a result of the horizontal separation of the boundary layer, acted like tornadoes, lifting the bed particles from behind the pier by suction action, resulting in sediment entrainment into the main flow. The entrained bed materials were then carried downstream due to the periodical vortex shedding behind the pier and deposited in the wake zone at the back of the pier, forming a dune. The size of the dune behind the pier increased with time as the total volume of scoured materials increased over time. The wake vortices were also responsible for the expansion of the scour hole behind the pier. During the scour process, the horseshoe vortex was the most dominant feature and acted at the front toe of the pier, and hence, maximum scour depth was observed near the front face of the pier. The wake vortices, due to their lower strength, acted mostly on displacing the entrained sediment, and consequently, scour depth was minimum at the back of the pier. Due to symmetry of the arrangement, the scour depths at the two sides of the pier were almost equal.

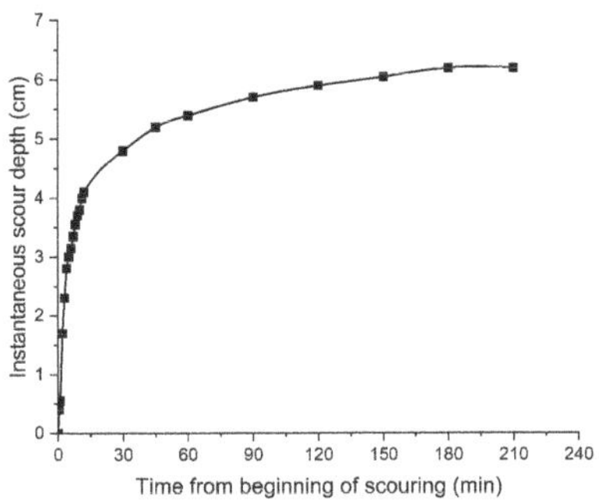

Figure 3. Temporal scour evolution of single pier at $V/Vc = 0.92$.

After the temporal scour experiment, a single-pier scour experiment of 3 h duration at flow intensity $V/Vc = 0.92$ was conducted as a control test for sacrificial pile scour experiments based on the result of the temporal scour evolution experiment. Additional single-pier tests were conducted at three additional flow intensities (0.68, 0.78, and 0.88) to study the effect of flow intensity on scour depth. From the experiment, it was found that the depth of local scour varied linearly with flow intensity. The higher the flow intensity, the greater the scour depth and longer the deposition length behind. This was because the strength of the horseshoe vortex depends directly on the inflow velocity. During the analysis, the maximum scour depths were determined by two methods: (a) from the graduated tape pasted directly on the pier face and (b) by analyzing the data obtained from the laser scanner. In general, the maximum scour depth obtained after processing the data obtained from the laser displacement meter was slightly higher than the graduated tape readings because maximum scour depth was usually found to be located at a small distance away from the pier face rather than exactly at the pier face. The higher value of the scour depth readings obtained from graduated tape and laser scanner was recorded as the maximum local scour depth. The maximum scour depth results are tabulated in Table 5.

Table 5. Measured scour depth values of single pier scour experiments.

Experimental Case Name	Flow Intensity (V/Vc)	Measured Value of Scour Depth [cm]			Maximum Scour Depth (d_{smax}) [cm]
		Front	Back	Sides (Averaged)	
S-0.92	0.92	6.20	3.90	5.35	6.20
S-0.88	0.88	5.64	3.10	4.70	5.64
S-0.78	0.78	4.80	2.60	4.05	4.80
S-0.68	0.68	3.40	1.80	3.30	3.40

The final scour depth readings from the graduated tape are shown in Figure 4.

The scour hole profile and cross-section data from the single-pier scour experiment were plotted and are presented in Figure 5.

The variation of maximum scour depth with flow intensity and the contour diagram of scour and deposition geometry are presented in Figure 6.

In Figure 6b, XX represents the streamwise direction (cm), YY represents the cross-stream direction (cm), and ZZ represents the vertical direction (cm). The dotted contour lines represent points of equal scour depth below the original bed level, while the solid contour lines represent points of equal deposition height above the original bed level. In addition, the size (length and width) of the scour hole as well as the length and width of the deposition can also be observed from the contour diagrams. The larger white circle represents the position of the pier model, while small black circles represent the position of sacrificial piles.

(**a**)

(**b**)

(**c**)

(**d**)

Figure 4. Final scour hole depth readings at the end of each test: (**a**) $V/Vc = 0.92$, (**b**) $V/Vc = 0.88$, (**c**) $V/Vc = 0.78$, and (**d**) $V/Vc = 0.68$.

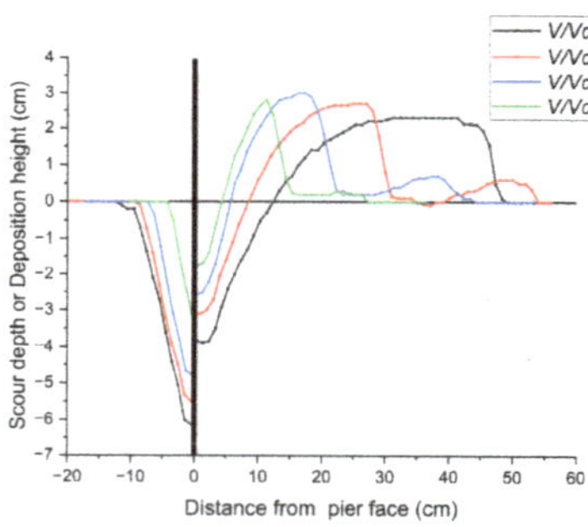

(**a**)

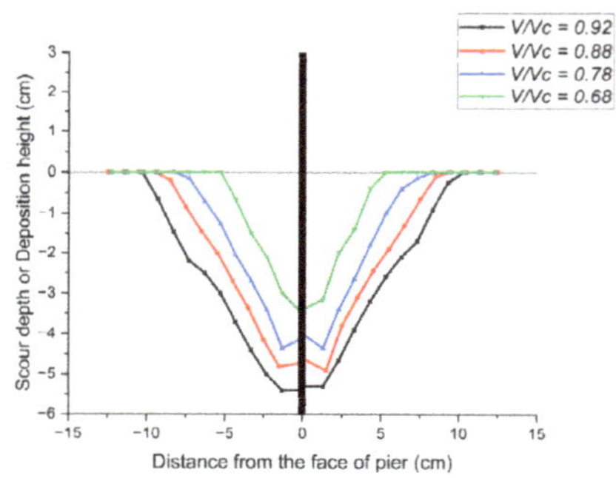

(**b**)

Figure 5. Scour and deposition plot from single pier scour experiment: (**a**) stream-wise centerline profile; (**b**) cross-section.

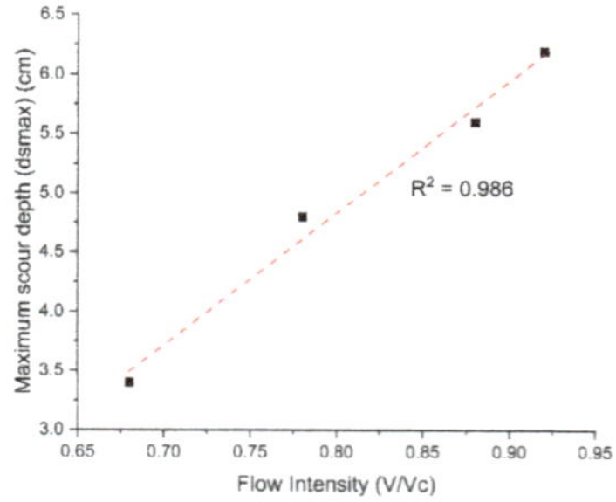

(**a**)

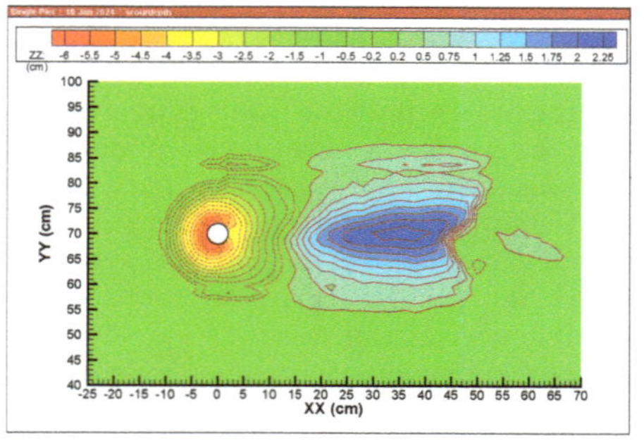

(**b**)

Figure 6. Results of single-pier scour experiments: (**a**) variation of maximum scour depth with flow intensity; (**b**) contour diagram for single-pier scour experiment at $V/Vc = 0.92$.

3.2. Linear Sacrificial Pile Scour Experiments

Experiments were conducted by placing sacrificial piles in a variety of configurations in front of the pier for protection against local scouring. Sacrificial pile configurations were made by varying the number of piles and the distance between piles and the pier, and the experimental cases are listed in Table 6. When the approach flow reached the sacrificial pile group, it was obstructed, and a local flow field was generated around the sacrificial pile group due to three-dimensional boundary-layer separation. The horseshoe vortex at the front toe of the piles and lee-wake vortices on the rear side of the piles acted together, forming a local scour hole around the pile group. The size of the scour hole around the piles increased with an increase in the number of piles as more flow was obstructed, and consequently, the size and strength of the horseshoe and wake vortices also increased. Due to the formation of a wake region behind the piles, the flow velocity was substantially reduced behind the piles. Due to this reduction in velocity, the strength of vortices (horseshoe and wake) forming at the pier located in the wake region behind the piles was also reduced. This is called the shielding effect of sacrificial piles. The shielding effect depended on the number of sacrificial piles as well as on the distance between the pier and the piles, and thus, a varying protection effect was observed for different configurations of piles.

Table 6. The efficiency of linear arrangements of sacrificial piles.

S.N.	Experimental Case Name	d_{smax} Observed [cm]	Reduction in d_{smax} Compared to Unprotected Pier [cm]	% Reduction in d_{smax}
1	L-P2-D	3.62	2.58	41.6
2	L-P2-2D	4.15	2.05	33.1
3	L-P2-3D	4.62	1.58	25.5
4	L-P3-D	3.85	2.35	37.9
5	L-P3-2D	4.05	2.15	34.7
6	L-P3-3D	4.4	1.8	29.0
7	L-P4-D	4.23	1.97	31.8
8	L-P4-2D	4.04	2.16	34.8
9	L-P4-3D	4.11	2.09	33.7

In addition to the shielding action, another action of piles was also observed during the experiment. The bed materials scoured from around the piles were carried downstream by the turbulent eddies shedding from the piles before being deposited at some distance behind the piles. When a pier was located behind the piles, the scoured bed materials reached the pier scour hole and filled it partially, due to which the scour depth around the pier was reduced. This sediment flux from piles depended on the number of piles installed. The higher the number of piles, the higher the sediment flux due to the increased scouring around the piles. The amount of sediment reaching the pier was affected by the gap width between piles and the pier. This was also a reason for the varying efficiency of sacrificial piles observed in the experiment under different configurations.

A schematic diagram showing the mechanism of local scour in the case of linear sacrificial piles scour experiment is presented in Figure 7.

The results of the sacrificial pile scour experiments were compared with the single-pier scour experiment at $V/Vc = 0.92$. In all sacrificial pile experiments, the maximum scour depth around the pier was less than that of an unprotected pier, thus justifying the protection provided by the sacrificial piles to the pier. However, it was observed that the linear arrangement of piles itself was subjected to substantial scour. The scour depth in front of the piles also exceeded that in front of the pier in many cases tested, particularly when the number of piles was high. The efficiency of each arrangement in reducing the maximum scour depth around the pier is presented in Table 6 in terms of % reduction in the maximum depth of local scour around an isolated bridge pier without the protection of sacrificial piles.

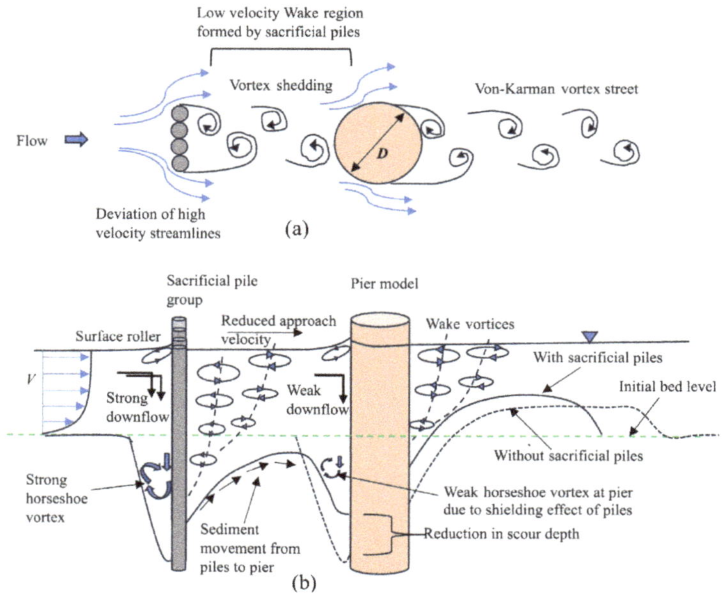

Figure 7. Flow structures and mechanism of local scour in the case of sacrificial piles with pier: (**a**) plan view; (**b**) side view.

The above calculations were performed based on the maximum scour depth of 6.2 cm recorded in the test of an unprotected, single pier. From the results, it was seen that the highest protection was obtained when two piles were linearly arranged in front of the pier at distance D (i.e., L-P2-D), while the least protection was obtained when the same was placed at distance $3D$ (i.e., L-P2-3D).

Scour and deposition profiles for all the experimental cases were plotted, and the variations in maximum scour depth with change in sheltered area (As) and spacing between piles and pier (X) were also plotted, which are presented in Figure 8.

Figure 8b shows the importance of the number of piles (or sheltering area) for a given X/D value. From the results, it can be concluded that when the sacrificial pile groups are placed close to the pier (at distance D), the maximum scour depth at the protected pier is likely to rise with a rise in the number of sacrificial cylinders (or sheltered area). This is because when the number of piles was increased, the size and strength of the wake vortices behind the piles increased significantly, which caused a deeper scour hole to be formed behind the piles due to the higher entrainment of the sediments into the flow. When the number of piles was high, and the piles were placed close to the pier (at D), the gap width was directly under the influence of those strong wake vortices, and the bed materials in the gap width were eroded by the recirculating flow, which were then transported downstream of the pier. In addition, the strong wake vortices caused high turbulence in the small gap width, which did not allow the scoured materials from around the piles be deposited in the pier scour hole; rather, they were transported downstream from the sides of the pier. Due to this action, the bed material in the gap width eroded over time, resulting in an increased scour depth in front of the pier. This shows the adverse effect of placing a large number of piles close to the pier. On the other hand, when a small number of piles (two piles) were placed at distance D, the wake vortices behind the pile group were small in size, which were not able to scour the vicinity of the pier and were also able to reduce the flow velocity behind them for the desired shielding of the pier. In addition, the materials scoured from the piles fed the scour hole around the pier properly. On the contrary, the scenario was opposite in the case of three and four piles, where the gap width D was insufficient to avoid the ill effect caused by the increased size of the wake vortices, as explained above. However, if the pile groups are placed far from the pier (at distances of 2*D* and 3*D*), the maximum scour depth around the pier decreases with an increase in the number of piles (or sheltered area). This is because when the distance between the piles and the pier was larger (2*D* and 3*D*), the wake vortices behind the piles could not affect the bed in the vicinity of the pier directly, as in the case when the gap width was *D*, and the scouring around the pier was primarily due to the horseshoe vortex formed in front of the pier. When the number of piles is increased, the size of the wake region behind the piles also increases, due to which the pier is subjected to a low-velocity near-wake region. As the flow velocity increases in the streamwise direction along a wake region, a pier located in a near-wake region will have a weaker horseshoe vortex. In addition, the volume of scouring around a higher number of piles is also higher. This increases the volume of scoured materials reaching the pier compared with a smaller number of piles. In the case of a small number of piles, the volume of scoured materials from around the piles is also small, which is mostly deposited in the gap width, and hence, fewer materials reach the pier scour hole. Due to this reason, the variation in scour depth with sheltering area showed an opposite trend at 2*D* and 3*D* compared to the case when the piles were located at distance *D*. From the slope of the plots, we can conclude that when the spacing between piles and the pier was 2*D*, the protection offered by the piles was the least sensitive to the change in sheltered area, while it was more sensitive to the sheltered area at lower and higher values of spacing (at *D* and 3*D*).

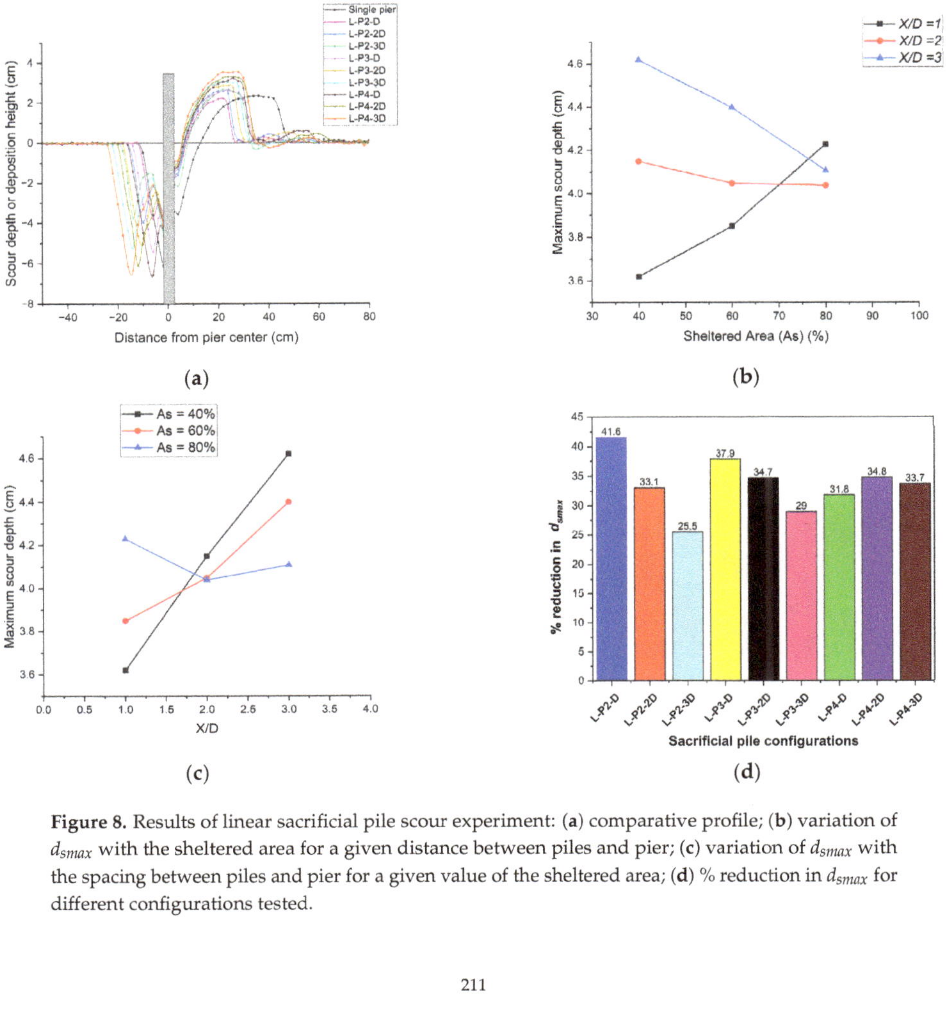

Figure 8. Results of linear sacrificial pile scour experiment: (**a**) comparative profile; (**b**) variation of d_{smax} with the sheltered area for a given distance between piles and pier; (**c**) variation of d_{smax} with the spacing between piles and pier for a given value of the sheltered area; (**d**) % reduction in d_{smax} for different configurations tested.

Figure 8c shows the importance of the distance between the pile group and the pier for a given value of the sheltered area (or pile numbers). For lower values of the sheltered area (40% and 60%), the protection provided by the pile group tended to decrease as the distance increased. This is because when the distance between piles and the pier increases while keeping the number of piles constant, the pier is subjected to a higher approach velocity as it shifts from the near-wake region to the far-wake region of the piles. In addition, when the distance is increased, materials scoured from around the piles are mostly deposited in the gap width, and hence, a smaller volume of scoured materials reach the pier scour hole to fill it. Due to this reason, the pier placed close to the piles (at distance D) experienced less local scouring in the current experiment. However, for 80% sheltering (four piles), the strong wake vortices directly influenced the pier at distance D, and hence, protection was better when the pier was placed outside the direct influence zone of the wake vortices. In this experiment, the optimum protection was observed at $X = 2D$. From the slope of the plots, it can be concluded that the efficiency of piles in reducing the depth of local scour was highly sensitive to the spacing between the piles and the pier for lower sheltering conditions, which decreased as the sheltered area was increased.

The scour and deposition contours for all sacrificial piles and pier arrangements are presented in Figure 9.

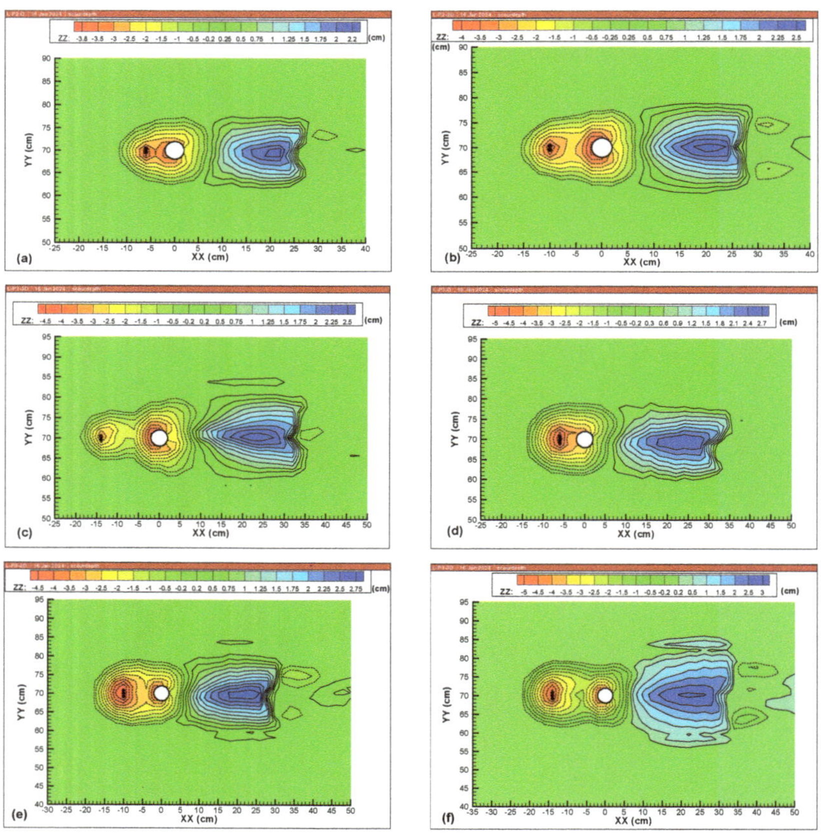

Figure 9. *Cont.*

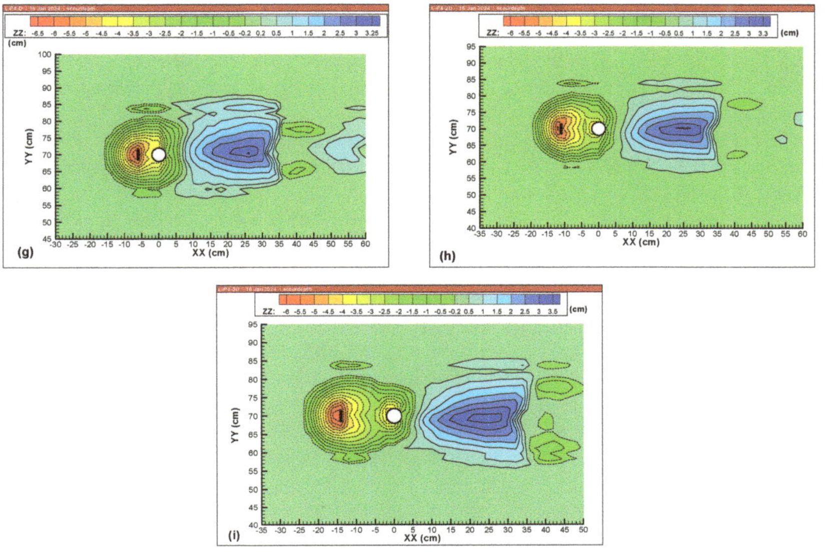

Figure 9. Contour plots of sacrificial piles experimental cases: (**a**) L-P2-D, (**b**) L-P2-2D, (**c**) L-P2-3D, (**d**) L-P3-D, (**e**) L-P3-2D, (**f**) L-P3-3D, (**g**) L-P4-D, (**h**) L-P4-2D, and (**i**) L-P4-3D.

The contour plots of linear sacrificial piles only (without placing the pier), i.e., cases L-P2, L-P3, and L-P4, are presented in Figure 10. Pier locations were placed on the resulting contour diagrams to understand the effect of sacrificial piles. The purpose of these cases was to understand the scouring and deposition patterns around the piles.

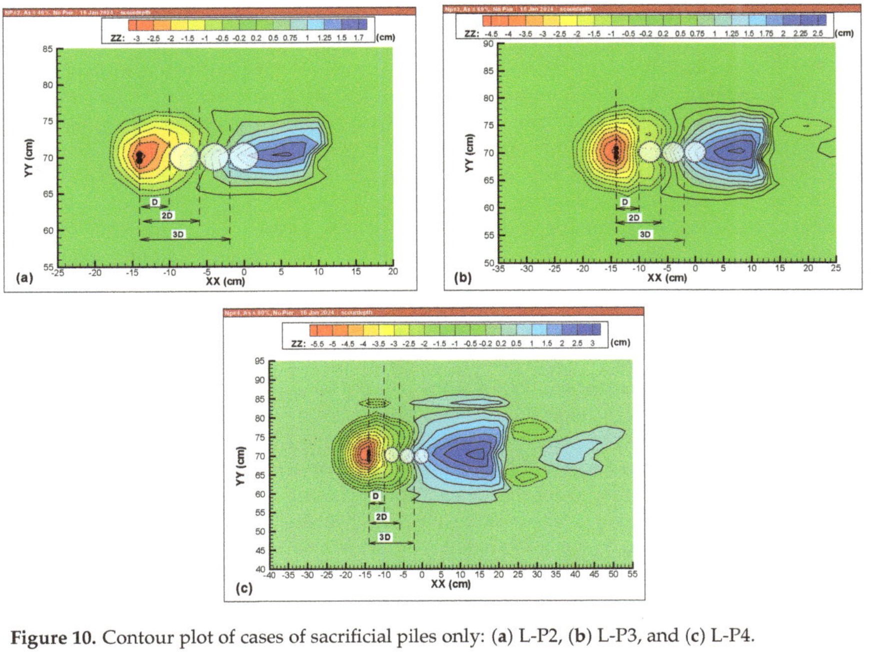

Figure 10. Contour plot of cases of sacrificial piles only: (**a**) L-P2, (**b**) L-P3, and (**c**) L-P4.

From Figure 10, it can be seen that when the pier is located in the outer region of the scour zone of sacrificial piles, the maximum scour depth around the pier tends to reduce, and when the pier is placed in the inner region of the scour zone or outside the scour region (or in the deposition zone) of the piles, the highest scour depth in the vicinity of the pier tends to increase.

The results obtained in this experiment for lower sheltering (40% and 60%) conditions are consistent with the results obtained by [6] in their scour experiments using a single sacrificial pile. Ref. [6], in their study, concluded that the optimal gap width for scour reduction using a single sacrificial pile (d = 1 cm) in front of a circular pier (D = 5.6 cm) as $1D$, for which the scour-reduction efficiency decreased as the gap width increased. In the present experiment, the highest protection was observed at $1D$ in the cases of two and three piles, and the efficiency was reduced as X increased. Ref. [20], in their numerical simulations, also concluded that the closer the sacrificial pile to the pier, the higher the protection efficiency. Ref. [7] also reported a higher percentage of scour depth reduction at the pier when a single sacrificial pile was close to the pier. In their experiment with the tandem arrangement of bridge piers, [21] also mentioned a reduction in the shielding effect when the spacing was increased beyond $1D$. These results conclude that the closer the sacrificial pile to the pier, the better the protection against local scouring. This is because when the pier was close to the sacrificial piles, the pier was in the low-velocity wake region formed by the piles, resulting in a weaker horseshoe vortex in front of the pier, while it was exposed to higher approach velocity when placed at larger distances. In addition, more scoured materials from around the piles reached the pier scour hole when the piles were placed close to the pier.

However, protection against local scouring is not only governed by the gap width but also by the size of the pile group (sheltering area). The scouring around the piles itself is another important factor, which depends on the overall size of the pile group. In the experiment of [6], when they increased the diameter of the sacrificial pile and kept the gap width equal to the optimum value $1D$, the protection decreased with increase in the sacrificial pile diameter. This result is consistent with the results obtained in the present experiment at $X = 1D$. The size and extent of scouring around the pile group increased with the increase in the number of piles. In the linear arrangement of piles in the present study, substantially high scouring around the piles was observed, particularly in cases with a larger number of piles. When the number of piles was increased while keeping the piles close to the pier (at $1D$), scour depth increased sharply around the pier, which was similar to the findings reported by the previous studies. This is because the stronger wake vortices directly affected the gap width region, resulting in scouring in front of the pier.

From the above discussions, it can be concluded that the spacing between sacrificial piles and piers should be small enough for the pier to be located within the low-velocity wake region formed by the piles (particularly in the near-wake region), while it should be large enough to avoid the high-turbulence recirculation zone behind the piles eroding the materials from the vicinity of the pier. Taking these into consideration, it can be concluded that the optimal gap width for a linear arrangement is $1D$ for sheltering areas below 60%, and the gap width should be increased for higher sheltering conditions to avoid reduced efficiency caused by the increased size of wake vortices behind the piles. In their study, [8] concluded that increasing the sheltering area from 60% to 100% in the case of linearly arranged sacrificial piles upstream of a circular cylinder at $2D$ yielded a sharp rise in the depth of the scour hole, suggesting that the gap width of $2D$ was insufficient for a 100% sheltering condition. Comparing this with the results of the present study, $2D$ can be considered the optimal gap width for more than 60% to around 80% sheltering conditions, and the gap width should be increased further for higher sheltering conditions (above 80%).

4. Conclusions

An experiment at model scale was carried out in this investigation for the purpose of studying the performance of sacrificial piles linearly arranged in a group ahead of a bridge pier on minimizing local scour phenomena in proximity to the pier. Sacrificial pile configurations with three different values of sheltering area to the pier were tested for three different values of distance between piles and the pier. A single pier without the protection of sacrificial piles was tested as a control experiment. Piles-only cases without placement of the pier were also tested. After conducting a number of experiments for the current investigation, the following results were drawn:

- Sacrificial piles linearly arranged in front of a circular bridge pier are effective in controlling the local scour around bridge piers. In the present research, their efficiency in controlling the maximum depth of local scour around the pier ranged between 25.5–41.6%;
- A configuration consisting of two piles placed at distance D was found to minimize the local scour depth in proximity of the bridge pier to the highest extent (41.6%), while the least protection was achieved when two piles were placed at a distance of $3D$ (25.5%);
- The closer the sacrificial piles to the pier, the better the protection. However, piles should be far enough apart to avoid the scouring effect of the recirculation zone behind the piles. The selection of the optimal distance depends on the sheltering area (or the number of piles);
- The efficiency of linearly arranged sacrificial piles depends on both the gap width (X) and the sheltering area (As). A gap width of $1D$ is suggested for sheltering areas below 60%, while $2D$ is suggested for sheltering areas around 80%, and a higher gap width is suggested for sheltering areas above 80%.

In this study, experiments were conducted using a circular pier and emergent sacrificial piles under steady, uniform, turbulent, and subcritical flow conditions on a uniform, cohesionless sediment bed. The sheltering area values of 40%, 60%, and 80% with the distance between piles and pier (X) values of D, $2D$, and $3D$ were adopted. Due to the limitations of available facilities and materials, the results of this experiment may have suffered some inherent scale/size effects (to some extent) associated with laboratory flume experiments of local scouring, as stated by [18]. However, the authors of this study tried to minimize this by following the suggestions provided by past researchers [4,8,14–16,18] during the selection of experimental parameters (explained in Table 1) and believe that the results are a closer representation of the actual phenomenon occurring in the field, thereby providing a sufficient idea about the efficiency of sacrificial piles. Considering the availability of experimental facilities and materials, future experiments using prototype-sized models, leading to more accurate geometric similitude between model and prototype, are recommended for higher accuracy in the assessment of local scour depths. In addition, the efficiency of sacrificial piles is also affected by the variation in thickness of the oncoming boundary-layer flow. In this experiment, due to the limitations of the experimental facility, measurement of boundary-layer thickness (δ) was not performed. Future studies are recommended with the measurement of flow velocity variation along the flow depth to determine the boundary-layer thickness at the locations of sacrificial piles and the pier and to understand the effect of boundary-layer thickness on the performance of sacrificial piles.

Future experiments are also suggested to be performed considering different pier shapes (e.g., rectangular, elliptical, sharp nose, etc.) to understand the performance of linear sacrificial piles on different pier shapes. In addition, submerged piles with different degrees of submergence can be studied to understand the effect of pile submergence. Likewise, a greater number of flow intensity values (low to high) and flow skewness conditions can be studied for understanding the performance under such flow conditions. A wider range of sheltering area values and wider range of values of distance between piles and the pier are recommended to be tested for a better understanding of the functioning of linear sacrificial piles. For each parameter investigated, statistical analysis should be conducted to confirm whether the difference is meaningful or not for the future study.

Author Contributions: Conceptualization, S.G. and N.T.; methodology, S.G.; model preparation S.G., investigation S.G., data processing and analysis, S.G.; writing—original draft, S.G.; visualization, S.G.; resources, N.T.; writing—review and editing, N.T.; supervision, N.T.; project administration, N.T.; funding acquisition, N.T. All authors have read and agreed to the published version of the manuscript.

Funding: This research received no external funding.

Institutional Review Board Statement: Not applicable.

Informed Consent Statement: Not applicable.

Data Availability Statement: Data are contained within the article.

Acknowledgments: The authors would like to acknowledge the support of the Japan Development Services-JDS Scholarship. The authors also acknowledge the anonymous reviewers for their valuable comments to improve this manuscript. I also sincerely thank Yoshiya Igarashi, Sai Seisin, Shohail Iqbal, and Dipak Negi for their continuous support of my studies.

Conflicts of Interest: The authors declare no conflicts of interest.

References

1. Hughes, D.; Ramey, G.E.; Hughes, M.L. Effects of Extreme Scour and Soil Subgrade Modulus on Bridge Pile Bent Buckling. *Pract. Period. Struct. Des. Constr.* **2007**, *12*, 96–108. [CrossRef]
2. Wardhana, K.; Hadipriono, F.C. Analysis of Recent Bridge Failures in the United States. *J. Perform. Constr. Facil.* **2003**, *17*, 144–150. [CrossRef]
3. Raudkivi, A.J. Functional Trends of Scour at Bridge Piers. *J. Hydraul. Eng.* **1986**, *112*, 1–13. [CrossRef]
4. Melville, B.W. The Physics of Local Scour at Bridge Piers. In Proceedings of the Fourth International Conference on Scour and Erosion 2008, Tokyo, Japan, 5 November 2008.
5. Melville, B.W.; Hadfield, A.C. Use of Sacrificial Piles as Pier Scour Countermeasures. *J. Hydraul. Eng.* **1999**, *125*, 1221–1224. [CrossRef]
6. Kharbeche, M.; Williams, P.; Balachandar, R. Effect of Gap Width and Pile Diameter on Efficacy of Sacrificial Piles for Scour Mitigation. In Proceedings of the International Conference on Scour and Erosion, Washington, DC, USA, 18–20 October 2021.
7. Kayatürk, Y.; Gö, M. Application of Sacrificial Piles to Control Scouring Around Bridge Piers. In Proceedings of the 3rd International Conference on Scour and Erosion (ICSE-3), Amsterdam, The Netherlands, 1–3 November 2006.
8. Tafarojnoruz, A.; Gaudio, R.; Calomino, F. Evaluation of Flow-Altering Countermeasures against Bridge Pier Scour. *J. Hydraul. Eng.* **2012**, *138*, 297–305. [CrossRef]
9. Elngar, M.A.; AboulAtta, N.M.; Saad, N.Y.; El-Molla, D.A. Effect of Sacrificial Piles on Local Scour around Complex Bridge Piers. *Ain Shams Eng. J.* **2024**, *15*, 102410. [CrossRef]
10. Gamboa, J.; Quintanilla, V.; Guzmán, E.; Santos, S. Evaluation of Diameter and Distance of Sacrificial Piles from the Bridge Pier in Reducing Local Scour. *Int. J. Struct. Civ. Eng. Res.* **2021**, *10*, 113–118. [CrossRef]
11. Qi, H.; Chen, G.; Zou, W.; Yuan, T.; Tian, W.; Li, J. Characteristics and Mechanism of Local Scour Reduction around Pier Using Permeable Sacrificial Pile in Clear Water. *Water* **2022**, *14*, 4051. [CrossRef]
12. Wang, C.; Liang, F.; Yu, X. Experimental and Numerical Investigations on the Performance of Sacrificial Piles in Reducing Local Scour around Pile Groups. *Nat. Hazards J. Int. Soc. Prev. Mitig. Nat. Hazards* **2017**, *85*, 1417–1435. [CrossRef]
13. Park, J.H.; Sok, C.; Park, C.K.; Kim, Y.D. A Study on the Effects of Debris Accumulation at Sacrificial Piles on Bridge Pier Scour: I. Experimental Results. *KSCE J. Civ. Eng.* **2016**, *20*, 1546–1551. [CrossRef]
14. Raudkivi, A.J.; Ettema, R. Clear-Water Scour at Cylindrical Piers. *J. Hydraul. Eng.* **1983**, *109*, 338–350. [CrossRef]
15. Melville, B.W.; Sutherland, A.J. Design Method for Local Scour at Bridge Piers. *J. Hydraul. Eng.* **1988**, *114*, 1210–1226. [CrossRef]
16. Chiew, Y.M.; Melville, B.W. Local Scour around Bridge Piers. *J. Hydraul. Res.* **1987**, *25*, 15–26. [CrossRef]
17. Sheppard, D.M.; Odeh, M.; Glasser, T. Large Scale Clear-Water Local Pier Scour Experiments. *J. Hydraul. Eng.* **2004**, *130*, 957–963. [CrossRef]
18. Ettema, R.; Melville, B.W.; Barkdoll, B. Scale Effect in Pier-Scour Experiments. *J. Hydraul. Eng.* **1998**, *124*, 639–642. [CrossRef]
19. Melville, B.W. Pier and Abutment Scour: Integrated Approach. *J. Hydraul. Eng.* **1997**, *123*, 125–136. [CrossRef]
20. Zhang, Z. Study on Local Scour Protection Countermeasures of Bridge Pier. *Highlights Sci. Eng. Technol.* **2022**, *28*, 464–470. [CrossRef]
21. Malik, R.; Setia, B.; Banik, A. Mechanism of Scouring Around Group of Bridge Piers in Tandem Arrangement. *J. Inst. Eng. India Ser. A* **2023**, *104*, 633–642. [CrossRef]

energies

MDPI

Article

Study on the Variation Law of Water Consumption Rate of Cascade Hydropower Station under Different Conditions

Xueshan Ai [1,2,3], Xiaoke Zhang [2], Renwei Guo [1], Xiao Li [1], Mian Wang [1], Fengshuo Liu [2,*] and Jiawei Zhang [2]

1 Hubei Key Laboratory of Intelligent Yangtze and Hydroelectric Science, China Yangtze Power Co., Ltd., Yichang 443000, China; xsai@whu.edu.cn (X.A.); guo_renwei@ctg.com.cn (R.G.); li_xiao2@ctg.com.cn (X.L.); wang_mian@ctg.com.cn (M.W.)
2 State Key Laboratory of Water Resources Engineering and Management, Wuhan University, Wuhan 430072, China; 2024282060102@whu.edu.cn (X.Z.); 2020302191178@whu.edu.cn (J.Z.)
3 Hubei Key Laboratory of Water System Science for Sponge City Construction, Wuhan University, Wuhan 430072, China
* Correspondence: 2023282060134@whu.edu.cn

Abstract: The water consumption rate for hydropower station power generation is an important index of the economic operation of hydropower stations, which has been widely considered and applied in the production practice of hydropower stations and has great significance for the full utilization of hydroelectric resources. This paper takes A and B cascade hydropower stations as the research objects to study the variation patterns of the water consumption rate of hydropower stations under different conditions, which mainly includes the following: (1) in this study, the relationship curve of output–gross head–flow ($NH_{gross}Q$) of each unit was established based on the relationship curve of output–net head–flow (NHQ) and the flow–head loss curve of each unit; (2) the mirror vertical shift method was used to calculate and plot the relationship curve of gross head–flow for combinations of multiple units; (3) the outflow–water–output characteristic curve of the hydropower station under the conditions of different reservoir water levels, different discharge flow, or different power generation output under the optimal operation of the hydropower station were calculated and drawn; (4) the outflow–water level–output curve of the hydropower station under various conditions was converted into an outflow–water level–water consumption curve; (5) the variation characteristics of the water consumption rate of the hydropower station under different conditions were analyzed; (6) on the basis of the definition of the water consumption rate of cascade hydropower stations, the research on the variation in the water consumption rate of cascade hydropower stations was carried out under the conditions of different water levels of the A reservoir, different water levels of the B reservoir, different discharge, and different total output of the hydropower stations. The variation pattern of the water consumption rate of hydropower stations under different conditions obtained in this paper can provide technical support for the real-time economic operation of cascade hydropower stations.

Keywords: the mirror vertical shift method; water consumption rate; unit combination; flow–gross head–output relationship; cascade hydropower station

1. Introduction

The water consumption rate of power generation is the ratio of a hydropower station's quotative flow Q to its output P in $m^3/(s \cdot kW)$, or the ratio of hydropower station's quotative water W to its power generation E in $m^3/(kWh)$, which is an important index of a hydropower station's economic operation and is the key concern of a hydropower station's water conservation and efficiency work. The study of the water consumption rate of power generation has important theoretical significance and practical application value for improving the operating efficiency of units, guiding the economic operation of hydropower stations and improving the utilization rate of hydropower resources in basins. Currently, there have been some studies on the definition and influencing factors of the water consumption rate of hydropower stations, and many scholars use this index to optimize the operation of hydropower stations.

The goal of a hydropower station is to produce more output with the same amount of water, that is, the minimum water consumption rate. However, the calculation methods of the water consumption rate are not unified. Different scholars have given specific calculation methods under different conditions. Bakke et al. [1] improved a conceptual framework of water consumption in order to calculate the water footprint from hydropower projects. Tu et al. [2] introduced the standard water consumption rate and proposed a method for calculating and assessing the long-term average water consumption rate of the cascade hydropower station in a study on improving the assessment method of the water consumption rate in hydropower stations. Wu [3] proposed to use the relative water consumption rate for power generation, i.e., the ratio of the actual water consumption rate during the assessment period to the assessment-calculated water consumption rate for power generation as the assessment index of the daily regulation of hydropower plants. However, the above studies give the calculation method of the water consumption rate based on the actual operating conditions and operating data. These studies can measure the operating efficiency in a certain period, but they did not find the rules of water the consumption rate under different conditions and cannot be used to guide the operation of hydropower stations.

In addition, the influencing factors of the water consumption rate are also the focus of research. For example, Sun et al. [4], Yu et al. [5], Hou [6], and Zhou et al. [7] all carried out some research and analyses on the influencing factors of the water consumption rate. Based on the actual operation data and actual operation conditions of hydropower stations, they summarized some influencing factors of the water consumption rate, but did not deeply explore how the water consumption rate changes with the changes in these influencing factors, that is, they did not systematically summarize the change law of the water consumption rate. And since some of the factors they summarize, such as upstream maintenance work, are based on specific hydropower stations and are under certain limitations, it is difficult to extend to other hydropower stations.

Based on the definition and influencing factors of the water consumption rate, the water consumption rate is used as an indicator of economic operation to solve the optimization problem. Zhao [8] established a dynamic planning model based on the rate of water consumption to carry out a study on the optimal scheduling of reservoirs of the Liangjiang hydropower station. Bu [9] studied the joint application of the water consumption rate curve and the water balance equation in the power generation plan of the Taoyuan hydropower station, and proposed measures to improve the water consumption rate. However, this paper uses the correlation between the water consumption rate and gross head to calculate the water consumption rate, and the calculation result is not rigorous enough. Shi [10], in the study of reservoir optimal dispatching based on the dynamic analysis of the water consumption rate in the dry period of the Ludila hydropower station, comprehensively analyzed the relationship between the water consumption rate and various variables, and transformed the problem of the optimal dispatching of reservoirs into a problem dealing with the correlation between power generation plans and the water consumption rate. Lei et al. [11] established the mathematical model of the water consumption rate of Longtan hydropower station to achieve the aim of energy saving and consumption reduction. Wu, Y. et al. [12] developed a short-term load distribution (SLD) model considering complicated hydraulic and electricity constraints. The results show that the developed model reduces water consumption by 4.02, 4.70, 5.79, and 5.28% compared to the constant proportional distribution model and the real operation. In the above studies, the water consumption rate is often used as an indicator; however, due to the complexity of the calculation process of the water consumption rate, it is often estimated, resulting in a lack of rigor.

The study of the water consumption rate is a vital work of hydropower stations but the water consumption rate calculated on the basis of actual operation data is difficult in systematically reflecting the change law of the water consumption rate, and it cannot be used to guide the real-time operation decisions of hydropower stations. In this paper, based on the unit flow characteristic curve, the flow–head loss curve of the unit, and the relationship curve of the hydropower station's flow–tail water level are used to deduce the water consumption change curve of the cascade hydropower station with the hydropower station's water level as the reference. The water consumption change law of the single unit, the multi-unit, the hydropower station, and the cascade hydropower station are found. These laws can be used to guide the economic operation of hydropower stations and improve the utilization rate of hydropower resources in basins.

2. Research Objects

Located in the lower reaches of the Jinsha River on the border between Yunnan and Sichuan Provinces in China, hydropower station A is a giant hydropower station that mainly generates electricity and also has the comprehensive benefits of sand blocking, flood control, and the improvement of downstream navigation, and it is the upstream hydropower station in the cascade hydropower stations of this study. With a normal water level of 600 m, the reservoir has a total capacity of 12.67 billion cubic meters, a regulated storage capacity of 6.46 billion cubic meters, and a flood control capacity of 4.65 billion cubic meters. The hub project consists of a barrage dam, a diversion power generation building, flood discharge, and an energy dissipation building. The barrage dam adopts a concrete double-curved arch dam with a crest elevation of 610 m, a crest length of 700 m, and a maximum dam height of 278 m. The left and right banks are arranged with underground plants, each installed with nine sets of hydroelectric generating units with a capacity of 700,000 kW, with a total installed capacity of 12.6 million kW and a multi-year average generating capacity of 57.12 billion kWh.

Hydropower station B is located about 196 km downstream of hydropower station A; it is a giant hydropower station that mainly generates electricity and takes flood control, sand retaining, irrigation, navigation, and other comprehensive benefits into consideration. The normal water level is 380 m, the total reservoir capacity is 5.185 billion cubic meters, and the regulated storage capacity is 905 million cubic meters; the hydropower station is installed with eight sets of hydroelectric turbine generators of 750,000 kW capacity. The total installed capacity is 6 million kW and the average power generation for many years is 30.13 billion kW per hour. The relative position map of the research hydropower stations A and B is shown in Figure 1.

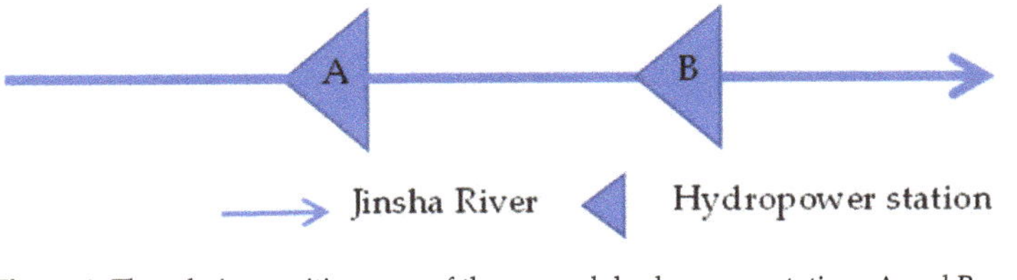

Figure 1. The relative position map of the research hydropower stations A and B.

Each unit of a hydropower station has a unit flow characteristic curve given by the manufacturer, which characterizes the relationship between the unit flow and output at each net head, also known as the NHQ relationship curve.

There are 18 hydro generator units installed in hydropower station A, with three types. Units no. 1–6 are of the same type, units no. 7–9 are of the same type, and units no. 10–18 are of the same type, and the flow characteristic curves of the three types of units are shown in Figure 2. The meaning of H in the figure is the net head.

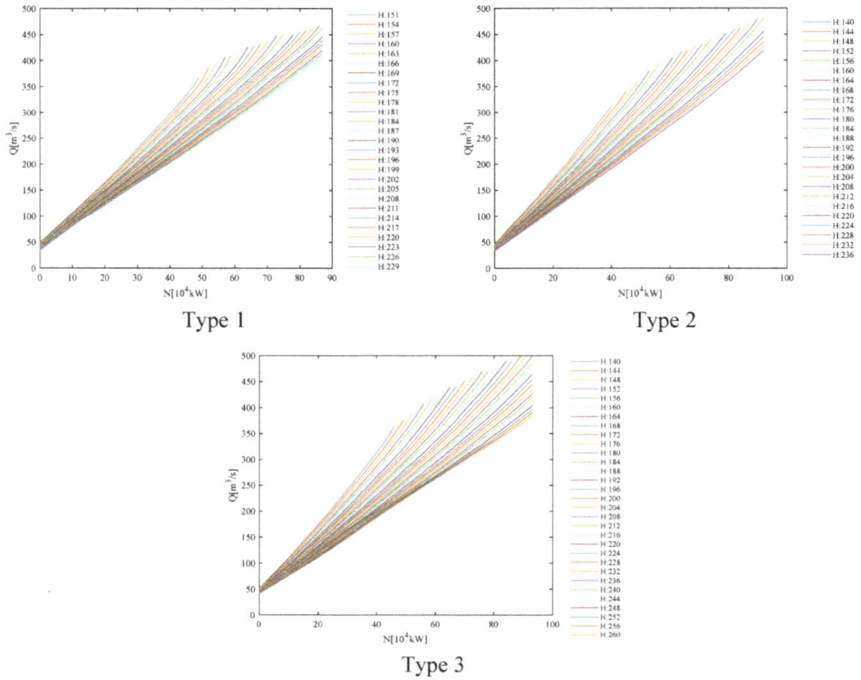

Figure 2. Flow characteristic curves of the units for hydropower station A.

Hydropower station B has eight units, of which units no. 1–4 are of the same type and units no. 5–8 are of the same type, and their flow characteristic curves are shown in Figure 3.

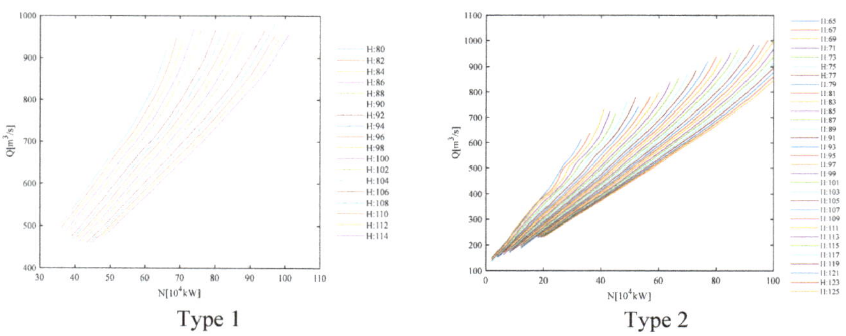

Figure 3. Flow characteristic curves of the units for hydropower station B.

3. Calculation of the Characteristic Curve of Water Consumption Rate for a Hydropower Station with Water Level as a Parameter

3.1. Calculation of the Characteristic Curve of Water Consumption Rate for a Single Unit

3.1.1. Plotting Relationship Curves of Flow–Gross Head–Output

Since the units of A and B hydropower stations are distributed side by side, in the process of power generation, the power generation flow of each unit of the same station may be different, i.e., the net head is different. While the gross head is the difference between the upstream water level and the downstream water level, which is the same, to better solve the output characteristics in the case of multi-unit stations, the net head of the unit characteristic curve is converted into the gross head to be calculated, in which the relationship between the unit's net head and the gross head is as follows:

$$H_{gross} = H + \Delta H \quad (1)$$

where H_{gross} is the hydropower station's gross head; H is the net head; and ΔH is the head loss in the diversion system.

Therefore, the net head of the unit is added to the head loss at the corresponding flow in the relationship curve output–net head–flow (NHQ) for a unit to obtain the relationship curve output–gross head–flow ($NH_{gross}Q$) for a unit.

3.1.2. Calculation of Characteristic Curve of Water Consumption Rate with the Gross Head as a Parameter

The water consumption rate is the amount of water consumed by a hydropower station for each unit of electricity generated, which is calculated as follows:

$$\mu = \frac{W}{E} = C\frac{QT}{NT} \quad (2)$$

where μ is the integrated water consumption rate, $m^3/(kWh)$; E is the total amount of power generation, kWh; W is the total amount of water generated, m^3; N is the power station output, kW; Q is the power generation flow, m^3/s; and C is the unit conversion factor.

From Equation (2), it can be seen that a set of determined flow and output data can determine a value of the water consumption rate, and for a single type of unit, the $QH_{gross}\,\mu$ characteristic curve or $NH_{gross}\mu$ characteristic curve can be derived from the unit's $NH_{gross}Q$ characteristic curve data, i.e., the flow–gross head–water consumption rate characteristic curve or the output–gross head–water consumption rate characteristic curve.

3.2. Plotting Characteristic Curve of Flow–Gross Head–Optimal Output for Multi-Unit Stations

3.2.1. Curve Plotting Method—The Mirror Vertical Shift Method

When two units are combined at a given gross head, a given generation flow is distributed among the two units to find the flow distribution scheme that maximizes the total generation output of the two units, i.e., the flow distribution scheme of the units at the optimal output. This process can be deduced by the mirror vertical shift method with the following steps:

1. Firstly, the relationship curves of output–gross head–flow ($NH_{gross}Q$) of the two units are organized into graphs whose horizontal coordinate is the value of output force, the vertical coordinate is the value of flow, and the scales of horizontal and vertical coordinate axes are the same.
2. Extract the graphs of the flow–output relationship between the two units at the same gross head separately and invert the graph of the second unit on the graph of the first unit.
3. Vertically move the power characteristic curve of the second unit, so that the longitudinal coordinates of the flow data are partially overlapped; it can be found that the sum of the flow values at each point is the same in the overlapped part. Plot several horizontal lines in the range of the overlapped part so that they intersect two curves, respectively. The sum of the coordinate values N_1 and N_2 corresponding to the intersections is the value of the total output of two units. As shown in Figure 4, the curve of the change in the total output can be plotted in the range of the overlap. The maximum output value on the curve is the optimal output $N_{i,max}$ that can be issued by the two units when the flow sum is Q. Record the flow and output value of each unit at the maximum output.

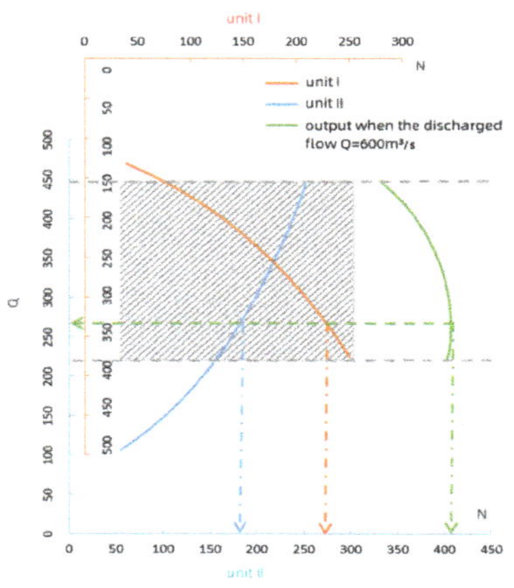

Figure 4. Schematic diagram of the optimal output of two units under a certain gross head.

4. Move the graph of the second unit up and down to obtain the optimal output $N_{1,max}$, $N_{2,max}$...$N_{n,max}$ corresponding to different flow sums Q_1, Q_2, Q_3,...Q_n of the two units, and then obtain the combination of the two units under the gross head. Then, obtain the optimal output–gross head–flow relationship curve for the combination of two units under this gross head.
5. Perform the same calculation for each gross head $H_{gross,1}$, $H_{gross,2}$...$H_{gross,n}$ to obtain the optimal output–gross head–flow relationship curve cluster in the case of the combination of two units.

3.2.2. Determination of Joint Operation Curve of Two Units Based on the Mirror Vertical Shift Method

The specific implementation idea of the multi-unit joint operation algorithm based on the mirror vertical shift method is as follows:

1. Based on the relationship curve of flow–net head–output for the unit and the relationship curve of flow–head loss for the unit, the head loss is calculated by interpolation. Then, add it to the net head to obtain the flow–gross head–output relationship data.
2. Fit the relationship curves of flow–output under an equal gross head H_m to obtain the flow–gross head–output relationship data for the two units.
3. Obtain the relationship curve of flow–gross-head–output when two units are combined. Firstly, determine the gross head H_{gross} and the combined flow Q_{zh} of the two units, then interpolate the total output N_{12} corresponding to different flow combinations to find the optimal output $N_{12,max}$. Secondly, change the combined flow Q_{zh} to obtain the corresponding optimal output. Thirdly, plot the data above into a curve. Finally, change the gross head H_{gross} and obtain the combined flow–optimal output curve under different gross head conditions, and then obtain the combination of optimal combined flow–gross head–output curve clusters.
4. Since the water consumption rate is uniquely determined by the flow and output, the relationship curve cluster of flow–gross head–optimal output for two units is converted into the relationship curve cluster of water consumption rate–gross head–combined flow for two units, i.e., the characteristic curve of the water consumption rate for the hydropower station.

The workflow of the two units' combined work operation model based on the mirror vertical shift method is shown in Figure 5.

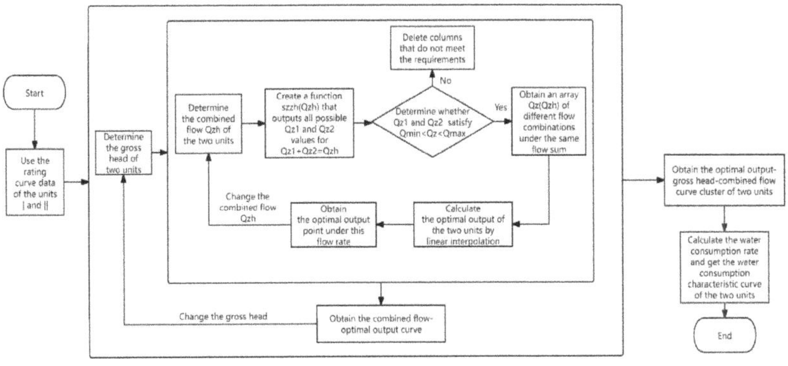

Figure 5. Operation model of two units' combined work based on the mirror vertical shift method.

3.2.3. Determination of Joint Operation Curve of Multi-Unit Stations Based on the Mirror Vertical Shift Method

First of all, merge the curve cluster of optimal output–gross head–flow for the combination of two units into one virtual unit's curve. Then, apply the mirror vertical shift method to calculate the curve cluster of flow–gross head–optimal output after adding one unit. After that, merge the curve cluster of flow–gross head–optimal output for the combination of three units into one virtual unit's curve, and apply the mirror vertical shift method to calculate the curve cluster of flow–gross head–optimal output after adding one unit, and so on until all available unit combinations of the hydropower station are obtained.

The units added to the calculation are all available units of the hydropower station. The use of this calculation method will not be affected when some units are repaired or broken down.

3.3. Calculation of Water Consumption Rate Characteristics of Hydropower Station Based on Station Water Level

3.3.1. Relationship between Gross Head and Water Level of Hydropower Station

The formula for calculating the gross head of a hydropower station is

$$H_{gross} = Z_{up} - Z_{down} \tag{3}$$

where H_{gross} is the gross head of the hydropower station, Z_{up} is the upstream water level of the reservoir, and Z_{down} is the tail water level of the reservoir. The tail water level of the reservoir can be obtained from the outflow Q_{out} by checking the relationship curve of reservoir outflow–tail water level. Therefore, when the outflow and the gross head of the reservoir are known, the upstream water level of the hydropower station can be determined.

3.3.2. Calculation of the Curve of Water Consumption Rate Characteristic for Hydropower Station

Applying the obtained curve cluster of outflow–gross head–optimal output relationship curves for hydropower stations, the curve cluster of outflow–water level–optimal output relationship curves for hydropower stations are obtained based on the property that the gross head is uniquely determined by the water level and the outflow of the hydropower station; then, as the water consumption rate is uniquely determined by the outflow and output, calculate the relationship curve cluster of outflow–water level–water consumption rate and the relationship curve cluster of output–station water level–water consumption rate.

4. Results of the Characteristic Curve of Water Consumption Rate of Hydropower Station with Water Level as a Covariant Variable

4.1. Hydropower Station A

The water consumption rate characteristic of hydropower station A was calculated by applying the previous calculation method. Figures 6 and 7 show the results of the characteristic curve of the water consumption rate for unit 1 and twin units 1 and 2 of hydropower station A, respectively. The legends in Figures 6 and 7 indicate the gross head in meters (note that the Hg in all the figures below represents the meaning of H_{gross} in this article—gross head).

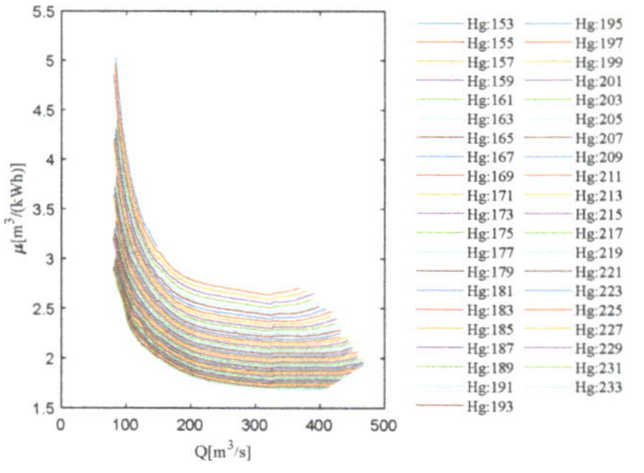

Figure 6. Relationship curve of flow–gross head–water consumption rate for unit 1.

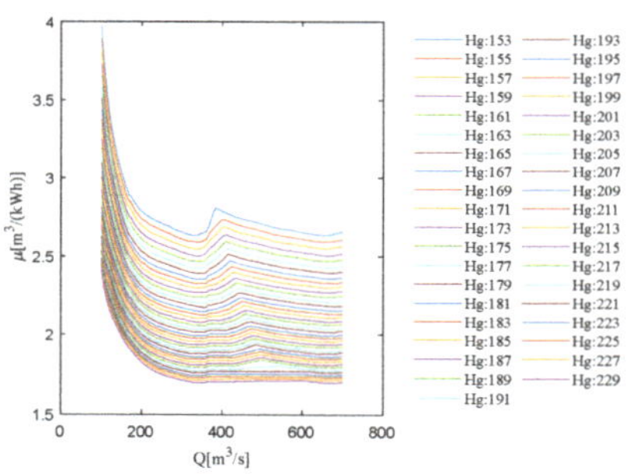

Figure 7. Relationship curve of flow–gross head–water consumption rate for the combination of units 1 and 2.

The changing pattern of the water consumption rate of each unit of hydropower station A is similar to that of Figure 6. The changing trend in the water consumption rate under each gross head is basically the same, decreasing with the increase in flow and decreasing with the increase in the gross head. When the flow is small, the water consumption rate decreases faster. With the increase in flow, the water consumption rate decreases slower. When the flow increases to a certain value and the gross head is small, the water consumption rate starts to increase slowly again.

As can be seen in Figure 7, the characteristic curve of the water consumption rate for the combination of two units shows a slow and smooth decreasing trend. When the gross head is certain, the water consumption rate firstly decreases with the increase in flow; when the flow is small, the water consumption rate decreases faster, and then the water consumption rate decreases slower. When the flow is certain, the water consumption rate decreases with the increase in the gross head. When the flow is around 350 m^3/s, there is a small fluctuation in the bottom curve, which is the result when converting from one unit of work to two units of work.

The characteristic curve and water consumption rate characteristic curve of hydropower station A are shown in Figures 8 and 9. Since the return water from reservoir B downstream of reservoir A may have a top-supporting effect on the tailwater level of reservoir A, the relationship curve of outflow–water level–optimal output and the relationship curve of outflow–water level–water consumption rate for hydropower station A when the water level of for hydropower station B is 370 m are shown in the figures. The legends in Figures 8 and 9 indicate the water level in meters.

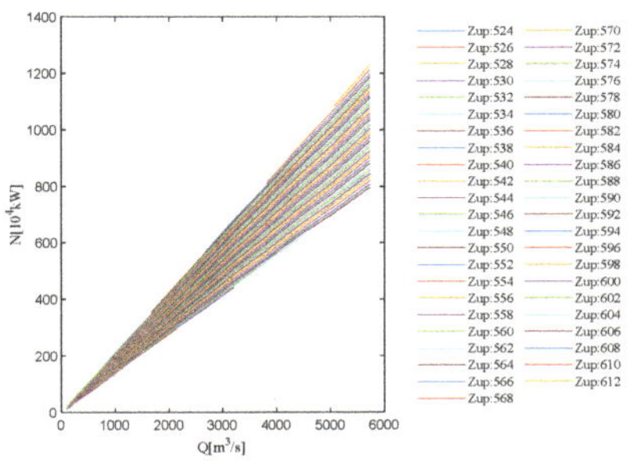

Figure 8. Relationship curve of outflow–water level–optimal output for hydropower station A.

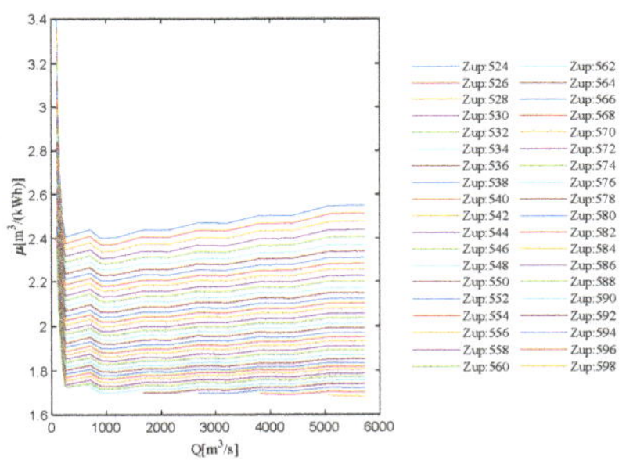

Figure 9. Relationship curve of outflow–water level–water consumption rate for hydropower station A.

As can be seen in Figure 8, the flow characteristic curve basically presents a linear upward trend. It can be seen from Figure 9 that the water consumption rate characteristic curve decreases rapidly when the flow is small and tends to be gentle when the flow is large, showing a stepwise upward trend.

4.2. Hydropower Station B

Figures 10 and 11 show the results of the characteristic curve of the water consumption rate characteristic curves for unit 1 and twin units 1 and 2 of hydropower station B, respectively. The legends in Figures 10 and 11 indicate the gross head in meters.

As can be seen in Figure 10, the characteristic curve of the water consumption rate for a single unit of hydropower station B presents V-shaped change characteristics. When the gross head is certain, the water consumption rate of a single unit first decreases with the increase in flow. When the flow increases to a certain value, the water consumption rate starts to increase again, and the increase trend is more obvious. From Figure 11, the characteristic curve of the water consumption rate for the combination of two units has an overall W-shaped trend, i.e., when the gross head is certain, the water consumption rate shows a decreasing–increasing–decreasing–increasing process with the increase in flow. When the flow is certain, the water consumption rate decreases with the increase in the gross head.

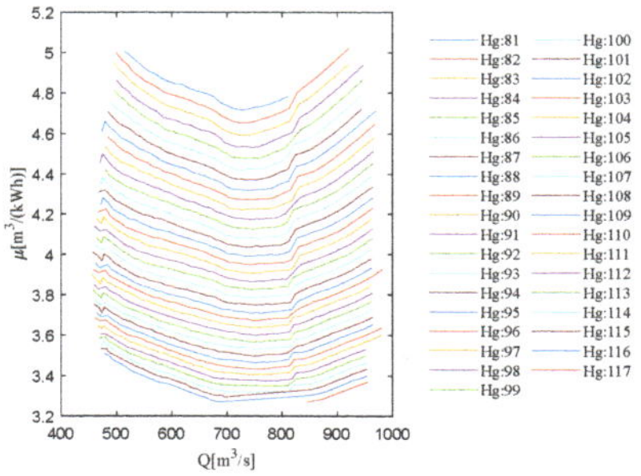

Figure 10. Relationship curve of flow–gross head–curve of water consumption rate for unit 1.

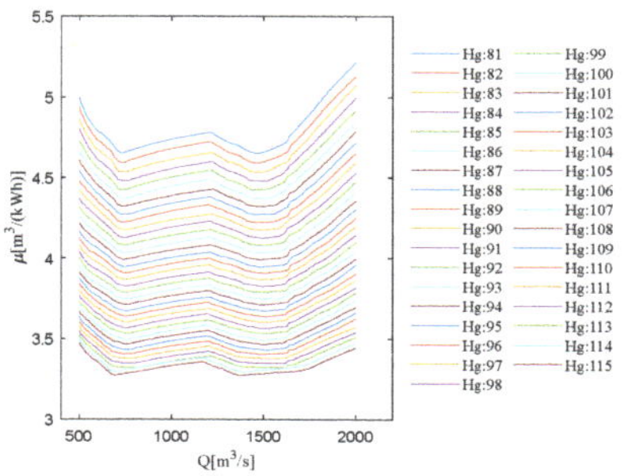

Figure 11. Relationship curve of flow–gross head–water consumption rate for the combination of units 1 and 2.

The characteristic curve and water consumption rate characteristic curve of hydropower station B are shown in Figures 12 and 13. The legends in Figures 12 and 13 indicate the water level in meters.

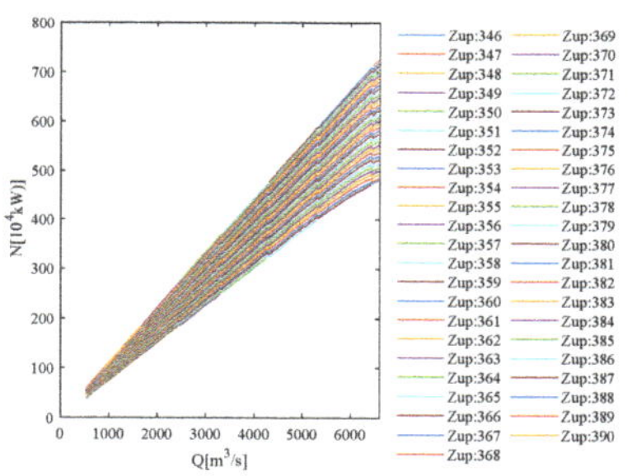

Figure 12. Curve of outflow–water level–optimal output for hydropower station B.

As can be seen in Figure 12, at a certain water level, the optimal output of the hydropower station has a linear growth trend with the outflow, and the growth rate slows down with the increase in the outflow. As can be seen in Figure 13, when the water level of the hydropower station is certain, the curve as a whole shows a stepwise upward trend with the increase in flow. When the flow is certain, the water consumption rate of the hydropower station decreases with the increase in the gross head.

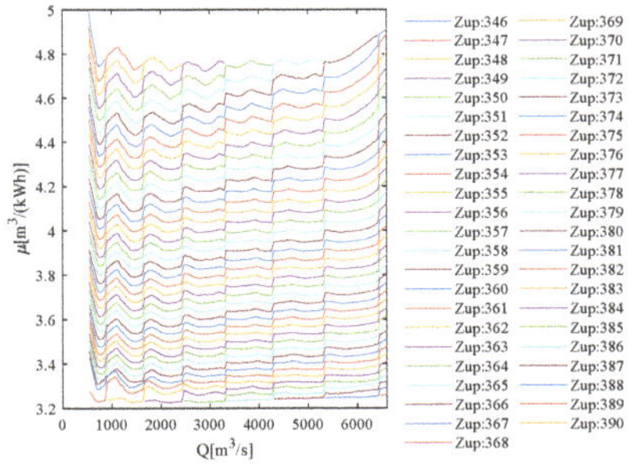

Figure 13. Curve of outflow–water level–water consumption rate relationship for hydropower station B.

5. Changing Pattern of Optimal Water Consumption Rate in Cascade Hydropower Station

5.1. Formula

Based on the characteristic curves of the water consumption rate of hydropower station A and hydropower station B, this section carries out the study on the changing pattern of the water consumption rate during the joint optimized operation of cascade hydropower stations.

When the water level of hydropower station A, the water level of hydropower station B, and the total output are given, it is possible to obtain the distribution of flow between the two hydropower stations at the time as the lowest rate of consumption, the distribution of output, and the rate of consumption at that time, thus realizing the economic operation of the cascade hydropower stations.

The water consumption rate of the cascade hydropower stations is characterized by the ratio of the total power generation flow of the cascade hydropower stations to the total output of power generation of the cascade hydropower stations [13], and the formula for calculating the water consumption rate of the cascade hydropower stations is as follows:

$$\mu_c = \frac{\sum_{i=1}^{n} W_i}{\sum_{i=1}^{n} E_i} = C\frac{\sum_{i=1}^{n} Q_i}{\sum_{i=1}^{n} N_i} \tag{4}$$

where μ_c is the total water consumption rate of the joint operation of cascade hydropower stations, $m^3/(kWh)$; E_i is the power generation of the ith hydropower station of the cascade hydropower stations, kWh; W_i is the power generation of water of the ith hydropower station of the cascade hydropower stations, m^3; N_i is the output of the ith hydropower station of the cascade hydropower stations, kW; Q_i is the power generation flow of the ith hydropower station of the cascade hydropower stations, m^3/s; and C is the unit conversion constant.

When the total load of the cascade hydropower stations is given, under the consideration of the minimum outflow of each hydropower station, i.e., the outflow of A is not less than 1200 m^3/s and the outflow of B is not less than 1700 m^3/s, the load can be distributed among each hydropower station of the cascade hydropower stations. Thus, the distribution combination that minimizes the water consumption rate of the cascade hydropower stations must be found, i.e., the optimal water consumption rate under the total load of the cascade hydropower stations.

5.2. Calculation Steps

The steps for calculating the water consumption rate for the joint operation of cascade hydropower stations are as follows:

1. Calculate the relationship curve cluster of outflow–water level–optimal output for each single station.

2. Under the condition of the given total output of the cascade hydropower stations and the initial water level of each hydropower station, the curve of outflow–water level–optimal output for each hydropower station can be used to obtain the corresponding outflow under different output distributions. When the sum of outflow is the smallest, the water consumption rate is the smallest. Record the output distribution and flow distribution in this case.
3. Change the total output, obtain the results of the output of each power station of the cascade under different total outputs, and plot the total output of the cascade hydropower stations to the optimal water consumption rate under the conditions of the combination of water level at this time.
4. Change the water level and repeat the above calculations to obtain the optimal water consumption rate–different water level combinations–the total output of the cascade hydropower stations relationship curve clusters under the conditions of different combinations of water levels.

5.3. Calculation Result

Taking A–B cascade hydropower stations as an example, the calculation of the characteristic curve of the optimal water consumption rate of the calculated cascade was carried out according to the above steps. When the water level of hydropower station A is 540 m and the water level of hydropower station B is 370 m, the relationship curve of total output–optimal water consumption rate for the cascade hydropower stations is obtained, as shown in Figure 14. The results of the optimal water consumption rate of the cascade hydropower stations in several typical cases are shown in Table 1.

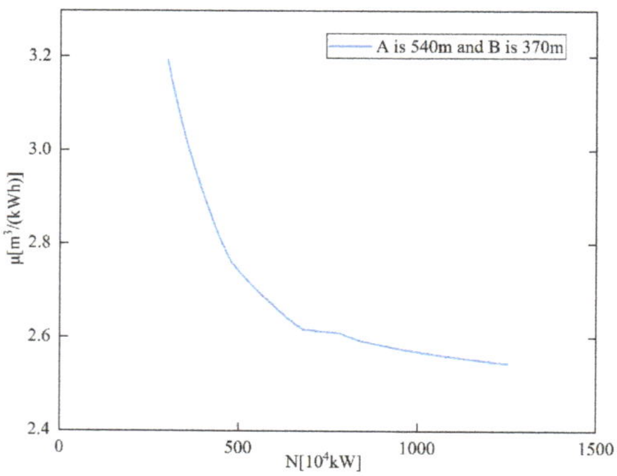

Figure 14. Relationship curve of total output–optimal water consumption rate for A–B cascade hydropower stations.

Table 1. Results of the optimal water consumption rate of cascade hydropower stations in several typical cases.

	A	B	A	B	A	B
Total output/10^4 kW	558		930		1116	
Output/10^4 kW	267	291	639	291	825	291
Flow/(m^3/s)	1721	3000	4246	3000	5703	3000
Gross head/m	165	102	161	102	158	102
Total water consumption rate (m^3/kWh)	2.7148		2.6606		2.6248	

It can be seen from the graph that as the output increases, the water consumption rate becomes lower and lower, and finally stabilizes. The specific output distribution is shown in Table 1. Three sets of examples are given in the table, meaning that when a total output value is determined, the output distribution scheme of the two hydropower stations can be output when the cascade water consumption rate is optimal, including the flow and gross head under the output. For example, when the total output is 558 × 10^4 kW, the optimal water consumption rate at this time is 2.7148 m^3/kWh, in which the output of hydropower station A is 267 × 10^4 kW and that of hydropower station B is 291 × 10^4 kW. Similarly, the corresponding flow and gross head under this scheme can be obtained.

Figure 15 shows the relationship curve cluster of total output–A–B joint optimal water consumption rate for hydropower station A with a water level of 580 m and for hydropower station B with different water levels. The legend in Figure 15 indicates hydropower station B's water level in meters.

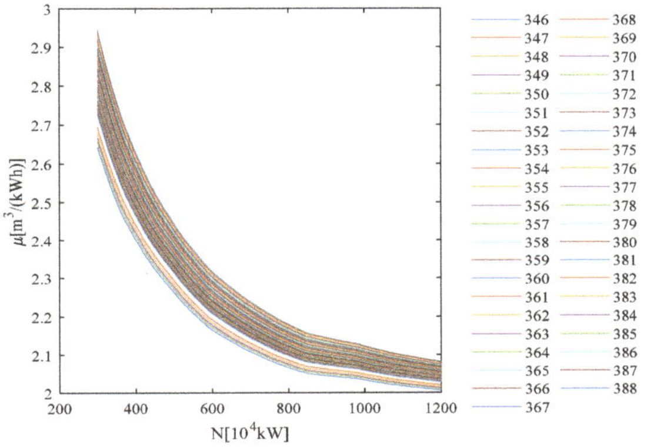

Figure 15. Relationship curve cluster of total output–A–B joint optimal water consumption rate.

From Figure 15, it can be seen that the optimal water consumption rate decreases with the increase in the total output of the cascade hydropower stations. The optimal water consumption rate of the cascade hydropower stations decreases with the increase in the water level of B in the case of the same water level of A and the total output.

6. Conclusions

In this paper, from the relationship curve of output–head–flow of the unit for hydropower stations, the relationship curve of output–gross head–flow is calculated and the power characteristic curve of unit combination with the gross head as the parameter is put forward. Then, the calculation method of the characteristic curve of the water consumption rate for hydropower stations with their water level as the parameter is established. Finally, the relationship between the change in the water consumption rate of hydropower stations under different conditions of water levels, flow, output, etc., was obtained. At the same time, the optimal water consumption rate changing pattern of cascade hydropower stations under the given various total output conditions was studied. The following conclusions were obtained:

The changing pattern of the water consumption rate of a single unit is as follows: when the gross head is certain, the water consumption rate decreases with the increase in flow when the flow is small; when the flow increases to a certain value, the water consumption rate starts to increase slowly. The changing pattern of the water consumption rate of the hydropower stations is as follows: when the gross head is certain, hydropower station 's water consumption rate with the increase in outflow shows slight fluctuations and an increasing trend. The changing pattern of the optimal water consumption rate of cascade hydropower stations is as follows: with the increase in the water level of each hydropower station, the optimal water consumption rate becomes smaller, and with the increase in total output, the optimal water consumption rate shows a smooth decline—first a fast decline, then the rate slows down.

The calculation method of the characteristic curve of the water consumption rate proposed in this research can be extended to other cascade hydropower stations. The results of the water consumption rate changing pattern are clear and intuitive, which is of great significance to understanding the characteristics of the water consumption rate of hydropower stations, applying the characteristics to carry out the economic operation of hydropower stations and promote the improvement of hydropower station power generation efficiency. It can be used as a reference for important decision-making in the economic operation of hydropower stations and cascade hydropower stations.

Author Contributions: Conceptualization, X.A. and X.Z.; methodology, F.L.; software, F.L. and J.Z.; validation, X.A., X.Z. and F.L.; formal analysis, X.A. and X.Z.; investigation, X.A.; resources, F.L.; data curation, X.A.; writing—original draft preparation, X.A. and X.Z.; writing—review and editing, X.A. and F.L.; visualization, X.Z.; supervision, F.L.; project administration, X.A.; funding acquisition, R.G., X.L. and M.W. All authors have read and agreed to the published version of the manuscript.

Funding: This research was funded by the Open Research Fund of Hubei Key Laboratory of Intelligent Yangtze and Hydroelectric Science, China Yangtze Power Co., Ltd., grant number 242202000916, the Natural Science Foundation of Hubei Province Innovation and Development Joint Fund Project, grant number 2022CFD169, and the National Natural Science Foundation of China, grant number 51979198, 52279024.

Data Availability Statement: Restrictions apply to the availability of these data. Data were obtained from [China Yangtze Power Co., Ltd.] and are available [from the authors] with the permission of [China Yangtze Power Co., Ltd.].

Conflicts of Interest: The authors declare that they have no known competing financial interests or personal relationships that could have appeared to influence the work reported in this paper. Renwei Guo, Xiao Li, Mian Wang are employees of China Yangtze Power Co., Ltd., who provided funding and technical support for the work. The funder had no role in the design of the study; in the collection, analysis, or interpretation of data, in the writing of the manuscript, or in the decision to publish the results.

References

1. Bakken, T.H.; Killingtveit, Å.; Engeland, K.; Alfredsen, K.; Harby, A. Water consumption from hydropower plants–review of published estimates and an assessment of the concept. *Hydrol. Earth Syst. Sci.* **2013**, *17*, 3983–4000. [CrossRef]
2. Tu, Q.; Cai, J. New and improved assessment method of water consumption rate of hydropower plant. *Water Resour. Power* **2013**, *31*, 32–34. (In Chinese)
3. Wu, S. Research on assessment method of water-saving generation increase of hydropower station with daily regulation performance. *Distrib. Energy* **2019**, *4*, 22–26. (In Chinese)
4. Sun, F.; Li, X. Research on optimal scheduling strategy of Pankou and Xiaoxuan reservoirs. *Hydropower New Energy* **2016**, *4*, 9–12. (In Chinese)
5. Yu, Y.; Wang, G. Analysis and improvement measures of increasing water consumption rate of large hydropower station during dry period. *Hydropower Pumped Storage* **2017**, *3*, 71–74. (In Chinese)
6. Hou, G. Analysis of Main influencing factors and countermeasures of comprehensive water consumption rate in Silin Power Plant. *Huadian Technol.* **2018**, *40*, 68–69+79–80. (In Chinese)
7. Zhou, J.; Feng, H. Analysis of influencing factors and countermeasures of water consumption rate of a cascade hydropower station in Wujiang River. *Hongshui River* **2021**, *40*, 135–137+147. (In Chinese)
8. Zhao, J. Reservoir optimization of hydropower station based on dynamic programming model of water consumption rate. *Jilin Water Resour.* **2014**, *8*, 24–26+29. (In Chinese)
9. Bai, F. Joint application of water consumption rate curve and water balance equation in power generation planning of Taoyuan Hydropower Station. *Hydropower Pumped Storage* **2020**, *6*, 73–76. (In Chinese)
10. Shi, J. Reservoir optimal operation based on dynamic analysis of water consumption rate of Ludila Hydropower Station during dry season. *Yunnan Water Power* **2022**, *38*, 12–15. (In Chinese)
11. Lei, X.; Yin, A. An Analysis on the Water Consumption Rate of LongTan Hydropower Station with Properties of Water Energy. *Adv. Mater. Res.* **2013**, *648*, 218–222. [CrossRef]

12. Wu, Y.; Liao, S.; Liu, B.; Cheng, C.; Zhao, H.; Fang, Z.; Lu, J. Short-term load distribution model for cascade giant hydropower stations with complex hydraulic and electrical connections. *Renew. Energy* **2024**, *232*, 121067. [CrossRef]
13. *Enterprise Standard Q/CSG1204020-2016*; Specifications for economic operation indicators of hydropower in China Southern Power Grid. China Southern Power Grid Co., Ltd.: Guangzhou, China, 2024. Available online: https://www.docin.com/p-2413212820.html (accessed on 3 April 2024).

water

MDPI

Article

A Monographic Experimental Investigation into Flood Discharge Atomized Raindrop Size Distributions Under Low Ambient Pressure Conditions

Dan Liu [1,2,3], Jijian Lian [1], Dongming Liu [1], Fang Liu [1,*], Bin Ma [1], Jizhong Shi [4], Linlin Yan [1], Yongsheng Zheng [5], Cundong Xu [2,3] and Jinxin Zhang [2,3]

1 State Key Laboratory of Hydraulic Engineering Intelligent Construction and Operation, Tianjin University, Tianjin 300350, China; liudan_hydraulic@tju.edu.cn (D.L.)
2 Nanxun Innovation Institute, Zhejiang University of Water Resources and Electric Power, Hangzhou 310018, China
3 Zhejiang Key Laboratory of River-Lake Water Network Health Restoration, Zhejiang University of Water Resources and Electric Power, Hangzhou 310018, China
4 Zhejiang East China Engineering Consulting Co., Ltd., PowerChina Huadong Engineering Corporation Limited, No. 201 Gaojiao Road, Yuhang District, Hangzhou 311122, China
5 Yalong River Hydropower Development Co., Ltd., Chengdu 610051, China
* Correspondence: fangliu@tju.edu.cn; Tel.: +86-155-1081-0907

Abstract: The construction and operation of high dam projects at high altitudes have led to concerns about the effectiveness of flood discharge security predictions resulting from the greater flood discharge atomized rain caused by ambient pressure reduction. In this study, self-similar characteristics and variation in atomized raindrop size distributions are analyzed to understand the phenomenon of increased atomized rain intensity under low ambient pressure from a mesoscopic scale. The monographic experiments are characterized by a low ambient pressure range ($0.66P_0$–$1.02P_0$) and a high waterjet velocity range (13.89–15.74 m/s). When the ambient pressure decreases by $0.10P_0$ (P_0 = 101.325 kPa) from the reference atmospheric pressure condition as the other conditions remain fixed, the total number concentration in a two-dimensional atomized raindrop spectrum (number/(54 cm^2)) and the peak value of the individual three-dimensional number concentration (number/(m^3·mm) increase, which can lead to the required industry standard protective level of atomized zones increasing by one level in some cases. In addition, the spectrum trend and typical particle size ranges of the atomized raindrop size distributions present self-similarity as the ambient pressure decreases. The above studies further confirm the effects of low-ambient pressure enhancement on flood discharge atomized rain intensity, which can provide a theoretical basis for the development of random splash simulation models characterized by low pressure for high-altitude hydropower stations.

Keywords: random splash experiments; low ambient pressure; atomized rain intensity; atomized raindrop spectrum; spectral characteristics application

1. Introduction

Rain hazards derived from extreme rain events (i.e., urban flooding [1], flash floods [2], soil erosion [3], building erosion [4], streambank erosion [5], and hydroelectric accidents [6,7]) have attracted interest as public safety problems at home and abroad. The direct disasters induced by heavy precipitation include confluence, erosion, infiltration, and other physical behaviors. Moreover, the absence of effective predictions and protective measures for these direct disasters often leads to a cascade of secondary disasters that inflict socioeconomic damage. In addition to natural rain, heavy precipitation at hydropower stations stems from another source: flood discharge atomization. This phenomenon has garnered considerable attention due to its undeniable threats not only to the hydropower station itself but also to the surrounding downstream environment [8–12]. As shown in Figure 1, the movement path of atomized rain mainly extends from splash zones downstream along the flow direction, which is unlike that of natural rain moving from cloud systems to the ground along the height direction [13]. Therefore, atomized rain is always concentrated in the downstream zone and varies in a dynamic changing process with obvious divisions. It is influenced by various factors, including hydraulic release conditions, design of release structures, regional topography, unique geological formations, and prevailing climatic characteristics [14]. In previous studies on atomized events, the influence of hydraulic factors has gradually been recognized, but the coupling analysis of environmental factors still remains a challenge because of a lack of thematic research. In particular, at high-altitude hydropower stations, low ambient pressure is a typical meteorological characteristic that affects the formation and diffusion of atomized rain. Therefore, it is important to mechanistically explore the influence of low ambient pressure on flood discharge atomization to adapt to the new demand for the development of hydropower projects.

Figure 1. Atomized rain fields induced by different flood discharge atomized sources at the Jinping-1 hydropower station. (**a**,**b**) Atomized source of a dam orifice discharge. (**c**) Atomized source of a spillway tunnel discharge [13].

Atomized areas can be classified as atomized rainstorm zones that require protection when their atomized rain intensity exceeds 10 mm/h. With the update of the engineering problem library induced by flood discharge atomized rain, the risk cognitions and protective technologies have improved. However, challenges persist in atomized prediction across various processes, locations, and scales. The risk perception of atomized rain-induced landslides has strengthened since one million cubic meters of landslides occurred at the Longyangxia hydropower station in 1988 and 1989 [12]. The relationship between discharge patterns and atomization intensities indicates that the problem of atomized rain induced by bottom-flow dissipation is weaker than that caused by ski-jump energy dissipation. Specifically, the prototypical atomized ranges of the Ertan hydropower station reached

0 + 2000 m along the flow direction and 1215 m in the elevation direction, which are greater than their corresponding design values of 1000 m and 75 m for ski-jump energy dissipation in its collision design [10,11]; however, those of the Xiangjiaba hydropower station are concentrated in the energy dissipation area and have almost no effect on the humidity of its downstream county [15]. The atomized predictions of ski-jump energy dissipation were subsequently introduced and developed into large-scale energy dissipation projects, such as the design projects of Wudongde, Baihetan, Lianghekou, Rumei, and the operation projects of Ertan, Laxiwa, Nazixia, Shuibuya, etc. [14]. Among the abovementioned hydropower stations, the maximum crest elevations of the Laxiwa, Longyangxia, and Nazixia hydropower stations are 2460.00 m, 2610.00 m, and 3204.60 m, corresponding to dam heights of 250 m, 178.0 m, and 121.5 m, respectively. Additionally, the maximum crest elevations of the Lianghekou and Rumei hydropower stations are 2875.00 m and 2902.00 m, corresponding to dam heights of 295.0 m and 315.0 m, respectively. The number of high-head hydropower stations at high altitudes is increasing. However, the impact of low ambient pressure is currently not taken into account in the atomized analysis of these projects. This oversight is likely to lead to an underestimation of the effects of atomized rain at high-altitude hydropower stations, due to the enhancement effect that low ambient pressure has on the downstream atomized rain field.

To mitigate and avoid atomized rain hazards, atomized scopes are evaluated in different regions with individual levels of engineering and nonengineering protective measures. In addition to prototypical and scale models, nonscale models, named "random splash experiments", are established with a major focus on atomized rain indices and factors during waterjet splashing processes [14,16]. The precipitation intensity and droplet spectrum are the principal macroscopic and microscopic characteristics of atomized rain fields, such as natural rain. In studies of atomized rain intensity, early random splash experiments controlled hydraulic conditions produced by the hydraulic impact of a single falling waterjet into a water cushion and the midair collision of two water jets under atmospheric ambient pressure [17–20]; however, a depressurized random splash experiment recently indicated that ambient pressure conditions are also important because lower ambient pressure can increase the area of an atomized storm [21]. The atomized raindrop size distributions can be statistically analyzed through a graph or digital signal acquisition. The relationships between rain intensity and raindrops information can be used for both the calculation of atomized light rain intensity according to the raindrop correlation parameters determined in advance and the exploration of atomized rain intensity laws characterized by dominant rainfall intensity and frequency under atmospheric pressure [22,23]. With statistical models of spectrum characteristics, atomized rain intensity distributions can be experimentally demonstrated or calculated by ignoring ambient pressure effects, among which experimental results of atomized raindrop size distributions are still limited [9,24]. Although atomized raindrop size distributions have been described and applied in random splash numerical models, the effects of low ambient pressure are still poorly understood [14]. To explore the enhancement mechanism of low ambient pressure on atomized rain, it is essential to gain insight into atomized raindrop size distributions in depressurized random splash experiments to further promote the development of atomization prediction methods that consider high-altitude flood discharge.

Depressurized experiments are widely applied and highly important in numerous fields, such as astrobiology models to predict the habitability of Mars [25], flame-geometry descriptors of fire jets and the diffusion characteristics of water mists to ensure fire safety in plateau areas [26,27], the splash form of liquid droplets to inhibit splashing action [28], the cracking properties of concrete to explore its plateau application [29], and the impact and splash characteristics of falling waterjets to guide flood discharge safety [21,30]. Analogously, these experiments require artificial low-pressure environments. Unlike other depressurized experiments, the ambient pressure stability of random splash experiments is an important challenge. This is primarily due to the involvement of two-stage multiphase coupling processes, which involve aerated waterjet movement based on a continuous phase and the diffusion of numerous droplets based on a discrete phase [31,32]. In general, the relationship between the splash and impact processes is characterized by negative compensation interactions without considering the influence of ambient pressure. However, the enhancement effects of the low ambient pressure not only appeared in the atomized rain field downstream of the water-collision locations but also in the impact pressure field near the water-collision locations. These phenomena were observed in depressurized experiments of model waterjets released from a pressure orifice with initial velocities of approximately 10 m/s and a surface orifice with initial velocities of approximately 5 m/s [21,30]. When these effects are applied synchronously to prototype engineering applications, the protection standards need to be improved in the flood discharge process. Precisely obtaining multiple factors, i.e., the water-entry velocity, water-entry angle, water-entry position, atomized source moisture, bucket nappe wind, and so on, is still limited, both experimentally and computationally [18,19,33,34]. Consequently, we have not explored the influence of independent factors but have focused on the terminal effect of the atomized field.

Although a clear relationship between atomized raindrops and ambient pressure is not available at present, other studies related to altitude have suggested that depressurized effects should be considered. In view of the droplet splash process, Xu et al. [35] and Latka et al. [36] reported that the ejected droplets splashed by a liquid drop and a solid surface can be impacted by ambient pressure and wall roughness. Xu et al. [35] reported that as the ambient pressure decreases, less liquid is ejected in the process of corona splashing on the smooth wall. Latka et al. [36] further reported that the ambient pressure needs to be considered during thin-sheet formation and prompt splashing. In view of the droplet diffusion process, the influence of ambient pressure on the natural raindrop spectrum is objective. Under climate conditions in the same area with different elevations, Song et al. [37] noted that a decrease in altitude can decrease the contribution rate of small raindrop particles to precipitation intensity. Li et al. [38] reported that the locations of raindrop diameter attenuation are in the order of mountain-side, mountain-bottom, and mountain-top, irrespective of rain type, and that the raindrop size distributions for convective precipitation are more sensitive to altitude than those for stratiform precipitation by comparing stratiform and convective precipitation raindrop size distribution datasets at Huangshan (vertical evolution at 1860 m, 1351 m, and 464 m). The raindrop size distribution (RSD) is the change in individual three-dimensional number concentration (number/(m^3·mm)) with raindrop particle size. Under climate conditions in different areas with different elevations, Sumesh et al. [39] reported that surface drop concentrations at a mid-altitude of 400 m show bimodal RSDs, whereas those at high altitudes of 1820 m represent monomodal RSDs regardless of rain class by characterizing the RSDs parameters of stratiform precipitation in the western Ghats Mountains. Wu et al. [40] detected large variability in RSDs in the Tibetan Plateau and southern China. This variability can be described by standardized gamma, whose parameters vary with changes in rain rate and precipitation type. Based on the above discussion, low ambient pressure should play an important role in the microdynamic mechanism (evaporation, collision, coalescence, breakup, etc.) of raindrops or raindrop groups. Thus, the ambient pressure is likely to affect the descriptive characteristics of raindrop size distributions with different ambient pressures.

Therefore, the study of raindrop size distributions under a low ambient pressure environment is not only the content of flood discharge atomization theory but also a practical necessity for numerical calculation of atomization at high-altitude hydropower stations [14]. To fill the above gaps, this study is a follow-up of the enhancement effect of low ambient pressure on downstream atomized rain intensity induced by depressurized random splash experiments. In this study, the atomized rain was generated by the collision process of a high-speed rectangular waterjet and a narrow water cushion under various low ambient pressure conditions. The observation device is an OTT Parsivel^2 (Beijing Sichuang Minda Measurement and Control Technology Co., LTD, Beijing, China) suspended in a horizontal guide rail with six measuring points along the overflow center line. In the following sections, the structure design and measurement method of the two-phase experimental platform will be introduced; the atomized quantitative precipitation estimation will be considered to access the protection level variation in atomized areas; the statistical characteristics of atomized raindrop size distribution will be analyzed and discussed; and the main conclusions and prospects will be summarized in the end to pave the way for a follow-up study.

2. Experimental Setup and Measurements

2.1. Design and Regulation of the Experimental Platform

A novel test control system chartered by high-speed waterjets and low ambient pressures was used to carry out the depressurized random splash experiments at the State Key Laboratory of Hydraulic Engineering Simulation and Safety (current name: State Key Laboratory of Hydraulic Engineering Intelligent Construction and Operation) of Tianjin University. The hydraulic depressurized chamber is intended for hydrodynamic simulations of high-altitude hydropower stations, not for traditional cavitation studies, so the vacuum degree is designed with reference to the absolute air pressure in the high-altitude area. The raindrop size distribution measurement methods are generally based on the principles of laser signals and image identification. It is reasonable to adopt various chemical test papers combined with postprocessing of graphics in the instantaneous atomized raindrop size distribution collection of random splash experiments under atmospheric pressure. However, this method is not suitable for the measurement of depressurized random splash experiments because measurement times are difficult to control, measurement conditions are difficult to switch, and instantaneous atomized raindrop size databases are not conducive to the statistical analysis of time–history characteristics. Therefore, an OTT Parsivel^2 with a 54 mm^2 rectangular area is introduced to collect the atomized raindrop size in real time, which can be moved horizontally along the flow direction via a remotely controlled horizontal slide rail. The experimental area is in the depressurized chamber that is connected with water-regulating and air-regulating systems, in which the atomized rain is splashed by a high-speed rectangular waterjet from a constricted nozzle that impacts a rectangular water cushion with an initial depth of 56 cm. The experimental coupling conditions between water flow and ambient pressure are adjusted with reasonable shunt designs and effective monitoring devices. In detail, the layout of the depressurized chamber is shown in Figure 2a,b, the layout of the measurement instrument and points is shown in Figure 2c, the air-regulating system is shown in Figure 3a, and the water-regulating system is shown in Figure 3b.

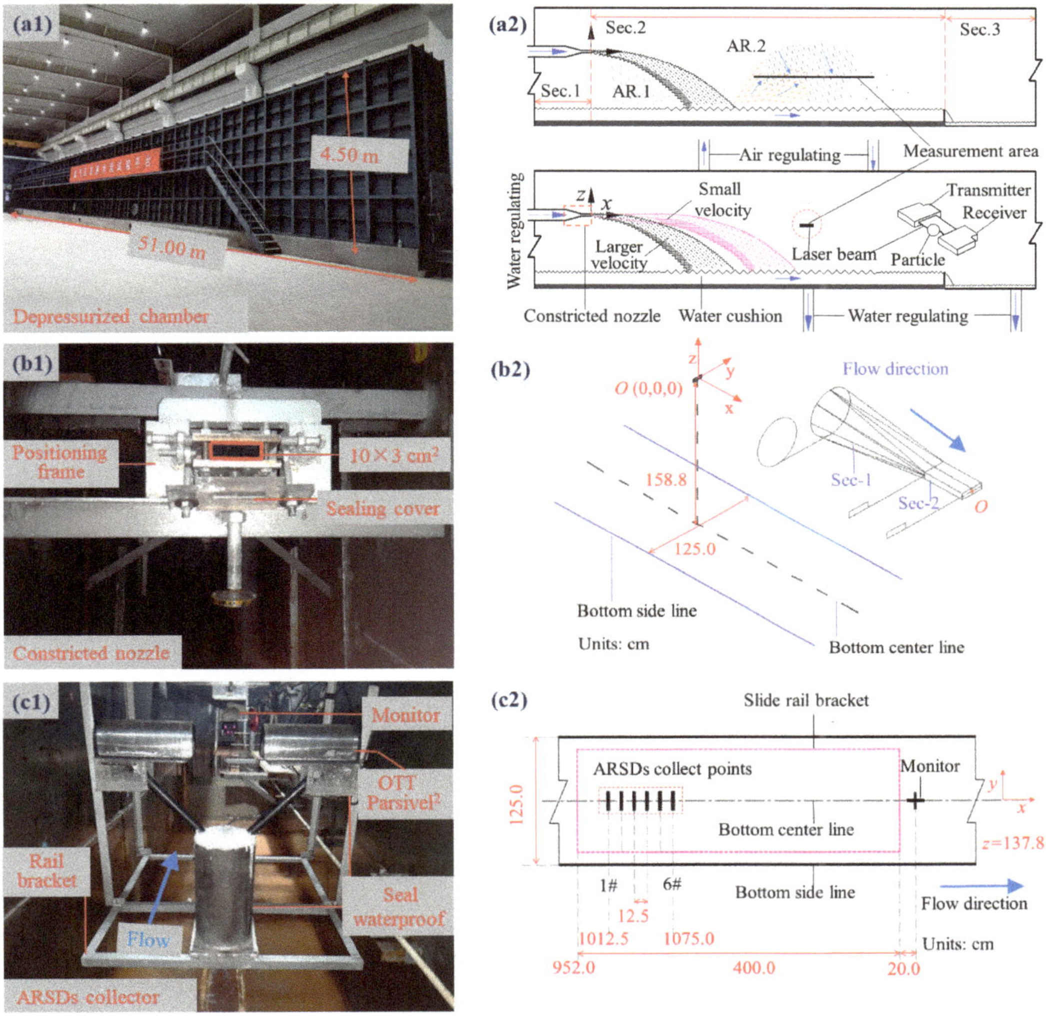

Figure 2. Physical diagram and partial design of the depressurized random splash experiments. (**a1**) Physical image of the depressurized chamber; (**a2**) Schematic diagram in the depressurized chamber; (**b1**) Physical image of the internal nozzle in the depressurized chamber; (**b2**) Schematic diagram of the internal nozzle in the depressurized chamber; (**c1**) Physical image of the measurement instrument and point layout ($ARSD_S$ is short for atomized raindrop distributions); (**c2**) Relative coordinates of measurement instrument and point layout ($ARSD_S$ is short for atomized raindrop distributions).

As shown in Figure 2a, the depressurized chamber has a sealable steel structure that is 51.00 m long, 4.50 m high, and 1.25 m wide in internal volume, in which three sections are divided along the flow direction, corresponding to the inflow Sec. I, the working Sec. II, and the outflow Sec. III, respectively. As rectangular waterjets are typically shaped as discharge waterjets, rectangular waterjets can also be considered component units of specially shaped waterjets in theoretical and numerical studies of flood discharge atomization. The main gathering parameters include the atomized raindrop distributions of six measurement points, the inflow discharge of the waterjet, and the ambient pressure in the depressurized chamber; two types of affiliated measurement indicators include the water level in the drainage area and the visual monitoring device in the measurement and drainage areas. The acquisition terminals of the above instruments are arranged on a measurement platform that is placed outside the depressurized chamber.

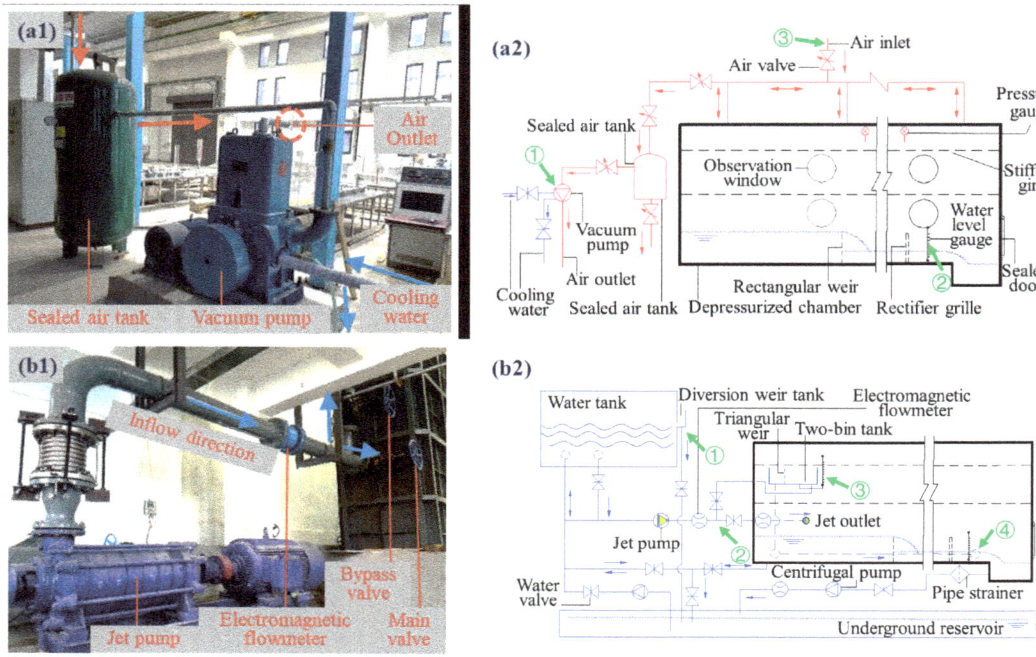

Figure 3. Physical and schematic diagram of the water- and air-regulating systems in the depressurized random splash experiments. (**a1**) Physical picture of air-regulating system layout; (**a2**) Schematic diagram of air-regulating system layout; (**b1**) Physical picture of water-regulating system layout; (**b2**) Schematic diagram of water-regulating system layout.

As shown in Figure 2b, to achieve stable high-speed rectangular waterjet discharge conditions, a constricted nozzle made of stainless steel was designed. This nozzle features a transition section measuring 55.0 cm in length and a rectifying section that is 20 cm long, with connection surfaces transitioning from an upstream circular cross-section (D = 209 cm) to a downstream rectangular cross-section (10.00 cm wide × 3.00 cm high), ultimately leading to the same rectangular surface at the exit. In the experiment, the streamlined nozzle effectively prevented water from detaching from the pipe wall, thereby mitigating the risk of local negative pressure and cavitation issues. Furthermore, no abnormal sounds or vibrations were detected at the nozzle during testing, and there were no indications of pitting or hollow erosion marks following the test. Consequently, we conclude that there are no cavitation-related problems at the nozzle that could potentially impact the test results, thus fulfilling all testing requirements. Furthermore, the center point (O) of the nozzle outlet surface is positioned 1.588 m above the bottom surface of the depressurized chamber, simulating an atomized rain field characterized by limited transverse diffusion while allowing for sufficient diffusion in both the flow and height directions. When the waterjet impacts the water cushion below, it generates a substantial number of atomized raindrops that are dispersed throughout space-time domains—particularly downstream from where collisions occur within the water body—which has been designated as the measurement area for atomized raindrop size distributions.

As shown in Figure 2c, to alleviate the diffusion interference caused by atomized rain, the instrument is secured firmly on a tray located beneath a straight slide rail. The four connecting rods are strategically positioned downstream of the instrument. To safeguard against water ingress, an appropriately designed partition protection mechanism has been implemented for the exterior of the instrument. To clearly define the spatial positions of the measurement points, we first establish the location of the center point within the designated measurement area as the coordinate reference for our measurement points. We subsequently designate point O—the geometric center of the nozzle outlet surface—as the origin of the coordinate system. We define the axes as follows: the x-axis represents the flow direction, the y-axis denotes the transverse direction, and the z-axis corresponds to the vertical direction. In analogous random splash experiments involving high-speed water jets with velocities of approximately 10 m/s, it has been demonstrated that there is a negligible transverse gradient in the two-dimensional distributions of downstream atomized rain (with rain intensities $\geq$ 2 mm/h). Consequently, six measurement points (labelled points #1 through 6) were arranged along the overflow centerline at a z-coordinate of 20 cm. To minimize errors arising from overlapping particles during heavy precipitation events, point #1 was positioned at a distance of 1012.5 cm from the coordinate origin, with an interval of 12.50 cm between adjacent measurement points.

As illustrated in Figure 3a, the air circulation within the depressurized chamber comprises two distinct pathways. The first pathway involves reducing the ambient pressure of the chamber through a vacuum pump connected to a sealed air tank. The second pathway facilitates the influx of external air into the depressurized chamber, driven by the existing differential pressure. Consequently, five suction air pipes located at the top of the depressurized chamber are individually linked to both exhaust and replenishment pipes, allowing for precise adjustment of the required airflow path by corresponding valves. During this adjustment process, real-time monitoring of ambient pressure within the depressurized chamber is conducted using an absolute pressure sensor installed at the top center front of the chamber. This sensor boasts an accuracy of 0.1% FS and operates within a range of 0 to 10 m. Additionally, water depth is measured utilizing a self-balancing manometer positioned downstream from the drainage area alongside a meter ruler. It is imperative that under enclosed conditions with atmospheric pressure, downstream water levels do not rise; conversely, these levels must not fall below the specified minimum design ambient pressures in similar enclosed environments. In addition, the testing was conducted in January, during which the recorded average temperature was 12 °C.

As illustrated in Figure 3b, high-speed waterjets derive energy from pumps, which increase velocity through a constricted nozzle and regulate waterjet inflow discharge via two branch pipes. The system is designed to recycle water through an underground reservoir and includes a centrifugal pump, an elevated water tank with a diversion weir tank, a jet pump, a #1 electromagnetic flowmeter, and two upstream branch pipes, along with another centrifugal pump downstream. Meanwhile, the rated head of 210 m for the jet pump provides a reliable assurance of water pressure within the supply system, ensuring the stability of the high-speed waterjet. In the inflow section of the depressurized chamber, one branch pipe connects to a #2 electromagnetic flowmeter and horizontally leads to a constricted nozzle for supplying the waterjet discharge. Another branch pipe connects to the inflow orifice of a dual-bin tank (with dimensions of 2.50 m long $\times$ 0.85 cm wide $\times$ 0.63 cm high for both the inflow and outflow sections) separated by a right-angled triangular weir (with right-angled lengths measuring 4 inches) to regulate the inflow discharge of the waterjet. During the measurement periods, the inflow discharge of the experimental events is recorded by the #2 electromagnetic flowmeter while being monitored synchronously through the discharge difference between the #1 electromagnetic flowmeter and the triangular weir. During non-measurement periods, it is possible to seal the branch pipe for waterjet discharge using a removable sealing cover to maintain air isolation within the piping system. In addition, the water tanks used for the experiment were thoroughly cleaned prior to testing, and the water utilized was sourced from tap water. Given that the quality of tap water is generally stable and no adverse factors affected its quality during the testing process, it is considered appropriate to regard the water quality as a constant variable.

Moreover, for both the waterproof design of testing instruments (specifically, an OTT Parsivel^2) within the depressurized chamber and the sealing design of instrument lines (utilizing a threading device) that pass through this chamber, achieving a perfect seal under differential pressure is crucial. In the sealing process, the use of TL990-KP800 cable refractory sealing material is feasible for anti-pressure packing due to its noncorrosive properties and its ease of removal.

2.2. Design and Optimization of the Test Conditions

In the depressurized random splash experiments, a total of 25 atomized events were conducted, corresponding to five inflow discharge rates V_m (m^3/h) and five ambient pressures P_n at six measurement points (#1–6) along the overflow centerline. Each event consisted of three sets lasting 10 min (total duration: 30 min), as detailed in Table 1. The average exit velocity is subsequently calculated by dividing the inflow discharge by the outlet area of the constricted nozzle (30.00 cm^2). The generation of a distinctly observable splash effect necessitates specific conditions regarding the Weber number of the spattering body. When the Weber number of the splashing droplets reaches 500, pronounced splashing is observed [41]. Furthermore, when the Weber number on the surface of the water jet utilized in an atomized physical model exceeds 500, it ensures that the water jet is fully dispersed and fragmented. The outcomes derived from this can be applied to mechanism research and related extensions [42]. However, it is important to note that the formula for calculating the Weber number at the surface of a water jet involves taking the square root of the traditional droplet's Weber number, with the radius of curvature along its longitudinal flow trajectory serving as a characteristic length. This value is smaller at the outlet of the waterjet but increases significantly as diffusion occurs within the waterjet. Consequently, achieving exit velocities for rectangular water jets pumped to approximately 15 m/s satisfies requirements for studying the atomization phenomena associated with these jets. Thus, such tests can be regarded as a generalized monomer test analysis at a 1:1 scale ratio. Additionally, conversions may be made based on similarities in scale modeling.

Table 1. Test parameters of the depressurized random splash experiments.

D_{mn}	Q_m m^3/h	V_m m/s	P_n/P_0 P_0 = 101.325 kPa				
$D_{1(1-5)}$	150.00	13.89	1.02	0.94	0.86	0.76	0.66
$D_{2(1-5)}$	155.00	14.35	1.02	0.94	0.86	0.76	0.66
$D_{3(1-5)}$	160.00	14.81	1.02	0.94	0.86	0.76	0.66
$D_{4(1-5)}$	165.00	15.28	1.02	0.94	0.86	0.76	0.66
$D_{5(1-5)}$	170.00	15.74	1.02	0.94	0.86	0.76	0.66

The initial step in data analysis involves calculating the atomized rainfall intensity RI_D (mm/h) (Formulas (1 and 2), see [43,44]) at the measurement point under normal ambient pressure. This corresponds to the actual atmospheric pressure within the laboratory setting (P_1 = 1.02P_0)) for further analysis. As illustrated in Figure 4, the atomized rain intensity directed downstream toward test surfaces 1# to 6# under various inflow conditions, with an ambient pressure of P_1 (1.02P_0), demonstrates that the atomized rain intensity increases as the inflow rate of the waterjet on the same test surface rises. Conversely, when the inflow rate of the waterjet remains constant, it exhibits a decreasing trend along the flow direction, which aligns with existing research findings [9]. Boundary values of 2, 10, 40, 200, and 600 mm/h have been selected for characteristic lines of rainfall intensity in accordance with industry standards outlined in the design specifications for concrete arch dams (SL282-2018) and spillways (SL253-2018). As illustrated in Figure 4, test surface #6 (flow direction coordinate of 10.750 m) under working condition $D_{(1-3)1}$ and test surface #5 (flow direction coordinate of 10.625 m) under working condition $D_{(3-5)1}$ belong to the above different protected zones. Therefore, these conditions can be considered baseline conditions for analysis. It can collaboratively reflect the impact of the inflow discharge from the waterjet and the measurement location on the test surface.

$$RI_D = \frac{(RA_D)_T}{\Delta t} \times 3600 = \frac{6\pi}{10^4} \times \sum_{i=3}^{32}\sum_{j=1}^{32} D_i^3 \times \frac{n_{ij}}{A_{eff}(D_i) \times \Delta t} \tag{1}$$

$$A_{\text{eff}}(D_i) = L \times (W - \frac{D_i \cdot 10^{-3}}{2}) \tag{2}$$

Here, Δt is the per-unit sampling period, 60 s; $(RA_D)_T$ is the precipitation measured during this sampling interval, mm; n_{ij} is the graded particle number, which refers to the quantity of particles per unit sampling period corresponding to a specific subchannel within a 30 × 32 coordinate matrix without two small particle size channels (Table A1); and $A_{eff}(D_i)$ is the effective sampling area estimated related to graded diameter sizes (D_i, mm), a flow length (W, m) of 0.03 m, and a transverse width (L, m) of 0.18 m (Equation (1)).

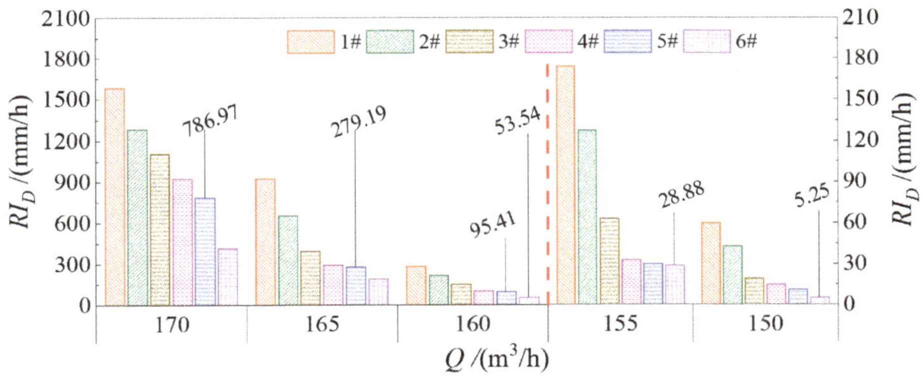

Figure 4. Atomized rain intensity distribution under experimental conditions of $D_{(1-5)1}$. (P ($D_{(1-5)1}$) = 1.02 P_0; $Q(D_{11})$–$Q(D_{51})$ = (150–170) m^3/h; ΔQ = 5 m^3/h).

3. Experimental Results and Discussion

3.1. Effects of Low Ambient Pressure on the Two-Dimensional Atomized Raindrop Spectrum

The two-dimensional atomized raindrop spectra of the above baseline conditions are compared under a normal ambient pressure environment (P_1 = 1.02P_0) and a low ambient pressure environment (P_5 = 0.66P_0), as shown in Figure 5. In total, 32 equidistant intervals are established along both the vertical and horizontal coordinates, with corresponding measurements of the particle equivalent diameter D_i (mm) and falling velocity V_{zj} (m/s) indicated. The one-minute average particle counts n_{ij} in each subchannel during the sampling period were categorized into nine distinct color levels. N_{st} represents the number concentration (number/(54 cm^2)) across all subchannels within a one-minute timeframe.

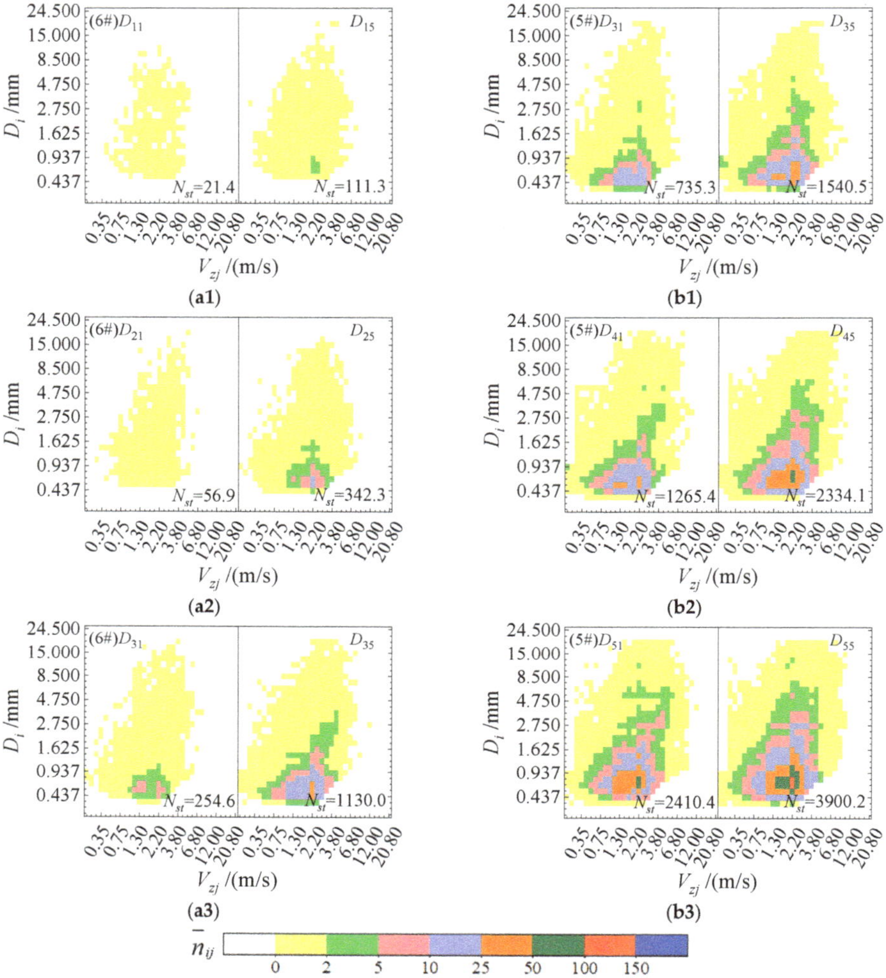

Figure 5. Two-dimensional atomized raindrop spectrum of typical atomizing rain intensity conditions under different ambient pressures ($P(D_{(1\text{–}3)1}$ and $D_{(3\text{–}5)1}) = 1.02\ P_0$, $P(D_{(1\text{–}3)5}$ and $D_{(3\text{–}5)5}) = 0.66\ P_0$; the flow direction coordinate of the test surface #5 and 6 is 10.625 m and 10.750 m, respectively). (**a1**) $RI_D(D_{11}) = 5.25$ mm/h, $Q(D_{1p}) = 150$ m^3/h; (**a2**) $RI_D(D_{21}) = 28.88$ mm/h, $Q(D_{2p}) = 155$ m^3/h; (**a3**) $RI_D(D_{31}) = 53.54$ mm/h, $Q(D_{3p}) = 160$ m^3/h; (**b1**) $RI_D(D_{31}) = 95.41$ mm/h, $Q(D_{3p}) = 160$ m^3/h; (**b2**) $RI_D(D_{41}) = 279.19$ mm/h, $Q(D_{4p}) = 165$ m^3/h; (**b3**) $RI_D(D_{51}) = 786.97$ mm/h, $Q(D_{5p}) = 170$ m^3/h.

Figure 5 shows that the atomized raindrop size two-dimensional sequence diagrams exhibit a continuous spectrum in both velocity and scale from the perspective of distribution and morphology. The distribution domain demonstrates geometric similarity, characterized by a "skewed pear shape" that lacks large droplet regions at low velocities and small droplet regions at high velocities. The total number of particles across all subchannels within a one-minute timeframe, N_{st} and the one-minute average particle counts in each subchannel n_{ij} are influenced not only by the inflow rate of the waterjet and the position of the measuring point but also by the environmental ambient pressure.

With increasing atomized rain intensity under baseline conditions, the influence rate of N_{st} on low ambient pressure initially increases, then decreases, and ultimately stabilizes.

When the environmental ambient pressure decreases by 0.10 P_0 relative to the baseline conditions, the increase rate of N_{st} is 0.51 times greater when the atomized rainfall intensity approaches or exceeds 100 mm/h. In contrast, when the atomized rainfall intensity is less than 100 mm/h, the influence rate of N_{st} increases by a factor of 1.45. Moreover, the variation in color gradients representing n_{ij} in each subchannel indicates that when other conditions are held constant, a decrease in ambient pressure results in a significant increase in n_{ij} during the sampling period within the small particle size range. In contrast, the changes observed within the larger particle size range are relatively minor.

Furthermore, by utilizing a particle diameter of 5 mm (the 20th particle size category in Table A1 of Appendix A) as the boundary, the number-dominated isovolumetric diameter of the rain intensity $(RI_D)_{nd}$ and the mass-dominated isovolumetric diameter of the rain intensity $(RI_D)_{md}$ can be calculated, as shown in Table 2. These metrics reflect the contribution of raindrop intensities from two distinct particle size segments. With increasing atomized rain intensity under baseline conditions, the growth rate of the atomized rain intensity initially decreases before fluctuating. When the environmental ambient pressure decreases by $0.10P_0$ relative to the baseline conditions, the fluctuation range of the RI_D increase rate is 0.38 to 0.87, corresponding to a rainfall intensity range of 28.88 mm/h to 1066.91 mm/h. The fluctuation range of the RI_D increase rate is 1.69 when the atomized rain intensity is 5.25 mm/h. In addition, when the atomized rain intensity under baseline conditions approaches or exceeds 20 mm/h, the impact of low ambient pressure on the ratios of $(RI_D)_{nd}$ and $(RI_D)_{md}$ becomes negligible. Consequently, their ratios remain relatively stable, with average values of 0.87 and 0.13, respectively.

Table 2. Variation characteristics of atomized rain intensity induced by low ambient pressure.

Test Condition / Atomized Rain Intensity (mm/h)	6#						5#					
	D_{11}	D_{15}	D_{21}	D_{25}	D_{31}	D_{35}	D_{31}	D_{35}	D_{41}	D_{45}	D_{51}	D_{55}
RI_D	5.25	31.94	28.88	57.89	53.54	167.64	95.41	180.99	279.19	601.98	786.97	1066.91
$(RI_D)_{nd}$	1.71	5.36	3.37	11.01	7.92	22.91	13.68	31.09	30.54	61.15	74.97	108.66
$(RI_D)_{md}$	3.54	26.59	25.51	46.88	45.62	144.73	81.73	149.9	248.65	540.83	712	958.24
$(RI_D)_{nd}/RI_D$	0.33	0.17	0.12	0.19	0.15	0.14	0.14	0.17	0.11	0.10	0.10	0.10
$(RI_D)_{md}/RI_D$	0.67	0.83	0.88	0.81	0.85	0.86	0.86	0.83	0.89	0.90	0.90	0.90

Note: P ($D_{(1-3)1}$ and $D_{(3-5)1}$) = $1.02P_0$, P ($D_{(1-3)5}$ and $D_{(3-5)5}$) = $0.66P_0$.

The significant increase in the number of particles within the small droplet size range during atomization clearly serves as a driving characteristic for the increase in atomized rainfall intensity under low ambient pressure. Additionally, an increase in the number of particles within the larger droplet size range can also contribute to the intensification of atomized rainfall under low ambient pressure conditions; however, this nonconstant increase may lead to greater randomness in the atomized rain intensity.

3.2. *Effects of Low Ambient Pressure on Three-Dimensional Atomized Raindrop Spectra*

The three-dimensional atomized raindrop spectra under the aforementioned baseline conditions are presented in single logarithmic coordinates. These spectra are compared in two different environments: normal ambient pressure ($P_1 = 1.02P_0$) and low ambient pressure ($P_5 = 0.66P_0$), as illustrated in Figure 6. The ordinate of Figure 6 represents the individual three-dimensional number concentration of atomized raindrops, denoted as $N(D_i)$ (particle counts/($m^3 \cdot mm$)), corresponding to 32 distinct raindrop diameter sizes with unequal spacing. $N(D_i)$ is calculated via Formulas (3) and (4), see [45]. ΔD_i (mm) is the channel width of each particle size class. The falling velocity of the particles at each velocity level V_j (m/s) is calculated as the average falling velocity V_{zj} (D_i) (m/s) measured at each particle size level. The fitted spectral pattern adheres to a gamma distribution, employing the M246 moment estimation method.

$$N(D_i) = \sum_{j=1}^{32} \frac{n_{ij}}{A_{eff}(D_i) \times \Delta t \times V_j \times \Delta D_i} \quad (3)$$

$$V_j = \frac{\sum\limits_{j=1}^{32} \overline{n}_{ij} \times V_{zj}(D_i)}{\sum\limits_{j=1}^{32} \overline{n}_{ij}} \quad (4)$$

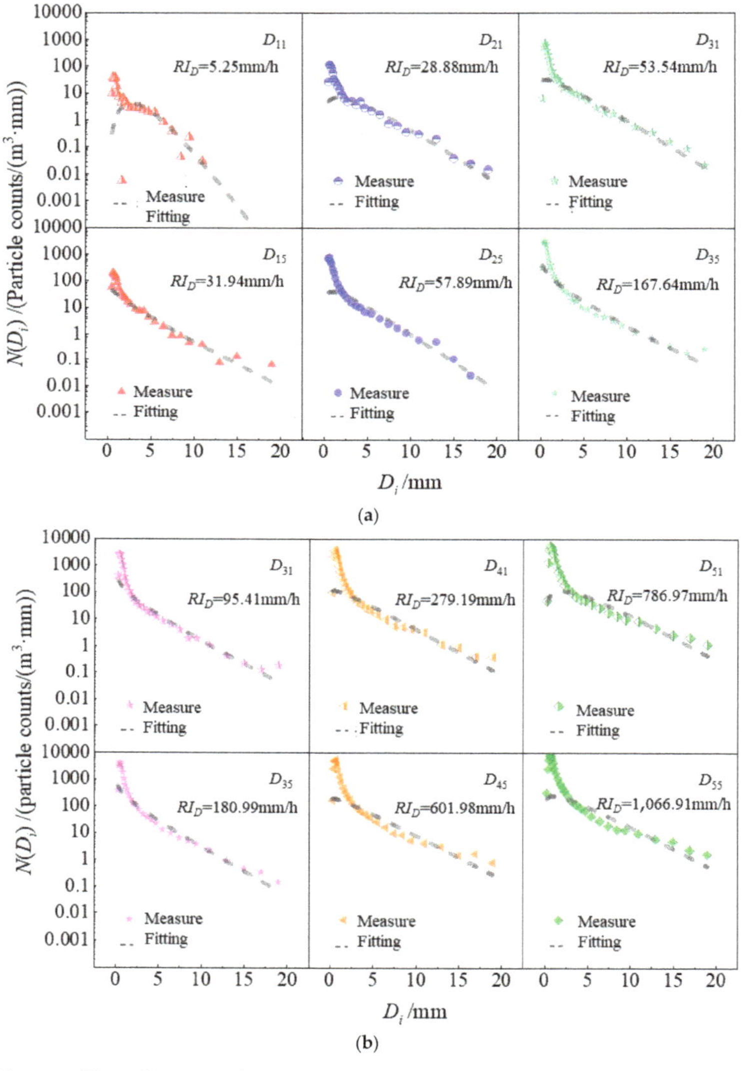

Figure 6. Three-dimensional atomized raindrop spectra of typical atomizing rain events intensity conditions under different ambient pressures: (**a**): Mearing points of #6 (Flow Direction coordinates are 10.750 m); (**b**): Mearing points #5 (the flow direction coordinates are 10.625 m) ($P(D_{q1}) = 1.02P_0$, $P(D_{q5}) = 0.66P_0$; $Q(D_{1p})$–$Q(D_{5p})$ = (150–170) m^3/h, ΔQ = 5 m^3/h).

As shown in Figure 6, the primary spectral type of these three-dimensional atomized raindrop spectra is unimodal. As the size of the atomized raindrop particles increases, their concentration initially rises before subsequently declining. The main peak is predominantly located within the small particle size range of less than 1 mm, while minor fluctuations are primarily observed in the tail end of the larger particle size segment. The peak value of the number concentration of atomized raindrops $N(D_i)$ is influenced not only by the inflow rate of the waterjet and the position of the measuring point but also by environmental ambient pressure.

With increasing atomized rain intensity under baseline conditions, the influence of the $N(D_i)$ peak value on low ambient pressure initially increases, then decreases, and ultimately stabilizes. When the environmental ambient pressure decreases by $0.10P_0$ relative to the baseline conditions, the rate of increase in the $N(D_i)$ peak value is 0.41 times greater when the atomized rainfall intensity approaches or exceeds 100 mm/h. In contrast, when the atomized rainfall intensity is less than 100 mm/h, the rate of increase in the $N(D_i)$ peak value increases by a factor of 1.48. Moreover, the three-dimensional atomized raindrop spectrum is self-similar and can be approximated using a gamma distribution curve. The fitted particle size in the mid-range aligns with the measured values; however, discrepancies are observed at both ends of the size spectrum, along with peak-shaving characteristics. This phenomenon is attributed to the broad spectral width of atomized raindrop distributions and the high concentration of small droplets, which results in significant increases and decreases in rates on either side of the main peak.

Figure 6 also shows that the ranges for the modal particle size, maximum particle size, and minimum particle size are relatively stable, measuring 0.437–0.687 mm, 0.312–0.562 mm, and 11.000–19.000 mm, respectively. When the atomized rain intensity is 5.25 mm/h, the corresponding maximum diameter is 11.000 mm, which is similar to the maximum particle size (10 mm) of atomized rain measured using the prototype at Wantangshui Power Station [22]. The range of the maximum diameter of atomized raindrops under different conditions is between 17.000 mm and 19.000 mm, with a concentration of raindrops at approximately 19 mm. This finding is consistent with the maximum size range of raindrops observed in the splash area as measured by Qiu Haijuan et al. during their model tests [46]. When the atomized rain intensity exceeds 10 mm/h, the resulting atomized raindrops can be compared with the mixed-phase precipitation raindrop spectrum characteristic of "heavy rainfall combined with hail or snowflakes" on a particle size scale [47–50].

It is evident that the impact of ambient pressure reduction on the three-dimensional atomized raindrop spectra is primarily observed in the spectral peak values, whereas it does not significantly affect the spectral type or range. The spectral types exhibit self-similarity, making it scientifically valid to fit them using a gamma approximation.

3.3. *Effects of Low Ambient Pressure on the Statistical Characteristics of the Atomized Raindrop Spectrum*

The reduction in ambient pressure, the migration upstream of measurement points, and the increase in inflow discharge are key factors that contribute to the heightened intensity of atomization rain. Based on the preceding analysis, it is imperative to conduct a more in-depth investigation into the influence of ambient pressure on key statistical particle size parameters of atomized raindrops under consistent rain intensities during atomization. Under the same atomized rain intensity, the arithmetic mean diameter of atomized raindrops (D_{ave}) and the relationship between the mass-weighted mean diameter of atomized raindrops ($(D_m)_{ave}$) in different ambient pressure environments are illustrated in Figure 7.

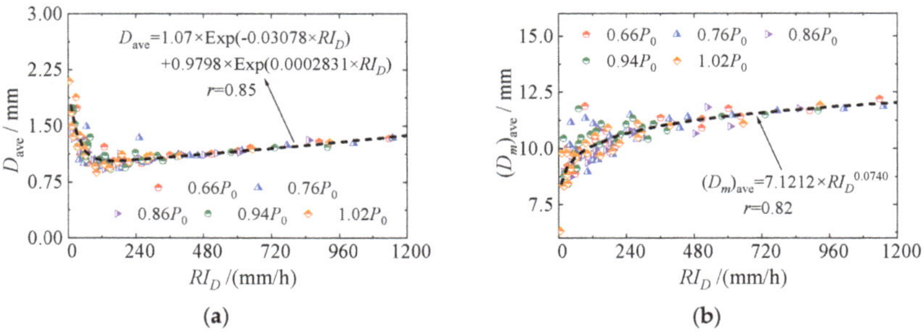

Figure 7. Statistical Characteristics of the Atomized Raindrop Spectrum. (**a**) Correlation analysis of RI_D–D_{ave}; (**b**) $(D_m)_{ave}$–RI_D under a series of test conditions.

As depicted in Figure 7, the D_{ave} and $(D_m)_{ave}$ of atomized raindrops fluctuate around the fitted curve under varying pressure conditions, while maintaining consistent atomized rain intensity. The observed trends are not distinctly pronounced when influenced by ambient pressure. As the atomized rain intensity increases, the D_{ave} of the atomized raindrops initially decreases rapidly before gradually increasing. The minimum value of D_{ave} corresponding to this atomized rain intensity is 153.78 mm/h, with a range of 0.88–1.88 mm. In contrast, $(D_m)_{ave}$ shows an exponential growth trend that eventually stabilizes, with an interval ranging from 6.32 to 12.03 mm. In addition, the $(D_m)_{ave}$ of atomized raindrops, along with the generalized intercept parameter N_0^*, can represent the synthesis of both the overall diameter and the number concentration within the three-dimensional spectrum of an atomized raindrop. A correlation analysis was performed between the logarithm of $\mathrm{Log}_{10}N_0^*$ and $(D_m)_{ave}$, N_0^*, and RI_D. The findings are presented in Figure 8.

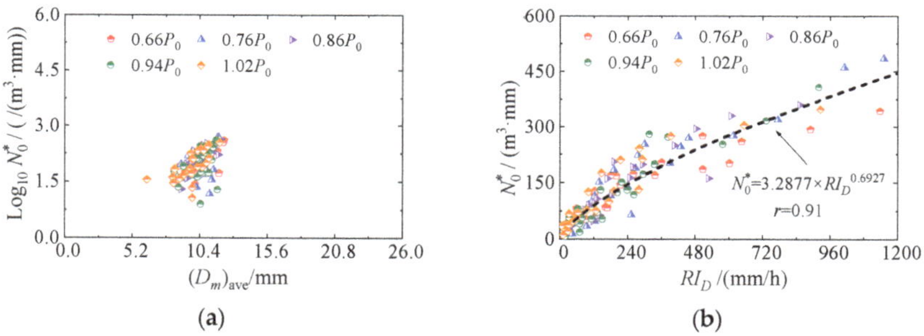

Figure 8. Statistical Characteristics of the Atomized Raindrop Spectrum. (**a**) Correlation analysis of $\log_{10}(N_0^*)$–$(D_m)_{ave}$; (**b**) N_0^*–RI_D under a series of test conditions.

As also depicted in Figure 8, the $\mathrm{Log}_{10}N_0^*$ of the atomized raindrop fitting spectrum corresponding to the same $(D_m)_{ave}$ and N_0^* of the atomized raindrop fitting spectrum related to identical RI_D are a multivalued discrete distribution and have no obvious correlation with the ambient pressure. The $\mathrm{Log}_{10}N_0^*$ of the fitting spectrum of atomized raindrops tends to increase with increasing $(D_m)_{ave}$. The N_0^* of the fitting spectrum of atomized raindrops is positively correlated with the RI_D. In the description of the natural raindrop spectrum, when $\mathrm{Log}_{10}N_0^*$ tends to be stable with increasing $(D_m)_{ave}$, the collision and fragmentation of natural raindrops tends to be balanced with increasing rain intensity [51,52]. Similarly, the atomized raindrops within the test area undergo dynamic evolution through processes of collision and fragmentation, with their activity increasing in correlation with the intensity of atomized rainfall.

4. Numerical Model Application Optimization

For existing flood discharge facilities, variations in reservoir depth can lead to corresponding changes in the upstream water level. Additionally, the dynamics of the associated waterjets are influenced by alterations in their inflow flow rate. This study presents a generalized model that is not derived from any specific project. While the high-speed water jet is supplied by a jet pump designed to regulate the inflow flow rate, the magnitude of this flow rate can qualitatively indicate the water level within the reservoir. In relation to prototype projects, it is important to note that ejection conditions are diverse. Consequently, employing a constant physical model presents challenges and makes it difficult to efficiently analyze the influence range of fogging rain within these projects. Therefore, a semi-empirical and semi-theoretical numerical prediction model for random splashing during jet atomization is considered valuable.

Random splash simulation models involving semiempirical and semitheoretical methods, which are developed based on the principles of Lagrangian mechanics, constitute a significant theory for the numerical prediction of flood discharge atomization. The Runge–Kutta method is employed to track the trajectories of both the water jet and the droplets, whereas the Monte Carlo method is utilized to assess the cumulative impact of the droplets on the terrain boundaries. Here, the application of experimental findings is examined to enhance the theoretical framework of stochastic numerical models.

In theory, a fundamental assumption is that the atomized water droplets are spherical and independent of the effective splashing volume on the outer surface of the waterjet dV. The quantity of atomized water droplets n_D is calculated using Formula (5), whereas the distribution of their diameters $f(D)$ is determined via Formula (6) [33]. The proportions of different atomized raindrop size counts are relatively stable in the numerical calculations.

$$f(D) = \frac{1}{{b_1}^{a_1} \times \Gamma(a_1)} \times D^{a_1 - 1} \times e^{-\frac{D}{b_1}} \qquad (a_1 = 2, b_1 = 0.5 \times \overline{D}) \tag{5}$$

$$n_D = \frac{6 \times dV}{\pi \times \overline{D}^3} \tag{6}$$

The probability density of the number concentration corresponding to the median value of each group of atomized raindrops with varying diameters is illustrated in Figure 9. Its predominant spectral type is unimodal, and its skewness trend increases rapidly and then decreases gradually as the particle size of the atomized raindrops increases, ultimately resulting in a left-skewed distribution. The positions of the main peaks are predominantly concentrated within the small particle size range, specifically for particles less than 1 mm in diameter. Furthermore, the impact of variations in the testing conditions on the probability density of the number concentration of spray droplets appears to be minimal, with a range between 1.13 and 1.93.

Therefore, when evaluating the impact of low ambient pressure on the atomization source, as the measurement area for the spectrum of atomized raindrops is located within a rainstorm zone, it can be classified as a downstream source of atomization. The distribution ratio of atomized raindrops across various particle sizes remains relatively stable.

In previous experiments, atomized rain was collected by an inclined plate positioned downstream of the water tongue impact area. The measured atomized rainfall can be considered the atomized source for the downstream region. When the ambient pressure is reduced by $0.10P_0$, the average rate of atomization on the inclined test surface ranges from 115 to 127.68 m^3/h. Correspondingly, there is a linear increase between 0.28 L/h and 2.83 L/h, with a relative growth rate ranging from 20.48% to 11.56% [21].

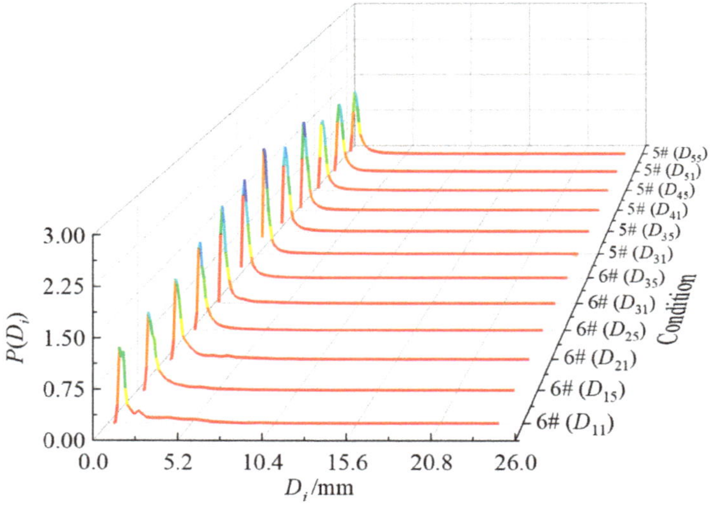

Figure 9. The probability density of the number concentration corresponding to the median value of each group of atomized raindrops with varying diameters ($P(D_{q1}) = 1.02P_0$,$P(D_{q5}) = 0.66P_0$; $Q(D_{1p})$ $-Q(D_{5p})$ = (150–170) m^3/h, $\Delta Q = 5$ m^3/h).

In this work, the atomized rain calculation is based on the raindrop spectrum at specific measurement points. The enhancement characteristics of rain intensity at various atomization planes under low pressure are illustrated in Figure 10a,b. The color code represents the atomized rain areas corresponding to specific rainfall intensity thresholds of 10 mm/h, 40 mm/h, 200 mm/h, and 600 mm/h. These categories are designated class ①, class ②, class ③, class ④, and class ⑤, respectively. Additionally, rainfall intensities of 1000 mm/h and above are classified as class ⑤+. Based on the test conditions of normal ambient pressure ($P_1 = 1.02P_0$), the average growth rate $\Delta\eta$ of the rain intensity ΔRI_D on the spray surface of the discharge flow was calculated when the ambient pressure was decreased by $0.1P_0$, as depicted in Figure 10c.

According to the analysis presented in Figure 10, as the environmental air pressure decreases ($1.02P_0$, $0.94\ P_0$, $0.86\ P_0$, $0.76\ P_0$, $0.66\ P_0$), the measurement area under working condition $D_{1(1\text{–}5)}$ ($Q_1 = 150$ m^3/h, $V_1 = 13.89$ m/s) shifts in atomized classification from levels ① to ③ to ② to ③; conversely, for working condition $D_{4(1\text{–}5)}$ ($Q_4 = 165$ m^3/h, $V_3 = 15.28$ m/s), the corresponding atomized classification changes from levels ③ to ⑤ to levels ④ to ⑤+.

Therefore, when evaluating the impact of low ambient pressure on the protection of the atomization field, the atomized classification of the testing surface is influenced by low air pressure environments. This may result in a one-level enhancement of the atomization protection level in certain areas, without the occurrence of multi-level increases, as all other conditions remain constant. This effect arises because as the reference rainfall intensity on the atomization surface increases, the average growth rate of the rainfall intensity affected by low air pressure initially declines sharply before gradually stabilizing.

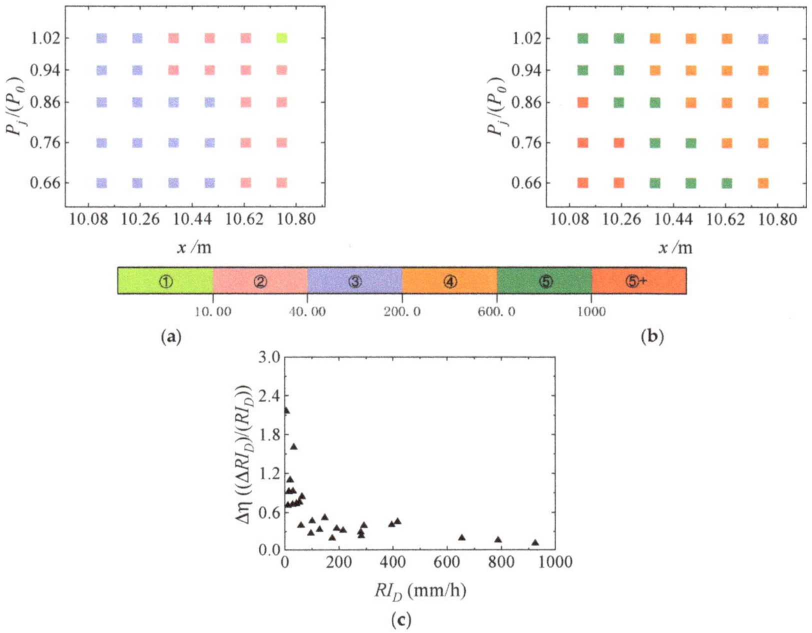

Figure 10. The enhancement characteristics of rain intensity at various atomization planes under low pressure. (**a**) Test conditions are $D_{1(1–5)}$ with an inflow discharge of 150 m^3/h; (**b**) Test conditions are $D_{4(1–5)}$ with an inflow discharge of 165 m^3/h; (**c**) Relative growth law of atomized rain intensity under low ambient pressure.

5. Conclusions

With the aim of decreasing the degree of aggravated risk of atomized rain at high-altitude hydropower stations, this study provides insights into the atomized raindrop size distributions of random splash experiments characterized by high-speed velocities and low-pressure environments. Random splash experiments were conducted under coupling conditions with high inflow discharge rates and low ambient pressures. The effects of low pressure on the atomized raindrop size distributions at six measuring points downstream along the overflow centerline collected by OTT Parsivel² can be summarized as follows:

(1) The two-dimensional atomized raindrop spectra reveal that the intensity of atomized rain increases with the increasing total number concentration of two-dimensional raindrop spectra at low ambient pressure. Notably, the diameter of the small particles within the subchannels significantly increases. In terms of the rain intensity components, the proportion of the number-dominated isovolumetric diameter of the rain intensity and the mass-dominated isovolumetric diameter of rain intensity remains relatively stable and is not influenced by variations in ambient pressure;

(2) According to the three-dimensional raindrop spectra, the increase in atomized rain intensity under low pressure is reflected in both the quantity and concentration of atomized rain, with the most pronounced effects occurring at peak levels. The maximum particle size and range are influenced by the intensity of atomized rain; however, the impact of low pressure appears to be less significant. The relationship between the atomized raindrop number concentration and characteristic particle size exhibits a unimodal distribution that skews toward smaller particles, which can be effectively modelled using the gamma distribution;

(3) Under the same atomized rain intensity, the atomized raindrops undergo a dynamic evolution characterized by collision and fragmentation, with their quantity increasing in response to heightened atomized rain intensity. Notably, there is no significant correlation between the statistical characteristics of three-dimensional raindrop spectra and low ambient pressure, although the relationships among the arithmetic mean diameter of atomized raindrops, the mass-weighted mean diameter of atomized raindrops, and the generalized intercept parameter concerning the atomized rain intensity exhibit a clear trend;
(4) Combined with the experimental results, this study provides a substantial reference for establishing random splash simulation models characterized by low pressure. It is advisable to appropriately increase the amount of the atomization source to accurately reflect the impact of low pressure on the atomized rain intensity within existing empirical formulas for the distribution of atomized raindrop diameters. When atomized protection is categorized into zones, enhancing the area adjacent to the characteristic rain intensity line is crucial, as its protection level may need to be elevated by one grade in accordance with current industry standards.

The statistical characteristics of atomized raindrop size distribution datasets under depressurized experimental conditions clearly demonstrated self-enhancing redistribution features. When investigating the enhancement effect of low ambient pressure on atomized rain intensity, it can be generalized that a proportional rise in the total number of particles generated by splashed atomized raindrops, without considering minor variations in particle size distribution, can establish a sufficient condition for advancements in numerical random splash models concerning low ambient pressure effects. To predicting rain protection scopes at a high-altitude hydropower station, future research should focus on how low ambient pressure affects the boundary conditions of atomization sources under identical discharge conditions, including trajectory distance of aerated waterjets; future research also should focus on the investigation of thematic mechanism experiments in conjunction with engineering challenges and explore the characteristics of atomized fields and sources from multiple perspectives, including water quality, water volume, and water temperature, so as to facilitate the development of advanced meteorological models for atomized protection in high-altitude areas by integrating both physical and numerical methods.

Author Contributions: Conceptualization, J.L.; methodology, J.L. and D.L. (Dan Liu); formal analysis, F.L. and D.L. (Dan Liu); data curation, D.L. (Dan Liu); writing—original draft preparation, D.L. (Dan Liu) and J.S.; writing—review and editing, D.L. (Dongming Liu), B.M., L.Y., Y.Z., C.X. and J.Z. All authors have read and agreed to the published version of the manuscript.

Funding: This research was funded by State Key Laboratory of Hydraulic Engineering Intelligent Construction and Operation, Tianjin University, grant number HESS-2216, National Natural Science Foundation of China, grant number U1765202, China Postdoctoral Science Foundation, grant number 2023M732603, the Joint Funds of the Zhejiang Provincial Natural Science Foundation of China, grant number LZJWY22E090003, University-Level Key Course of Zhejiang University of Water Resources and Electric Power, grant number ZDKC202319, Scientific research foundation of Zhejiang University of Water Resources and Electric Power, grant number xky2022002 and Nanxun scholars program of ZJWEU, grant number RC2025011034, and Science and Technology Projects of Xizang Autonomous Region, China, grant number XZ202501ZY0109. All workers from the State Key Laboratory of Hydraulic Engineering, Intelligent Construction, and Operation of Tianjin University are acknowledged. The authors are also grateful for the assistance of the anonymous reviewers.

Data Availability Statement: The original contributions presented in this study are included in the article. Further inquiries can be directed to the corresponding author.

Acknowledgments: The authors would like to express their sincere gratitude to all those who have offered selfless help during the course of this research.

Conflicts of Interest: Jizhong Shi was employed by the company Zhejiang East China Engineering Consulting Co., Ltd. of PowerChina Huadong Engineering Corporation Limited. Yongsheng Zheng was employed by the company Yalong River Hydropower Development Co., Ltd. The remaining authors declare that the research was conducted in the absence of any commercial or financial relationships that could be construed as a potential conflict of interest.

Appendix A

Table A1. Graded diameter size and fall velocity of atomized raindrop size distributions databases collected by the OTT Parsivel^2.

Classes Number	Raindrop Diameter Sizes		Raindrop Fall Velocities	
	Mid-Value of Class (mm)	Class Width (mm)	Mid-Value of Class (m/s)	Class Width (m/s)
1	0.062 *	0.125	0.050	0.100
2	0.187 *	0.125	0.150	0.100
3	0.312	0.125	0.250	0.100
4	0.437	0.125	0.350	0.100
5	0.562	0.125	0.450	0.100
6	0.687	0.125	0.550	0.100
7	0.812	0.125	0.650	0.100
8	0.937	0.125	0.750	0.100
9	1.062	0.125	0.850	0.100
10	1.187	0.125	0.950	0.100
11	1.375	0.250	1.100	0.200
12	1.625	0.250	1.300	0.200
13	1.875	0.250	1.500	0.200
14	2.215	0.250	1.700	0.200
15	2.375	0.250	1.900	0.200
16	2.750	0.500	2.200	0.400
17	3.250	0.500	2.600	0.400
18	3.750	0.500	3.000	0.400
19	4.250	0.500	3.400	0.400
20	4.750	0.500	3.800	0.400
21	5.500	1.000	4.400	0.800
22	6.500	1.000	5.200	0.800
23	7.500	1.000	6.000	0.800
24	8.500	1.000	6.800	0.800
25	9.500	1.000	7.600	0.800
26	11.000	2.000	8.800	1.600
27	13.000	2.000	10.400	1.600
28	15.000	2.000	12.000	1.600
29	17.000	2.000	13.600	1.600
30	19.000	2.000	15.200	1.600
31	21.500	3.000	17.600	3.200
32	24.500	3.000	20.800	3.200

Notes: The first and second particle sizes represented by * are discarded in the database analysis due to the low signal-to-noise ratios of the instrument. The median of the subsequent class is determined by adding the class width at the level of the current median to the existing median, whether in terms of raindrop diameter sizes or raindrop fall velocities.

References

1. Ward, P.J.; Jongman, B.; Aerts, J.C.J.H.; Bates, P.D.; Botzen, W.J.W.; Diaz Loaiza, A.; Hallegatte, S.; Kind, J.M.; Kwadijk, J.; Scussolini, P.; et al. A global framework for future costs and benefits of river-flood protection in urban areas. *Nat. Clim. Change* **2017**, *7*, 642–646. [CrossRef]
2. Ballesteros-Canovas, J.A.; Czajka, B.; Janecka, K.; Lempa, M.; Kaczka, R.J.; Stoffel, M. Flash floods in the Tatra Mountain streams: Frequency and triggers. *Sci. Total Environ.* **2015**, *511*, 639–648. [CrossRef]
3. Angulo-Martínez, M.; Beguería, S. Trends in rainfall erosivity in NE Spain at annual, seasonal and daily scales, 1955–2006. *Hydrol. Earth Syst. Sci. Discuss.* **2012**, *9*, 6285–6309. [CrossRef]
4. Blocken, B.; Carmeliet, J. A review of wind-driven rain research in building science. *J. Wind Eng. Ind. Aerodyn.* **2004**, *92*, 1079–1130. [CrossRef]
5. Dave, N.; Mittelstet, A.; Korus, J.; Waszgis, M. Impact of an extreme flood event on streambank retreat: Cedar River, Nebraska, USA. *J Am Water Resour Assoc.* **2020**, *56*, 528–541. [CrossRef]
6. Vahedifard, F.; AghaKouchak, A.; Ragno, E.; Shahrokhabadi, S.; Mallakpour, I. Lessons from the Oroville dam. *Science* **2017**, *355*, 1139–1140. [CrossRef]
7. Liu, G.; Tong, F.; Tian, B.; Gong, J. Finite element analysis of flood discharge atomization based on water–air two-phase flow. *Appl. Math. Model.* **2020**, *81*, 473–486. [CrossRef]
8. Gong, J.B.; Pi, J.H. Reflection and countermeasures of "7·19" flood event of Shuibuya dam. *Dam Saf.* **2018**, *2*, 20–26.
9. Lian, J.J.; He, J.L.; Liu, F.; Ran, D.J.; Wang, X.Q.; Wang, C. An improved empirical model for flood discharge atomization and its application to optimize the flip bucket of the Nazixia Project. *Int. J. Environ. Res. Public Health* **2019**, *16*, 316. [CrossRef]
10. Wang, F.H. Influence of discharging foggy and its protection at Ertan hydropower station. *Dam Saf.* **2011**, *5*, 42–46.
11. Wu, W.Y.; Yang, Y.H.; Min, S.H. Analysis on collapse and sliding control of high slope in atomized area of right bank of Ertan hydropower station. *Dam Saf.* **2018**, *4*, 49–52.
12. Li, Z. On landslide induced by water-fog from ski-jump energy dissipation of Longyangxia Hydropower Station. *Dam Saf.* **2001**, *3*, 17–20+29–56.
13. Lian, J.J. *Hydraulic Prototype Observation Test at Jinping Hydropower Station and Its Early Warning System Development with Real-Time Monitoring for Flood Discharge Security*; Research report; Tianjin University: Tianjin, China, 2016.
14. Lian, J.J.; Liu, D.; Liu, F. Research progress and frontiers on flood discharge atomization of Chinese high dam projects. *J. Hydraul. Eng.* **2019**, *50*, 283–293.
15. Zeng, S.Y.; Zhang, Y.T.; Zhang, B.C.; Gu, L.; Dai, X.B.; Miao, B.G.; Wang, L.J. Discharge atomization and its affection analysis of Xiangjiaba Hydropower Station. *Water Power* **2019**, *45*, 54–58.
16. Zhou, H.; Wu, S.Q.; Chen, H.L.; Zhou, J.; Wu, X.F. Similarity criterion of flood discharge atomization. *Water Sci. Eng.* **2008**, *1*, 59–65.
17. Liu, H.T.; Sun, S.K.; Wang, X.S.; Xia, Q.F. Study on the distribution of splash intensity during nappe impingement. *J. Hydrodyn.* **2009**, *24*, 217–223.
18. Yuan, H.; Xu, W.L.; Li, R.; Feng, Y.Z.; Hao, Y.F. Spatial distribution characteristics of rainfall for two-jet collisions in air. *Water* **2018**, *10*, 1600. [CrossRef]
19. Duan, H.D.; Liu, S.H.; Luo, Q.S.; Huang, W. Rain intensity distribution in the splash region of atomized flow. *J. Hydrodyn. Ser.* **2006**, *18*, 362–366. [CrossRef]
20. Liu, F.; Lian, J.J.; Zhang, X.J.; Li, C.Y. Experimental study of atomization and splashing caused by a ski-jump jet into scour pool. *J. Hydroelectr. Eng.* **2010**, *29*, 113–117.
21. Liu, D.; Lian, J.J.; Liu, F.; Liu, D.; Ma, B.; Shi, J. An experimental study on the effects of atomized rain of a high velocity waterjet to downstream area in low ambient pressure environment. *Water* **2020**, *12*, 397. [CrossRef]
22. Wu, X.F.; Wu, S.Q.; Zhou, H.; Chen, H.L. Atomization measurement for Wantang Hydroplant. *Hydro Sci. Eng.* **2001**, *4*, 71–74.
23. Chen, D.; Jin, F.; Li, J. Study on model law for intensity of rainfall from atomization of flood discharging flow for high dam. *Water Resour. Hydr. Eng.* **2005**, *36*, 47–49.
24. Liu, S.H.; Ran, Q.S.; Luo, Q.S.; Tai, W.; Fan, M. An in-depth study of partical size of atomization rainfall and generated flow on a slope. *Eng. J. Wuhan Univ.* **2013**, *46*, 1–5.
25. Schwendner, P.; Schuerger, A.C. Exploring microbial activity in low-pressure environments. *Curr. Issues Mol. Biol.* **2020**, *38*, 163–196. [CrossRef]
26. Rengel, B.; Àgueda, A.; Pastor, E.; Casal, J.; Planas, E.; Hu, L.; Palacios, A. Experimental and computational analysis of vertical jet fires of methane in normal and sub-atmospheric pressures. *Fuel* **2020**, *265*, 116878. [CrossRef]
27. Cai, X.; Wang, X.S.; Liao, G.X. Experimental study on the effects of ambient pressure on spray characteristics of water mist. *J. Disas. Prev. Mitig. Eng.* **2012**, *32*, 235–241.
28. Ersoy, N.E.; Eslamian, M. Phenomenological study and comparison of droplet impact dynamics on a dry surface, thin liquid film, liquid film and shallow pool. *Exp. Therm. Fluid Sci.* **2020**, *112*, 109977. [CrossRef]

29. Ge, X.; Ge, Y.; Li, Q.; Cai, X.; Yang, W.; Du, Y. Effect of low air pressure on the durability of concrete. *Constr. Build. Mater.* **2018**, *187*, 830–838. [CrossRef]
30. Lian, J.J.; Dong, Z.; Liu, F.; Liu, D. Experimental study on the influence of low atmospheric pressure on the dynamic pressure. *J. Hydroelectr. Eng.* **2019**, *38*, 1–13.
31. Liu, X.L.; An, G.; Yao, Z.D. The investigation on the mechabism and sphere of influence of atomization by discharge flow. *J. Tianjin Univ.* **1991**, *24*, 30–36.
32. Liang, Z.C. A analysis and computation for droplets-splashing region of atomization water flow. *J. Yangtze River Sci. Res. Inst.* **1996**, *13*, 9–13.
33. Lian, J.J.; Li, C.Y.; Liu, F.; Wu, S.Q. A prediction method of flood discharge atomization for high dams. *J. Hydraul. Res.* **2014**, *52*, 274–282. [CrossRef]
34. Liu, H.T.; Liu, Z.P.; Xia, Q.F.; Sun, S.K. Computational model of flood discharge splash in large hydropower stations. *J. Hydraul. Res.* **2015**, *53*, 576–687. [CrossRef]
35. Xu, L.; Barcos, L.; Nagel, S.R. Splashing of Liquids: Interplay of surface roughness with surrounding gas. *Phys. Rev. Stat. Nonlin. Soft Matter Phys.* **2007**, *76*, 066311. [CrossRef]
36. Latka, A.; Strandburg-Peshkin, A.; Driscoll, M.M.; Stevens, C.S.; Nagel, S.R. Creation of prompt and thin-sheet splashing by varying surface roughness or increasing air pressure. *Phys. Rev. Lett.* **2012**, *109*, 054501. [CrossRef]
37. Song, C.; Zhou, Y.Q.; Wu, Z.H. Vertical profils of raindrop size distribution observed by Micro Rain Radar. *J. Appl. Meteor. Sci.* **2019**, *30*, 479–490.
38. Li, H.; Yin, Y.; San, Y.P.; Jin, Q. Statistical characteristics of raindrop size distribution for stratiform and convective precipitation at different altitudes in Mt. Huangshan. *Chin. J. Atmos. Sci.* **2018**, *42*, 268–280.
39. Sumesh, R.K.; Resmi, E.A.; Unnikrishnan, C.K.; Jash, D.; Sreekanth, T.S.; Resmi, M.C.M.; Rajeevan, K.; Nita, S.; Ramachandran, K.K. Microphysical aspects of tropical rainfall during Bright Band events at mid and high-altitude regions over Southern Western Ghats, India. *Atmos. Res.* **2019**, *227*, 178–197. [CrossRef]
40. Wu, Y.; Liu, L. Statistical characteristics of raindrop size distribution in the Tibetan Plateau and southern China. *Adv. Atmos. Sci.* **2017**, *34*, 727–736. [CrossRef]
41. Deegan, R.D.; Brunet, P.; Eggers, J. Complexities of splashing. *Nonlinearity* **2008**, *21*, C1–C11. [CrossRef]
42. Chen, H.L. Study of flood discharge atomization for Xiaowan Hydropower Project. *Yunnan Water Power* **1998**, *14*, 51–55.
43. Alessandro, B.; Elke, R.; Ali, T.; Ulrich, B.; Clemens, S. PARSIVEL snow observations: A critical assessment. *J. Atmos. Ocean. Technol.* **2010**, *27*, 333–344.
44. Jaffrain, J.; Berne, A. Experimental quantification of the sampling uncertainty associated with measurements from Parsivel disdrometers. *J. Hydrometeorol.* **2011**, *12*, 352–370. [CrossRef]
45. Serio, M.A.; Carollo, F.G.; Ferro, V. Raindrop size distribution and terminal velocity for rainfall erosivity studies. A review. *J. Hydrol.* **2019**, *576*, 210–228. [CrossRef]
46. Qiu, H.J. Hydraulic Characteristics of Splash of Aerated Jet. Master's Thesis, Hohai University, Nanjing, China, 2011.
47. Löffler-Mang, M.; Joss, J. An optical disdrometer for measuring size and velocity of hydrometeors. *J. Atmos. Ocean. Technol.* **2000**, *17*, 130–139. [CrossRef]
48. Luo, L.; Xiao, H.; Yang, H.L.; Chen, H.N.; Guo, J.; Sun, Y.; Feng, L. Raindrop size distribution and microphysical characteristics of a great rainstorm in 2016 in Beijing, China. *Atmos. Res.* **2020**, *239*, 104895. [CrossRef]
49. Hill, M.; Grieser, J. How to express hail intensity-modeling the hailstone size distribution. *J. Appl. Meteorol. Climatol.* **2019**, *58*, 2329–2345.
50. Huang, G.J.; Kleinkort, C.; Bringi, V.N.; Notaros, B.M. Winter precipitation particle size distribution measurement by Multi-Angle Snowflake Camera. *Atmos. Res.* **2017**, *198*, 81–96. [CrossRef]
51. Zhang, A.; Hu, J.; Chen, S.; Hu, D.; Li, H. Statistical characteristics of raindrop size distribution in the monsoon season observed in southern China. *Remote Sens.* **2019**, *11*, 432. [CrossRef]
52. Islam, T.; Rico-Ramirez, M.A.; Thurai, M.; Han, D. Characteristics of raindrop spectra as normalized gamma distribution from a Joss–Waldvogel disdrometer. *Atmos. Res.* **2012**, *108*, 57–73. [CrossRef]

Journal of
Marine Science and Engineering

MDPI

Article

A Comparative Study on Generation and Propagation of Nonlinear Waves in Shallow Waters

Jiaqi Liu [1], Masoud Hayatdavoodi [1,2,*] and R. Cengiz Ertekin [1,3]

1 College of Shipbuilding Engineering, Harbin Engineering University, Harbin 150001, China
2 Civil Engineering Department, School of Science and Engineering, University of Dundee, Dundee DD1 4HN, UK
3 Ocean & Resources Engineering Department, University of Hawaii, Honolulu, HI 96822, USA
* Correspondence: mhayatdavoodi@dundee.ac.uk; Tel.: +44-(0)1382-385720

Abstract: This study is concerned with the generation and propagation of strongly nonlinear waves in shallow water. A numerical wave flume is developed where nonlinear waves of solitary and cnoidal types are generated by use of the Level I Green-Naghdi (GN) equations by a piston-type wavemaker. Waves generated by the GN theory enter the domain where the fluid motion is governed by the Navier–Stokes equations to achieve the highest accuracy for wave propagation. The computations are performed in two dimensions, and by an open source computational fluid dynamics package, namely OpenFoam. Comparisons are made between the characteristics of the waves generated in this wave tank and by use of the GN equations and the waves generated by Boussinesq equations, Laitone's 1st and 2nd order equations, and KdV equations. We also consider a numerical wave tank where waves generated by the GN equations enter a domain in which the fluid motion is governed by the GN equations. Discussion is provided on the limitations and applicability of the GN equations in generating accurate, nonlinear, shallow-water waves. The results, including surface elevation, velocity field, and wave celerity, are compared with laboratory experiments and other theories. It is found that the nonlinear waves generated by the GN equations are highly stable and in close agreement with laboratory measurements.

Keywords: Green–Naghdi equations; Navier–Stokes equations; Boussinesq equations; KdV equation; solitary wave; cnoidal wave; numerical wave tank

1. Introduction

Solitary wave is a nonlinear wave with an isolated elevation of finite amplitude and permanent form over a flat seafloor. The presence of the solitary wave was first reported by [1], who conducted extensive experiments on the generation and propagation of the wave in 1834 and 1835, see [2]. At the time, there was a conflict between Russell's observations and [3] prediction (shallow-water-theory) that a wave of finite amplitude cannot propagate without a change of form. This conflict was resolved independently by [4–8]. Boussinesq and Rayleigh showed that appropriate allowance for vertical acceleration and finite amplitude leads to an explicit solution of the solitary wave. Ref. [6]'s theory of solitary wave was accepted although the solution is only first-order accurate. Ref. [9] suggested a theoretical solution whose mathematical formulations are provided by Lee et al. (1982). As [9]'s equations are implicit, an iterative method, such as Newton–Raphson, is needed to solve the equations. An equation only admitting right-running solutions was derived by [10], named the Korteweg–de Vries (KdV) equation.

The periodic form of a solitary wave with the same height is named a cnoidal wave by [7]. A cnoidal wave is a periodic wave with isolated wave crests and flat wave troughs. A cnoidal wave without a nonlinear and asymmetrical wave profile about the still-water level (SWL) could become a sinusoidal wave. Ref. [10] obtained the solutions for cnoidal waves by means of the Jacobian elliptic cosine function Cn contained in the wave profile formula. Following the idea of [10] about cnoidal waves, Ref. [11] established the Keulegan–Patterson theory for cnoidal waves. Ref. [12] proved the existence of cnoidal waves with a complex potential function. By then, the equations were complicated to be used in practical application for the incompleteness of second-order solutions. Based on the work of [13,14] a second-order approximation to the cnoidal wave theory was developed by using the expansion method, and [15] added a nonzero term to the equations.

Ref. [16] developed the Green–Naghdi (GN) equations for shallow-water wave motions based on the directed fluid sheet theory. The GN theory satisfies exactly the nonlinear boundary conditions and conservation of mass, postulates the integrated form of the conservation of momentum, and does not require the irrotationality of the flow. This set of equations contains solutions for solitary wave (see [17]) and cnoidal wave (see [18]). The GN equations are used successfully for a wide range of problems involving water wave dynamics; for example, nonlinear wave scattering by submerged objects ([19]), and diffraction and refraction of solitary and cnoidal waves ([20]).

Although there are several equations that can be used to describe the solitary and cnoidal waves, it is difficult to solve the equations analytically due to the nonlinear nature of these waves, especially the implicit part of solutions of the cnoidal wave, see [18]. Thanks to the increase in computational power, the computational fluid dynamics (CFD) approach is used in the study of solitary and cnoidal waves. The waves are generated by use of different equations, see, e.g., Refs. [21,22] for the use of the KdV equations to generate solitary waves, Ref. [23] for the use of the GN equations to generate solitary waves, and propagate in the domain governed by the Navier–Stokes (NS) equations, Ref. [24] for the use of cnoidal waves by use of the GN equations, and [25] for the use of the Boussinesq equations to generate solitary waves. It is desirable to generate accurate nonlinear waves when studying wave dynamics in coastal zones, for example for wave propagation over an uneven seafloor, or in wave-structure interaction problems. While various theories are used to generate nonlinear waves of solitary and cnoidal types in computational wave tanks, there has not been any systematic investigation about the accuracy and stability of the waves when compared to each other. It is desirable to find out how the solitary wave and cnoidal wave generated based on the existing equations propagate in the computational fluid dynamics-based numerical wave tanks.

In this study, we create a numerical tank to investigate the accuracy and stability of solitary and cnoidal waves of different theories. We develop numerical wave makers capable of generating solitary and cnoidal waves of different theories. Waves are generated by a piston-type wave maker and enter a domain of constant water depth, where the flow is governed by the NS equations. We then investigate the characteristic of the solitary and cnoidal waves of different theories within the domain, and compare the results with available laboratory measurements. To further assess the performance of the GN equations, we also create a numerical wave tank based entirely on the GN equations, i.e., the waves generated by the GN theory enter a domain where the flow is governed by the GN equations. The results are then compared with the laboratory measurements, and with the results of the case of the NS domain.

In this paper, we first present the theories used for wave generation and propagation, followed by the introduction of the set-up of the numerical tank. Steady-state results are given first and unsteady results of solitary waves and cnoidal waves are presented and discussed thereafter. The conclusions then follow.

2. The Theories and Numerical Set Up

We create a numerical tank, in which solitary and cnoidal waves generated by different theories are investigated. The governing equations of the domain and the wave theories used are described in this section.

We adopt a right-handed two-dimensional (2D) Cartesian coordinate system, with x_1 pointing to the right, and x_2 pointing vertically opposite to the direction of the gravitational acceleration such that $x_2 = 0$ is the SWL. Indicial notation and Einstein's summation convention are used. Subscripts after the comma indicate differentiation. The equations are given in dimensionless forms using density of water, ρ, gravitational acceleration, g, and water depth, h, as a dimensionally independent set. Hence, the dimensionless form of time is $t' = t \times \sqrt{g/h}$, surface elevation is $\eta' = \eta/h$, velocity is $u' = u/\sqrt{gh}$ and pressure is $p' = p/\rho gh$. For simplicity, we drop primes from all dimensionless variables in the rest of this work unless otherwise indicated.

2.1. The Navier–Stokes Equations

For a homogeneous, Newtonian and incompressible fluid, the 2D NS equations are given by the following conservation of mass and momentum equations:

$$u_{i,i} = 0, \quad i = 1,2 \tag{1}$$

$$u_{j,t} + (u_i u_j)_{,i} = \frac{1}{Fr^2} g_j - p_{,j} + \frac{1}{Re} u_{j,ii}, \quad i,j = 1,2 \tag{2}$$

where $\vec{u} = u_i \vec{e_i}$ is the velocity vector, and $\vec{e_i}$ is the unit normal vector in the i-th direction, $Fr = \frac{U}{\sqrt{gl}}$ is the Froude number, with U and l as the characteristic velocity and length, respectively, $Re = \frac{Ul}{\nu}$ is the Reynolds number, ν is kinematic viscosity, $\vec{g} = (0, -1)$ is the gravitational acceleration vector, and p is the pressure.

A computational tank is created, in which waves are generated by the use of different wave theories and propagate over a flat tank floor. A schematic of the tank is shown in Figure 1. The tank is separated by dash lines into three parts, namely the wave generation zone, the computational domain and the wave absorption zone. Within the computational domain, the governing equations are solved for both water and air above. The volume of fluid method, originally introduced by [26], is used to determine the free surface between air and water.

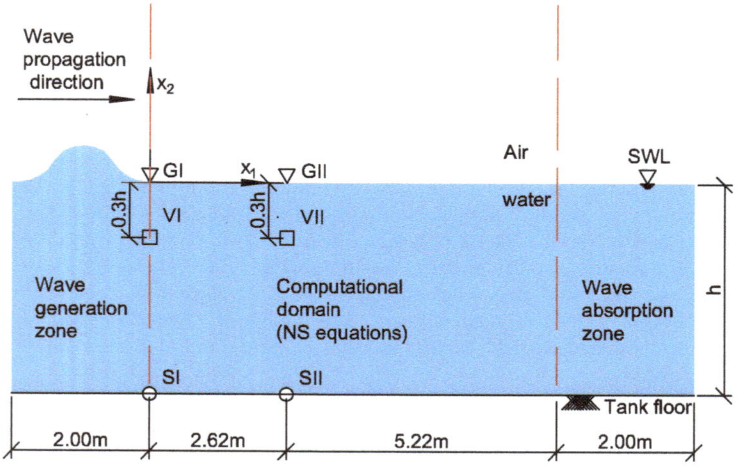

Figure 1. Schematic of the NS computational domain used in this study. Solitary and cnoidal waves are created at the generation zone, using different theories.

The Finite Volume approach is used to discretize the equations. The boundary conditions used in this study are the same as that of [27]. The details of these boundary conditions can be found in, e.g., [28,29].

2.2. The Green–Naghdi Equations

Considering the theory of directed fluid sheets of an incompressible fluid and from the laws of conservation of mass and linear momentum, [30] derived a set of equations, the Green–Naghdi (GN) equations, describing the nonlinear motion of inviscid fluids. The dimensionless GN equations read as

$$\eta_{,t} + [(1+\eta-\alpha)u_1]_{,1} = \alpha_{,t}, \tag{3}$$

$$\dot{u}_1 + \eta_{,1} + \hat{p}_{,1} = -\frac{1}{6}[2\eta+\alpha]_{,1}\ddot{\alpha} + [4\eta-\alpha]_{,1}\ddot{\eta} + (1+\eta-\alpha)[\ddot{\alpha}+2\ddot{\eta}]_{,1}, \tag{4}$$

$$u_2 = \dot{\alpha} + \frac{x_2+1-\alpha}{\eta+1-\alpha}(\dot{\eta}-\dot{\alpha}), \tag{5}$$

$$\bar{p} = \frac{1}{2}(1+\eta-\alpha)(\ddot{\alpha}+\ddot{\eta}+2) + \hat{p}, \tag{6}$$

where η is the free surface elevation measured from the SWL, $\alpha(x_1,t)$ describes the bottom surface of the fluid sheet measured from a fixed reference point, $\bar{p}(x_1,t)$ is the pressure on α, and $\hat{p}(x_1,t)$ is the pressure on top of the fluid sheet. The superposed dot denotes the material time derivative, and double dot is the second material derivative.

The pressure distribution in shallow water, p, is given as ([31,32]),

$$p(x_1,x_2,t) = \frac{1}{2}(1+\eta-x_2)(\ddot{\alpha}+\ddot{\eta}+2) + \hat{p}. \tag{7}$$

In the cases studied here, the bottom surface is flat and stationary, hence $\alpha = 0$ and $\ddot{\alpha} = 0$, and we assume $\hat{p} = 0$, i.e., pressure is atmospheric on the top surface. Hence, Equations (5) and (7) can be written as,

$$u_2(x_1,x_2,t) = \frac{x_2+1}{\eta+1}\dot{\eta}, \tag{8}$$

and

$$p(x_1,x_2,t) = \frac{1}{2}(1+\eta-x_2)(\ddot{\eta}+2). \tag{9}$$

The above solution belongs to the Level I GN equations, in which the vertical velocity varies linearly along the water column and horizontal velocity is invariant in the vertical direction. The Level I GN equations, hence, are best applicable to the propagation of long waves in shallow water. High level GN equations can be obtained by considering higher order functions for the distribution of the velocities in the vertical direction, see, e.g., [33,34].

2.3. Solitary Wave Solution

The solitary wave surface elevation of the Level I GN equations is given by [16] as

$$\eta(x_1,t) = A\, sech^2(\epsilon x_{C1}), \tag{10}$$

where A is the amplitude of the solitary wave, $x_{C1} = (x_1 - Ct)$, $\epsilon = \sqrt{\frac{3A}{4(A+1)}}$, and C is the wave speed, which is

$$C = \sqrt{(A+1)}. \tag{11}$$

The horizontal particle velocity is given by

$$u_1(x_1,t) = C\frac{\eta}{1+\eta}. \tag{12}$$

Equations (8)–(10) can be used to obtain analytical solutions for the solitary wave velocity and pressure fields. $\dot{\eta}$ is given by

$$\dot{\eta} = \eta_{,t} + u_1\eta_{,1} + u_2\eta_{,2}. \tag{13}$$

The first part of the right-hand side of Equation (13) is obtained as

$$\eta_{,t} = [A\ sech^2(\epsilon x_{C1})]_{,t} = 2A\epsilon C\ cosh^{-2}(\epsilon x_{C1}) tanh(\epsilon x_{C1}), \tag{14}$$

The second part of the right-hand side of Equation (13) is given as

$$u_1\eta_{,1} = u_1[A\ sech^2(\epsilon x_{C1})]_{,1} = -2u_1 A\epsilon\ cosh^{-2}(\epsilon x_{C1})\ tanh(\epsilon x_{C1}). \tag{15}$$

As $\eta(x_1, t)$ is a function of x_1 and t, but not x_2, the third part on the right-hand side of Equation (13) is given as

$$u_2\eta_{,2} = u_2[A\ sech^2(\epsilon x_{C1})]_{,2} = 0. \tag{16}$$

Substituting Equations (14)–(16) into Equation (13) gives

$$\dot{\eta} = 2A\epsilon(C - u_1)\ sech^2(\epsilon x_{C1})\ tanh(\epsilon x_{C1}), \tag{17}$$

and, substituting Equation (17) into Equation (8) gives

$$u_2(x_1, x_2, t) = 2\frac{x_2 + 1}{\eta + 1}\epsilon\eta\ tanh(\epsilon x_{C1})(C - u_1). \tag{18}$$

Next, we obtain $\ddot{\eta}$, given as

$$\ddot{\eta} = \dot{\eta}_{,t} + u_1\dot{\eta}_{,1} + u_2\dot{\eta}_{,2}. \tag{19}$$

For the first part of the right-hand side of Equation (19), we can write

$$\dot{\eta}_{,t} = 2A\epsilon^2(C - u_1)C(3\ sech^2(\epsilon x_{C1})\ tanh^2(\epsilon x_{C1}) - 1). \tag{20}$$

For the second part of the right-hand side of Equation (19)

$$u_1\dot{\eta}_{,1} = -2A\epsilon^2(C - u_1)u_1(3\ sech^2(\epsilon x_{C1})\ tanh^2(\epsilon x_{C1}) - 1). \tag{21}$$

As in the derivation of Equation (16), $\eta(x_1, t)$ is a function of x_1 and t, but not x_2; therefore, the third part of the right-side of Equation (19) can be written as

$$u_2\dot{\eta}_{,2} = 0. \tag{22}$$

Hence, substituting Equations (20)–(22) into Equation (19) gives

$$\ddot{\eta} = 2A\epsilon^2(C - u_1)^2(3\ sech^2(\epsilon x_{C1})\ tanh^2(\epsilon x_{C1}) - 1), \tag{23}$$

Substituting Equation (23) into Equation (7) gives

$$p(x_1, x_2, t) = (\eta - x_2)(2A\epsilon^2(C - u_1)^2(3\ sech^2(\epsilon x_{C1})\ tanh^2(\epsilon x_{C1}) - 1) + 1). \tag{24}$$

Equations (10), (12), (18) and (24) provide the surface elevation, horizontal and vertical velocities and the pressure under a steady-state solitary wave solution of the GN equations.

There are several solutions of a solitary wave provided by, e.g., Boussinesq (1872), [14] and KdV equations given by [10]. The analytical solutions (used in this study) and the GN solution derived above are listed in Table 1.

Table 1. Solutions of solitary wave by different theories.

	Boussinesq	KdV	Laitone's 1st Order	Laitone's 2nd Order	GN
η	$A\ sech^2(\epsilon_{BL}x_{C1})$	$A\ sech^2(\epsilon_{BL}x_{C1})$	$Asech^2(\epsilon_{BL}x_{C1})$	$A\ sech^2\chi_2$ $-3A^2\ sech^2\chi_2$ $\times(1-\ sech^2\chi_2)/4$	$A\ sesh^2(\epsilon x_{C1})$
C	$\sqrt{1+A}$	$\sqrt{1+A}$	$\sqrt{1+A/2}$	$\sqrt{1+A/2-3A^2/20}$	$\sqrt{1+A}$
u_1	$C\eta/(1+\eta)$	$A[a_{KdV}$ $+b_{KdV}$ $\times cosh(\sqrt{3A}x_{C1})]$ $\times sech^4(\epsilon_{BL}x_{C1})/4$	$A\ sech^2(\epsilon_{BL}x_{C1})$	$A(1+A/4-3Ax_2^2/2)$ $\times sech^2\chi_2$ $+\epsilon_{L1}\ sech^4\chi_2$	$C\eta/(1+\eta)$
u_2		$\sqrt{3A^3}(1+x_2)$ $\times[c_{KdV}+d_{KdV}$ $\times cosh(\sqrt{3A}x_{C1})]$ $\times sech(\epsilon_{BL}x_{C1})/4$	κx_2 $\times sech^2\chi$ $\times tanh\chi$	$\kappa(x_2+1)\ sech^2\chi_2$ $\times tanh[(1-3A/8)\chi_2]$ $-Ax_2^2/2+A\ tan\epsilon_{L2}$ $\times sech^2\chi_2$	$2\epsilon\eta$ $\times tanh(\epsilon x_{C1})$ $\times(C-u_1)$ $\times(x_2+1)$ $/(\eta+1)$

where $\epsilon=\sqrt{3A/4(A+1)}$, $\epsilon_{BL}=\sqrt{3A/4}$, $\kappa=\sqrt{3A^3}$, $\chi=\sqrt{3/4}x_{C1}$, $\chi_2=x_1\sqrt{3A/(4(1-5A/8)}-Ct$, $\epsilon_{L1}=A^2(-1+9x_2^2/4)$, $\epsilon_{L2}=-2+3x_2^2$, $a_{KdV}=2+A+12Ax_2+6Ax_2^2$, $b_{KdV}=2-A-6Ax_2-3Ax_2^2$, $c_{KdV}=2-7A+10Ax_2+5Ax_2^2$, $d_{KdV}=2+A-2Ax_2-Ax_2^2$.

2.4. Cnoidal Wave Solution

Similar to the solitary wave, there are several solutions for the cnoidal wave, such as through the Boussinesq equations, KP equations and KdV equation. The cnoidal waves generated according to the GN equations, KP equations, and KdV equation are discussed in this study.

2.4.1. The Green-Naghdi Equations

The cnoidal wave solution of the GN equations provided by [18] gives the water surface elevation, the wave phase speed, C, and the water particle velocity as

$$C^2=(1+\eta_1)(1+\eta_2)(1+\eta_3), \tag{25}$$

where $\eta_1=\frac{H}{k^2}(1-\frac{E}{K})$, $\eta_2=\frac{H}{k^2}(1-k^2-\frac{E}{K})$, and $\eta_3=\frac{H}{k^2}\frac{E}{K}$ are three wave parameters,

$$\eta=\eta_2+HC_n^2=\frac{H}{k^2}(1-k^2-\frac{E}{K})+HC_n^2, \tag{26}$$

$$u_1=\frac{\eta}{1+\eta}C, \tag{27}$$

$$u_2=\frac{1+x_2}{(1+\eta)}\frac{\partial\eta}{\partial x_1}C, \tag{28}$$

where H is the wave height of the cnoidal wave, K is the complete elliptic integral of the first kind, E is the complete elliptic integral of the second kind, C_n is the Jacobian elliptic function and k is the modulus of the respective elliptic integral function. For details, see [18].

2.4.2. The Keulegan and Patterson Equations

The solution of cnoidal wave derived by [11] share Equations (25) and (26) for C and η with the solution of the GN theory. The equations for u_1 and u_2 are:

$$u_1=(\eta-\frac{1}{4}\eta^2+(\frac{1}{3}-\frac{1}{2}(1+x_2)^2)\frac{\partial^2\eta}{\partial x_1^2})C, \tag{29}$$

$$u_2=-(1+x_2)((1-\frac{1}{2}\eta)\frac{\partial\eta}{\partial x_1}+\frac{1}{3}(1+\frac{1}{2}(1+x_2)^2)\frac{\partial^3\eta}{\partial x_1^3})C. \tag{30}$$

2.4.3. The Korteweg-de Vries Equations

The KdV solution of the cnoidal wave is given by [35]:

$$C^2 = 1 + \frac{1}{k}(2 - k - 3\frac{E}{K}), \tag{31}$$

$$\eta = \frac{H}{k}(1 - k - \frac{E}{K}) + HC_n^2, \tag{32}$$

$$u_1 = (\eta - \eta^2 + \frac{1}{2}(\frac{1}{3} - (1 + x_2)^2)\frac{\partial^2 \eta}{\partial x_1^2})C, \tag{33}$$

$$u_2 = -(1 + x_2)((1 - 2\eta)\frac{\partial \eta}{\partial x_1} + \frac{1}{6}(1 - (1 + x_2)^2)\frac{\partial^3 \eta}{\partial x_1^3})C. \tag{34}$$

2.5. The Numerical Tank

A numerical tank is created in this study, in which solitary and cnoidal waves of different theories can be generated. Inside the domain, the fluid is governed by the NS equations. The computational domain of the numerical tank used in this study is 11.84 m long and 0.21 m deep, as shown in Figure 1. The wave generation and absorption zones in this study are shown in Figure 1, follow the same principle as those given by [36], within the open source computational fluid dynamics package, OpenFOAM. Waves are generated in a zone on the left of the tank. On the right-hand side of the domain, a relaxation zone is used to minimize the wave reflection back into the domain. The generation zone is set from $x_1 = -2$ m to $x_1 = 0$ and the relaxation zone is located at $x_1 = 7.84$ m to $x_1 = 9.84$ m. There are two wave gauges, GI located at the first node outside of the relaxation zone at $x_1 = 0$, and GII located at $x_1 - 2.62$ m from the end of the relaxation zone. Two pressure sensors, SI and SII, are placed on the tank bottom, at $x_2 = -1$, and two velocimeters, VI and VII, are placed at $x_2 = -0.3$. The horizontal location of the gauges are shown in Figure 1.

For the NS computations, two Intel Xeon E5-2697A v4 processors (16 cores, 40 M Cache, 3.00 GHz) are used. The maximum Courant Number is 0.25 and average Courant Number is 0.0086. All the cases are performed in 2D. The mesh sizes are $\Delta x_1/h = 0.067$ and $\Delta x_2/h = 0.167$, following the grid convergence study of [27].

We also use a tank that utilizes the GN equations both at the wavemaker and inside the domain, see [37] for details of the theoretical tank of the GN equations. See [38–41] for some studies on nonlinear wave dynamics and wave-structure interaction utilizing GN numerical tanks. For internal wave solutions of the GN equations, see, e.g., [42].

3. Results and Discussion

The results, including the surface elevations, velocities and pressure fields of solitary and cnoidal waves of different theories, are presented in this section. The results are compared with laboratory measurements where available and also with the results of the GN equations, (GN-GN, where both the wave maker and the domain are governed by the GN theory). The steady-state results generated by different theories are discussed first, followed by the unsteady results obtained from the computational tank. All results are given in a dimensionless form using ρ, g and h as a dimensionally independent set.

3.1. Steady-State Solution

The surface elevation of solitary waves generated by the five equations listed in Table 1 are presented first, followed by discussion on the surface elevations, horizontal velocities, and vertical velocities of cnoidal waves generated by the GN equations, the KP equations and the KdV equations.

3.1.1. Solitary Wave

The profiles of solitary waves with $A = 0.4$ generated by the Boussinesq equations, Laitone's 1st equations, Laitone's 2nd equations, the GN equations, and the KdV equations are shown in Figure 2.

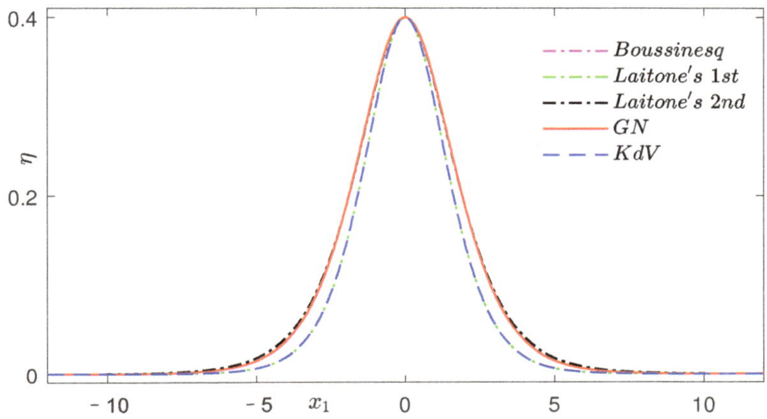

Figure 2. Comparisons of steady-state surface elevations of a solitary wave with $A = 0.4$, generated by various theories.

In Figure 2, the surface elevation of the Boussinesq equations and Laitone's 1st equations are covered by that of the KdV equations, as they share the same equation for η. The GN solution is wider, and that of Laitone's 2st theory is the widest.

In these cases, the theoretical wave speeds of the Boussinesq equations ($C = 1.1832$), the KdV equations ($C = 1.1832$), and the GN equations ($C = 1.1832$) are the same, followed by that of Laitone's 1st equations ($C = 1.2000$) is the fastest, and that of Laitone's 2nd equations ($C = 1.1760$) is the slowest.

3.1.2. Cnoidal Wave

The theoretical cnoidal wave surface elevation and velocity field, for $\lambda = 29.5$ and $H = 0.156,\ 0.311$, generated by the GN equations and the KP equations are shown in Figures 3–5, where λ is the wave length.

Figures 3–5 show that the results of the three equations are almost on top of each other, except there are small differences between the peak of the results. The peak of the surface elevation, horizontal and vertical velocities of the three equations are listed in Tables 2 and 3, together with the ratio, $R = \frac{\theta_M - \theta_{GN}}{\theta_{GN}} \times 100\%$, where θ is an arbitrary variable (η, u_1 or u_2), and M refers to different models.

Table 2. Comparisons of the peak values of the surface elevation, horizontal and vertical velocities of cnoidal wave ($H = 0.156$) of the GN equations and the KP equations.

	η	u_1	u_2
GN	0.1251	0.1145	0.0252
KP	0.1251	0.1219	0.0250
R	0.00%	6.46%	−0.79%
KdV	0.1251	0.1162	0.0247
R	0.00%	1.48%	−1.98%

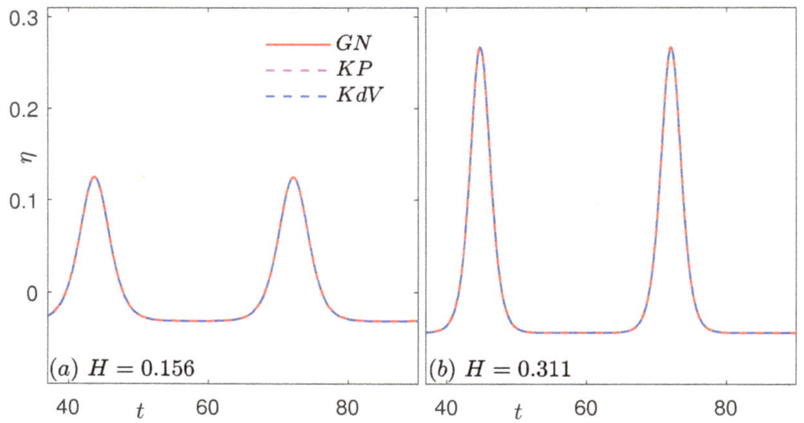

Figure 3. Comparisons of steady-state surface elevation of cnoidal waves generated by various theories. $\lambda = 29.6$.

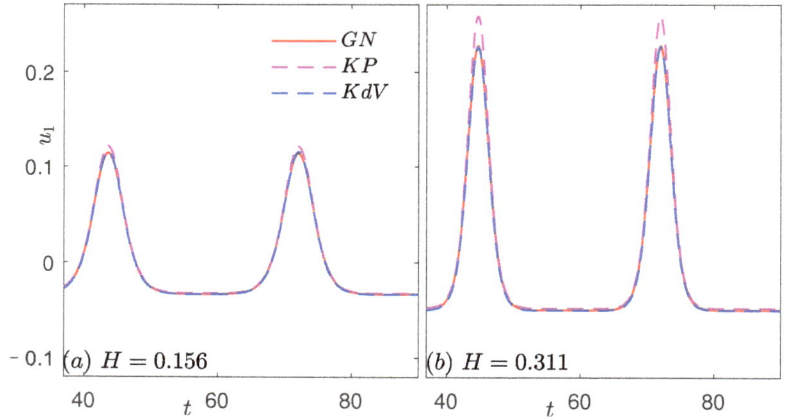

Figure 4. Comparisons of steady-state horizontal particle velocity of cnoidal waves generated by various theories. $\lambda = 29.6$.

Table 3. Comparisons of the peak values of the surface elevation, horizontal and vertical velocities of cnoidal wave ($H = 0.311$) of the GN equations and the KP equations.

	η	u_1	u_2
GN	0.2675	0.2273	0.0618
KP	0.2675	0.2580	0.0635
R	0.00%	13.51%	2.75%
KdV	0.2674	0.2265	0.0603
R	−0.04%	−0.35%	−2.43%

Shown in Figure 3, the peak values of the surface elevation of the three theories are the same, except that of the KdV equations with $H = 0.311$, which is 0.04% less than that of the other two equations. The same results are found for the vertical velocity, where the peak values are almost the same. The peak value of the horizontal velocity of the KP theory is significantly larger, where R is 6.46% and 13.51% for the two cases, than that of the other two theories.

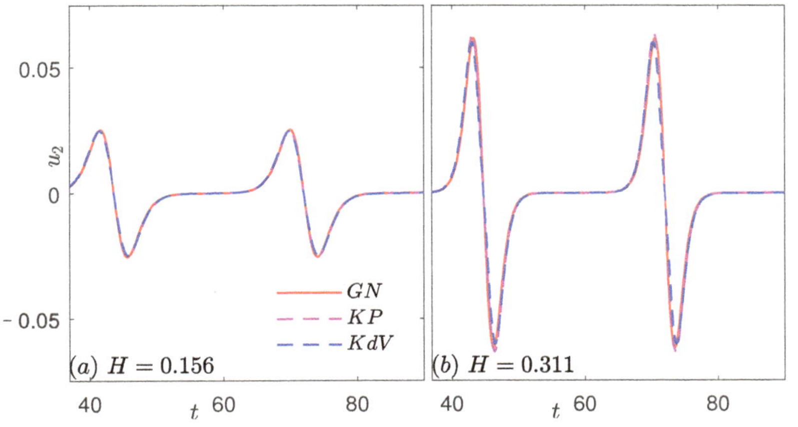

Figure 5. Comparisons of steady-state vertical particle velocity of cnoidal waves generated by various theories. $\lambda = 29.6$.

4. Unsteady Results

In this section, the propagation of solitary and cnoidal waves generated by a wave maker (using different theories) over a flat seafloor is studied. The wave maker is placed at $x_1 = 0$. In all cases, waves are generated by wave makers using different theories, and enter the domain governed by the NS equations. We also consider waves generated by the GN theory, entering a domain also governed by the GN equations. These combinations are listed in Table 4. The results for solitary waves are shown first, followed by the cnoidal wave results.

Table 4. List of the combination of wave makers and domain equations of the unsteady cases considered in this study.

Case	Wave Maker Theory	Domain Governing Equations	Type of Waves
KdV-NS	KdV	NS	solitary & cnoidal
KP-NS	KP	NS	cnoidal
GN-NS	GN	NS	solitary & cnoidal
GN-GN	GN	GN	solitary & cnoidal

4.1. Solitary Wave

Ref. [21] conducted laboratory experiments on the generation of solitary waves. They report the surface elevations of four cases with the same water depth $h = 0.114$ m. In all the cases presented in this subsection for cnoidal waves, water depth is fixed at $h = 0.114$ m. These four cases are listed in Table 5. Waves are measured in the wave flume in the absence of any obstruction by GII, located at $x_1 = 2.62$ m. These laboratory measurements are used for comparison here.

Table 5. Solitary wave cases considered in this study.

Case	A
1	0.192
2	0.301
3	0.408
4	0.525

Snapshots at $t = 0$ and surface elevations gathered by the GII of solitary waves are compared between GN-NS, KdV-NS, GN-GN, and laboratory experiments of [21], and are shown in Figures 6–9.

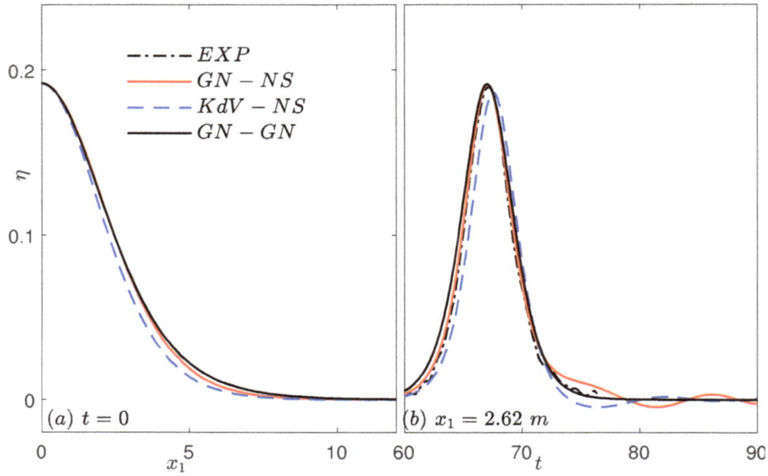

Figure 6. Comparisons of surface elevations of a solitary wave with $A = 0.192$, generated by the KdV and GN equations, propagating in domains of the NS and GN equations.

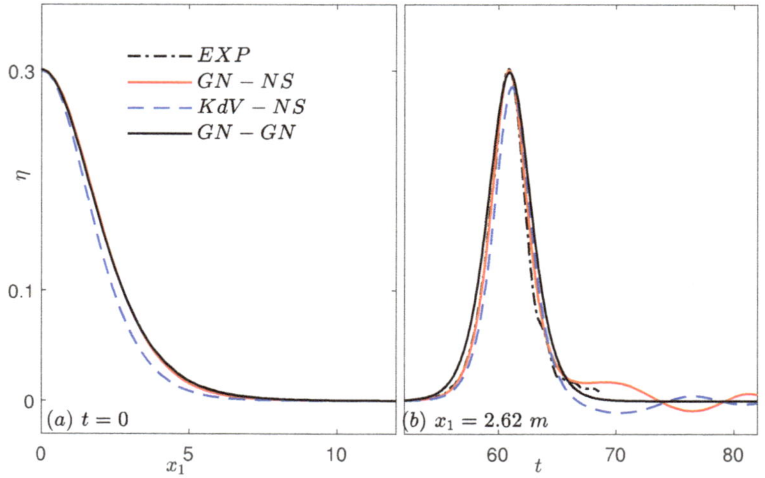

Figure 7. Comparisons of surface elevations of a solitary wave with $A = 0.301$, generated by the KdV and GN equations, propagating in domains of the NS and GN equations.

At $t = 0$, the surface elevations of GN-NS and GN-GN are almost identical (as expected), and the KdV-NS solitary wave is slimmer, see sub-figure (a) of Figures 6–9.

Table 6 shows the comparisons between the steady-state propagation speeds of solitary waves calculated by the equations listed in the table and unsteady ones measured from Figure 8 ($A = 0.4$). The wave speeds calculated by Laitone's 1st theory are the fastest, followed by those of the Boussinesq theory and the GN analytical solution, whose wave speeds are equal to each other. After these are the GN-GN, GN-NS and Laitone's 2nd theory. When compared to the other models, the KdV-NS predicts the slowest wave propagation.

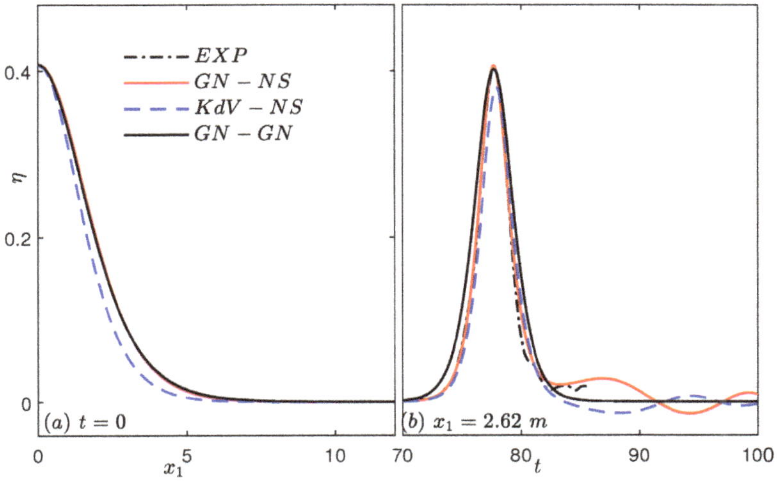

Figure 8. Comparisons of surface elevations of a solitary wave with $A = 0.408$, generated by the KdV and GN equations, propagating in domains of the NS and GN equations.

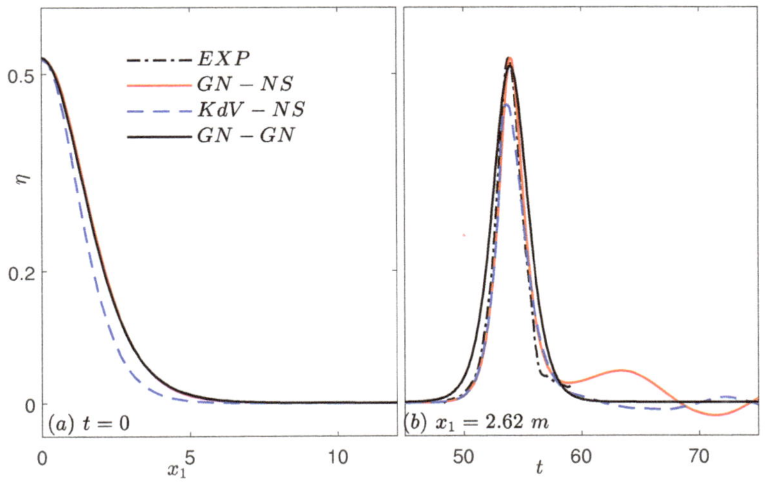

Figure 9. Comparisons of surface elevations of a solitary wave with $A = 0.525$, generated by the KdV and GN equations, propagating in domains of the NS and GN equations.

For the surface elevations recorded at x_1 = 2.62 m, the peak value of the GN-GN and the GN-NS drops slightly from the initial amplitude and are in good agreement with the laboratory experimental data of [21], while that of the KdV-NS drops more remarkably and is smaller than the laboratory measurements, particularly where the wave amplitude is larger, shown in sub-figures (b) of Figures 6–9. It shows the solitary wave generated by the GN equations are more stable than that of the KdV equations, especially in the cases of larger wave amplitudes.

Table 6. The analytical and numerical wave speeds of various theories ($A = 0.408$).

		Analytical	Value of C
Steady-state	Boussinesq Theory	$\sqrt{1+A}$	1.1832
	Laitone's 1st Theory	$\sqrt{1+A/2}$	1.2000
	Laitone's 2st Theory	$\sqrt{1+A/2-3A^2/20}$	1.1760
	GN Theory	$\sqrt{1+A}$	1.1832
	KdV equations	$\sqrt{1+A}$	1.1832
Unsteady	KdV-NS		1.1767
	GN-NS		1.1769
	GN-GN		1.1769

4.2. Cnoidal Wave

Ref. [43] conducted laboratory experiments on the generation of cnoidal waves in a tank with a flat floor. They report the surface elevations of cnoidal waves at water depth $h = 0.071$ m. In all the cases presented in this subsection for cnoidal waves, the water depth is fixed at $h = 0.071$ m. Details of two cases are listed in Table 7 used here for comparison. Waves are measured in the wave flume in the absence of any obstruction by only one gauge located at 2.62 m in the laboratory experiments.

Table 7. Cnoidal wave cases considered in this study.

Case	λ	H
1	29.6	0.156
2	29.6	0.311

The surface elevations recorded by the two wave gauges are presented in Figure 10. The results at $x_1 = 2.62$ m are compared between the experimental data of [43], and the numerical results of GN-NS, KP-NS, KdV-NS and GN-GN, in Figure 10.

Figure 10 shows that the results of the GN-GN model is the closest to the laboratory measurements. The wave speed of KP-NS is the fastest, followed by GN-NS, GN-GN, and KdV-NS.

The surface elevation differences between the experimental data and numerical results are presented in Table 8. In the table, $R = (\frac{P_M}{P} - 1) \times 100\%$, where P is the peak of the surface elevation in experiments and P_M is that of the equations. The average percentage difference ratios of the peak of the surface elevation of GN-NS (3.99%) and KdV-NS (4.36%) are close to each other, while that of KP-NS (5.71%) is the largest, and that of GN-GN (3.03%) is the smallest.

Table 8. The percentage differences between the peak of the surface elevation of cnoidal waves of different models and the laboratory measurements of [43]. $\lambda = 29.6$.

Case	$H = 0.156$	$H = 0.311$
Experiments	0.132	0.262
GN-NS	0.123	0.259
R	−6.82%	−1.15%
KP-NS	0.126	0.280
R	−4.55%	6.87%
KdV-NS	0.123	0.257
R	−6.82%	−1.90%
GN-GN	0.125	0.264
R	−5.30%	0.76%

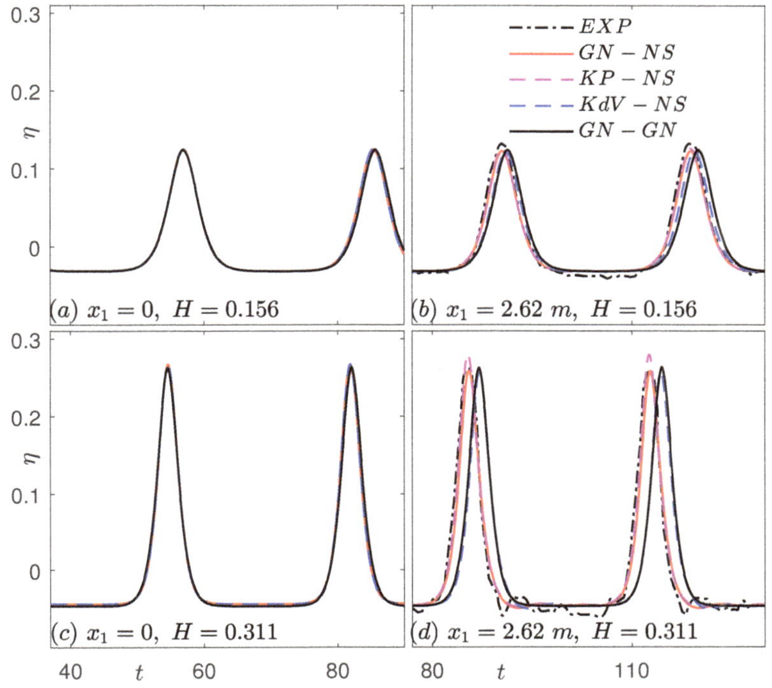

Figure 10. Comparisons of surface elevation of cnoidal waves of GN-NS, KP-NS, KdV-NS and GN-GN and the laboratory measurements of [43]. $\lambda = 29.6$.

Comparisons of pressures, horizontal and vertical velocities of the GN-NS, KP-NS, KdV-NS, and GN-GN models, recorded at pressure sensors, SI and SII, velocimeters, VI and VII, under GI and GII, respectively, of the two cases are presented in Figures 11–13, respectively. In the absence of laboratory measurements of the pressure and velocity, and given that the surface elevations of the GN-GN is the closest to the laboratory experimental data, the GN-GN model is used as a benchmark and differences between the results of the models, $R = \frac{\theta - \theta_{GNGN}}{\theta_{GNGN}}$, are presented in Tables 9–11.

Table 9. The differences between the peak pressure of cnoidal waves generated by various theories.

Case	$H = 0.156$	$H = 0.311$
GN-GN	1.124	1.263
GN-NS	1.100	1.201
R	−2.14%	−4.91%
KP-NS	1.108	1.216
R	−1.42%	−3.72%
KdV-NS	1.105	1.200
R	−1.69%	−4.99%

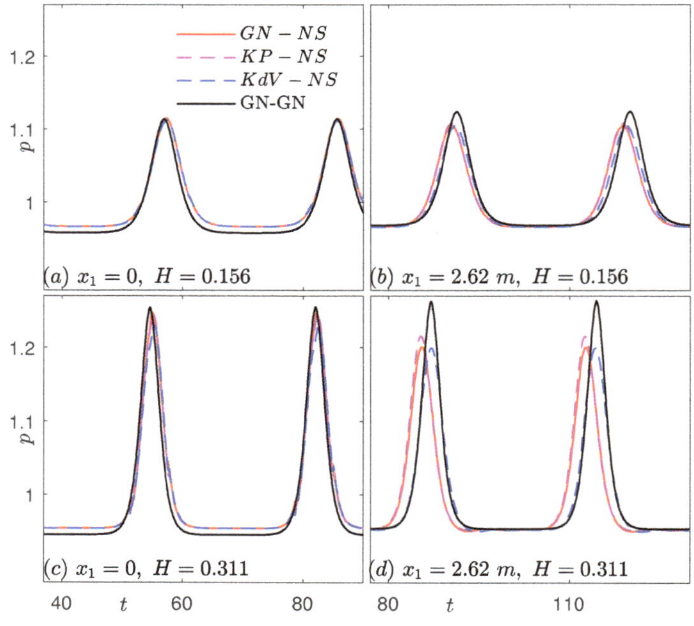

Figure 11. Comparisons of pressure of cnoidal waves of various theories, propagating in domain of the NS and GN equations. $\lambda = 29.6$.

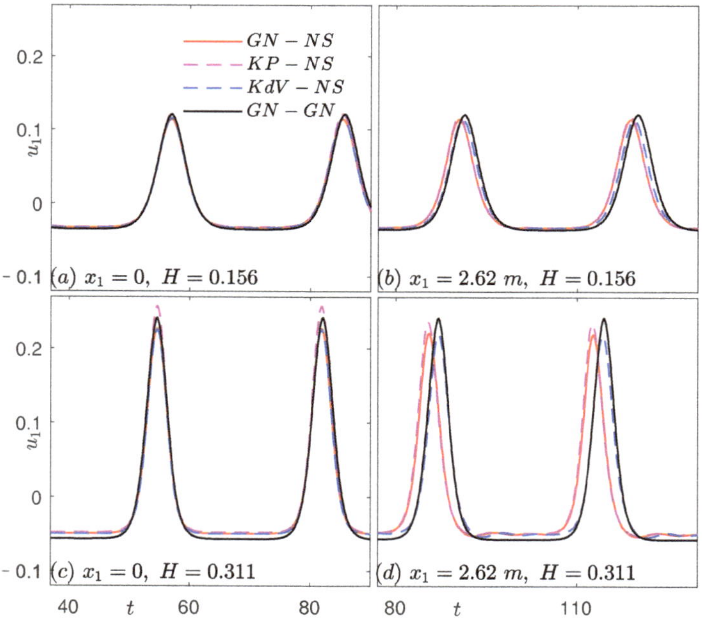

Figure 12. Comparisons of horizontal particle velocity of cnoidal waves generated by various theories, propagating in domain of the NS equations. $\lambda = 29.6$.

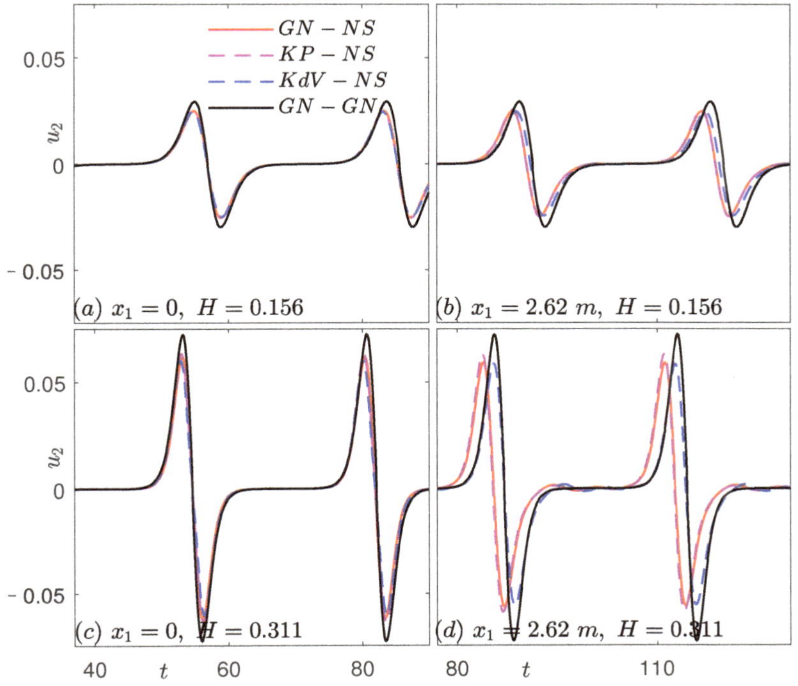

Figure 13. Comparisons of vertical particle velocity of cnoidal waves generated by various theories, propagating in domain of the NS equations. $\lambda = 29.6$.

Table 10. The differences between the peak horizontal velocity of cnoidal waves generated by various theories.

Case	$H = 0.156$	$H = 0.311$
GN-GN	0.121	0.242
GN-NS	0.114	0.220
R	−5.79%	−9.09%
KP-NS	0.117	0.235
R	−3.31%	−2.89%
KdV-NS	0.114	0.219
R	−5.79%	−9.50%

Table 11. The differences between the peak vertical velocity of cnoidal waves generated by various theories.

Case	$H = 0.156$	$H = 0.311$
GN-GN	0.0296	0.0726
GN-NS	0.0250	0.0592
R	−15.54%	−18.46%
KP-NS	0.0255	0.0632
R	−6.93%	−12.95%
KdV-NS	0.0249	0.0588
R	−15.88%	−19.00%

Figures 11–13 and Tables 9–11 show that p, u_1, u_2 of KP-NS are closer to that of GN-GN, while that of the GN-NS and KdV-NS are close to each other and slightly smaller than that of KP-NS.

We recall that in the Level I GN equations, we assume linear distributions of the vertical velocity. It is interesting to compare the results of GN-GN model with that of GN-NS and KP-NS to find out how close the assumption is to the computational results. Distribution of the horizontal and vertical velocity along the vertical direction at $x_1 = 0$ and $x_1 = 2.62$ m of the two cases are shown in Figures 14–17.

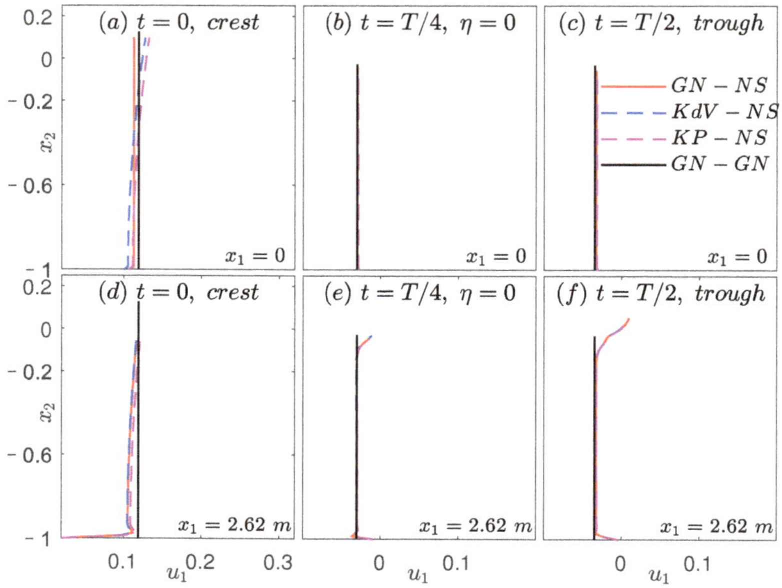

Figure 14. Comparisons of horizontal velocity distribution of cnoidal waves of theories, propagating in domains of the NS and GN equations in Case 1. $H = 0.156$, $\lambda = 29.6$.

For horizontal velocity distribution of cnoidal wave, GN-GN shows constant velocity along the vertical direction and this agrees with Equation (27). Although the horizontal velocity of GN-NS is constant at $x_1 = 0$, which is the same as GN-GN, the velocity becomes closer to that of KP-NS and KdV-NS as the wave propagates in the NS domain (at $x_1 = 2.62$ m). At $x_2 = 2.62$ m, the horizontal velocity of the cnoidal wave propagating in the NS domain shows a slight difference with that of GN-GN near the tank bottom and free surface. This is due to the effect of viscosity and the atmospheric boundary effects. Away from these boundaries, the horizontal velocities of the GN-NS, KP-NS and KdV-NS are constant and very close to that of the GN-GN model.

For vertical velocity distribution of cnoidal wave, that of GN-GN, GN-NS, KP-NS, and KdV-NS are almost linear along the vertical direction (agree with Equation (28) for the GN-GN model). Nonlinear distributions are found at $t = 0$, where the distributions under the crest is observed, see sub-figs (a) and (d) of Figures 16 and 17.

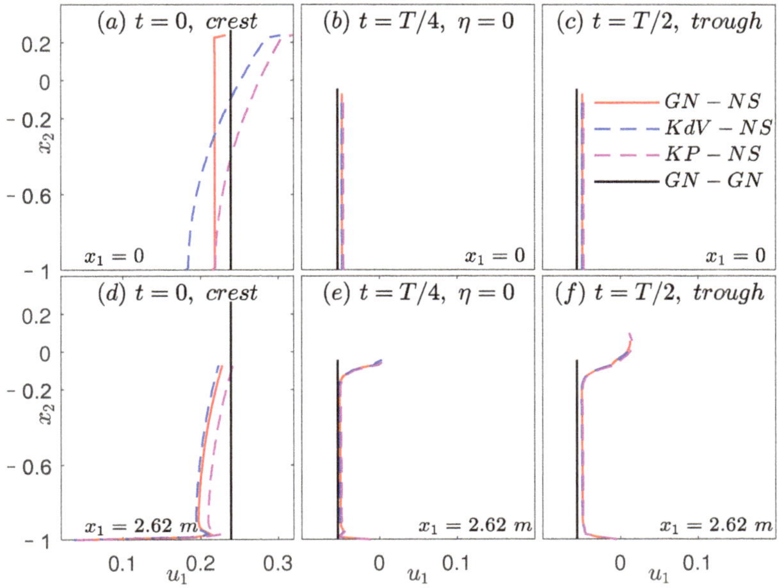

Figure 15. Comparisons of horizontal velocity distribution of cnoidal waves of theories, propagating in domains of the NS and GN equations in Case 2. $H = 0.311$, $\lambda = 29.6$.

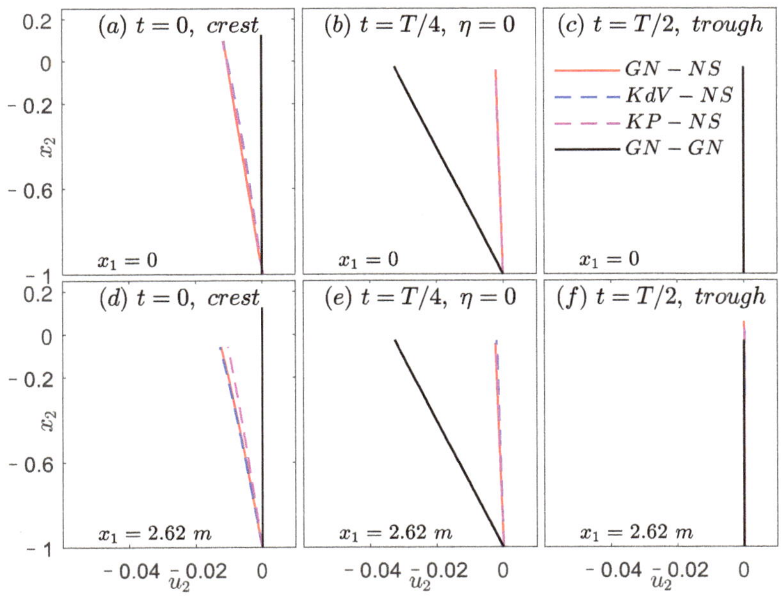

Figure 16. Comparisons of vertical velocity distribution of cnoidal waves of theories, propagating in domains of the NS and GN equations in Case 1. $H = 0.156$, $\lambda = 29.6$.

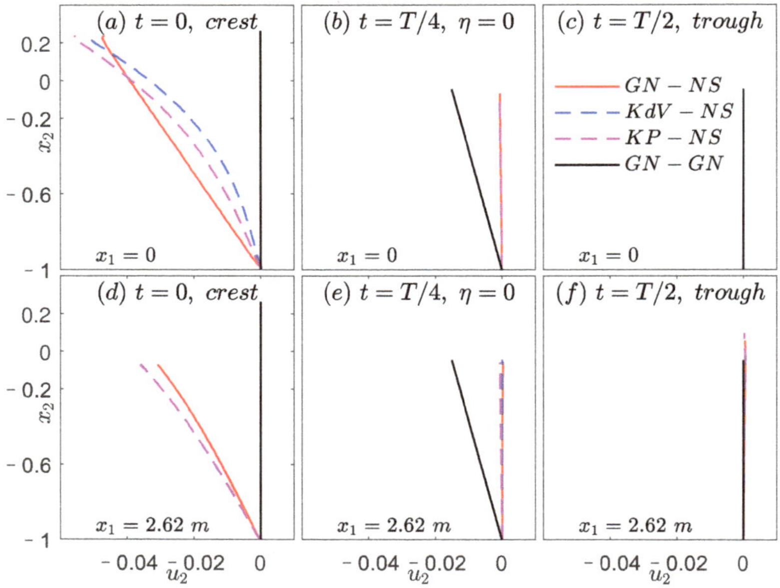

Figure 17. Comparisons of vertical velocity distribution of cnoidal waves of theories, propagating in domains of the NS and GN equations in Case 2. $H = 0.311$, $\lambda = 29.6$.

5. Concluding Remarks

The surface elevation, pressure and velocities of solitary and cnoidal waves generated by different theories are investigated. The stability and accuracy of the waves generated by different theories, propagating in a domain governed by the NS equations, are assessed. We have also considered waves generated by the GN equations, entering the domain of the GN equations.

For a steady-state solitary wave, the surface elevation of the wave generated by Laitone's 2nd is close to that of the GN equations, while those of the Boussinesq equations and Laitone's 1st are close to each other and slimmer than the previous two equations. The wave speed of Laitone's 1st equation is the largest, followed by the Boussinesq equations and the GN equations, and that of Laitone's 2nd is the slowest. Although the GN equations and the KdV equations share the same expression of wave speed for solitary waves, the speed of the wave in the GN-NS model is faster and closer to the theoretical values than that of KdV-NS, while that of GN-GN shows close agreement with the GN-NS model.

For the cnoidal wave, the surface elevation of the GN-GN model is closer to the experimental data, followed by KP-NS, GN-NS, and KdV-NS. Similar results are found for pressure, horizontal and vertical velocities. The wave speed of KP-NS is the fastest, GN-NS is the second, GN-GN is the third, and KdV-NS is the last. The main difference in the horizontal and vertical velocity distribution between GN-NS, KdV-NS, KP-NS, and GN-GN is found under the crest of the cnoidal wave, while the distributions are all the same when observed under the trough.

Overall, it is concluded that nonlinear waves generated by the GN equation are stable and show the least difference when compared with the laboratory measurements.

Author Contributions: Conceptualization, M.H. and R.C.E.; Methodology, M.H. and R.C.E.; Software, J.L., M.H. and R.C.E.; Validation, J.L. and M.H.; Formal analysis, J.L.; Investigation, J.L. and M.H.; Resources, M.H.; Writing—original draft, J.L.; Writing—review & editing, M.H. and R.C.E.;

Visualization, J.L.; Supervision, M.H. and R.C.E.; Project administration, M.H. All authors have read and agreed to the published version of the manuscript.

Funding: This research received no external funding.

Data Availability Statement: The data that support the findings of this study are available within the article.

Conflicts of Interest: The authors declare no conflict of interest.

References

1. Robison, J.; Scott, R.J. Report of the Committee on Waves. In Proceedings of the 7th Meeting of the British Association for the Advancement of Science, Liverpool, UK, September 1837; pp. 417–496.
2. Scott, R.J. Report on waves. In Proceedings of the Fourteenth Meeting of the British Association for the Advancement of Science, York, UK, September 1844; John Murray: London, UK, 1845; pp. 311–390.
3. Airy, G.B. *Tides and Waves, Section 192*; Encycyclopedia Metropolitan: London, UK, 1845; Volume 5, pp. 241–396.
4. Boussinesq, M.J. Théorié de l'intumescence liquide, appelée onde solitaire ou de translation, se propageant dans un canal rectangulaire. *Acad. Sci. Comptes Rendus* **1871**, *73*, 755–759.
5. Boussinesq, M.J. Théorié générale des mouvements qui sont propagés dans un canal rectangulaire horizontal. *Acad. Sci.* **1871**, *73*, 256–260.
6. Boussinesq, M.J. Théorié des ondes et des remous qui se propagent Ie long d'un canal rectangulaire horizontal. en communiquant au liquide contenu dans ce canal des vitesses sensiblement pareilles de la surface au fond. *J. MathéMatiques Pures Appliquées* **1872**, *17*, 55–108.
7. Boussinesq, M.J. Essai sur la théorié des eaux courantes. *Mémoires Présentés Divers Savants L'Académie Sci. (Séries 2)* **1877**, *23*, 1–680.
8. Rayleigh, L. On waves. *Philos. Mag.* **1876**, *1*, 257–259.
9. McCowan, J. VII. On the solitary wave. *Lond. Edinb. Dublin Philos. Mag. J. Sci.* **1891**, *32*, 45–58. [CrossRef]
10. Korteweg, D.J.; de Vries, G. On the change of form of long waves advancing in a rectangular canal and on a new type of long stationary waves. *Philos. Mag.* **1895**, *39*, 422–443. [CrossRef]
11. Keulegan, G.H.; Patterson, J.A. Mathematical Theory of Irrotational Translation Waves. *J. Res. Natl. Bur. Stand.* **1940**, *24*, 47–101. [CrossRef]
12. Littman, W. On the existence of periodic waves near critical speed. *Commun. Pure Appl. Math.* **1957**, *10*, 241–269. [CrossRef]
13. Friedrichs, K.O. On the Derivation of the Shallow Water Theory, Appendix to the Formation of Breakers and Bores by J. J. Stoker. *Commun. Pure Appl. Math.* **1948**, *1*, 81–85.
14. Laitone, E.V. The second approximation to cnoidal and solitary waves. *J. Fluid Mech.* **1960**, *9*, 430–444. [CrossRef]
15. Batchelor, C.K.; Batchelor, G.K. *Cnoidal Waves over a Gently Sloping Bottom*; Series Paper No. 6; Institute of Hydrodynamics and Hydraulic Engineering, Technical University of Denmark: Lyngby, Denmark, 1974.
16. Green, A.E.; Naghdi, P.M. A derivation of equations for wave propagation in water of variable depth. *J. Fluid Mech.* **1976**, *78*, 237–246. [CrossRef]
17. Green, A.E.; Naghdi, P.M. Directed fluid sheets. *Proc. R. Soc. Lond.* **1976**, *347*, 447–473.
18. Sun, X. Some Theoretical and Numerical Studies on Two-Dimensional Cnoidal-Wave-Diffraction Problems. Ph.D. Thesis, University of Hawaii, Honolulu, HI, USA, 1991.
19. Hayatdavoodi, M.; Ertekin, R.C.; Valentine, B.D. Solitary and cnoidal wave scattering by a submerged horizontal plate in shallow water. *AIP Adv.* **2017**, *7*, 065212. [CrossRef]
20. Hayatdavoodi, M.; Ertekin, R.C. Diffraction and Refraction of Nonlinear Waves by the Green–Naghdi Equations. *J. Offshore Mech. Arct. Eng.* **2022**, *145*, 021201. [CrossRef]
21. Seiffert, B.; Hayatdavoodi, M.; Ertekin, R.C. Experiments and computations of solitary-wave forces on a coastal-bridge deck. Part I: Flat plate. *Coast. Eng.* **2014**, *88*, 194–209. [CrossRef]
22. Hayatdavoodi, M.; Seiffert, B.; Ertekin, R.C. Experiments and Computations of Solitary-Wave Forces on a Coastal-Bridge Deck. Part II: Deck with Girders. *Coast. Eng.* **2014**, *88*, 210–228. [CrossRef]
23. Hayatdavoodi, M.; Ertekin, R.C. Nonlinear wave loads on a submerged deck by the Green–Naghdi equations. *J. Offshore Mech. Arct. Eng.* **2015**, *137*, 011102. [CrossRef]
24. Hayatdavoodi, M.; Seiffert, B.; Ertekin, R.C. Experiments and calculations of cnoidal wave loads on a flat plate in shallow-water. *J. Ocean Eng. Mar. Energy* **2015**, *1*, 77–99. [CrossRef]
25. Hallak, T.S.; Islam, H.; Mohapatra, S.C.; Soares, C.G. Comparing Numerical and Analytical Solutions of Solitary Water Waves Over Finite and Variable Depth. In Proceedings of the 40th International Conference on Ocean, Offshore and Arctic Engineering, Online, 21–30 June 2021; American Society of Mechanical Engineers: New York, NY, USA, 2021.
26. Hirt, C.W.; Nichols, B.D. Volume of fluid (VOF) method for the dynamics of free boundaries. *J. Comput. Phys.* **1981**, *39*, 201–225. [CrossRef]
27. Hayatdavoodi, M.; Liu, J.; Ertekin, R.C. Bore Impact on Decks of Coastal Structures. *J. Waterw. Port Coast. Ocean Eng.* **2021**, *148*, 04021051. [CrossRef]

28. Higuera, P.; Lara, J.L.; Losada, I.J. Realistic wave generation and active wave absorption for Navier–Stokes models: Application to OpenFOAM®. *Coast. Eng.* **2013**, *71*, 102–118. [CrossRef]
29. Greenshields, C.J. *OpenFOAM User Guide, Version 6*; OpenFOAM Found. Ltd.: London, UK, 2018.
30. Green, A.E.; Naghdi, P.M. Water waves in a nonhomogeneous incompressible fluid. *J. Appl. Mech.* **1977**, *44*, 523–528. [CrossRef]
31. Neill, D.R.; Hayatdavoodi, M.; Ertekin, R.C. On solitary wave diffraction by multiple, in-line vertical cylinders. *Nonlinear Dyn.* **2018**, *91*, 975–994. [CrossRef]
32. Hayatdavoodi, M.; Neill, D.R.; Ertekin, R.C. On Nonlinear Wave Diffraction by Vertical Cylinders in Shallow Water. *Theor. Comput. Fluid Dyn.* **2018**, *32*, 561–591. [CrossRef]
33. Zhao, B.B.; Ertekin, R.C.; Duan, W.Y.; Hayatdavoodi, M.H. On the steady solitary-wave solution of the Green-Naghdi equations of different levels. *Wave Motion* **2014**, *51*, 1382–1395. [CrossRef]
34. Zhao, B.B.; Duan, W.Y.; Ertekin, R.C.; Hayatdavoodi, M. High-level Green–Naghdi wave models for nonlinear wave transformation in three dimensions. *J. Ocean Eng. Mar. Energy* **2015**, *1*, 121–132. [CrossRef]
35. Dingemans, M.W. Water wave propagation over uneven bottoms: Linear wave propagation. *Adv. Ser. Ocean Eng.* **1997**, *13*, 708–715.
36. Jacobsen, N.G.; Fuhrman, D.R.; Fredsøe, J. A Wave Generation Toolbox for the Open-Source CFD Library: OpenFoam®. *Int. J. Numer. Methods Fluids* **2012**, *70*, 1073–1088. [CrossRef]
37. Ertekin, R.C.; Hayatdavoodi, M.; Kim, J.W. On some solitary and cnoidal wave diffraction solutions of the Green-Naghdi equations. *Appl. Ocean Res.* **2014**, *47*, 125–137. [CrossRef]
38. Hayatdavoodi, M.; Ertekin, R.C. Wave Forces on a Submerged Horizontal Plate. Part I: Theory and Modelling. *J. Fluids Struct.* **2015**, *54*, 566–579. [CrossRef]
39. Hayatdavoodi, M.; Ertekin, R.C. Wave Forces on a Submerged Horizontal Plate. Part II: Solitary and Cnoidal Waves. *J. Fluids Struct.* **2015**, *54*, 580–596. [CrossRef]
40. Hayatdavoodi, M.; Treichel, K.; Ertekin, R.C. Parametric study of nonlinear wave loads on submerged decks in shallow water. *J. Fluids Struct.* **2019**, *86*, 266–289. [CrossRef]
41. Hayatdavoodi, M.; Chen, Y.B.; Zhao, B.B.; Ertekin, R.C. Experiments and computations of wave-induced oscillations of submerged horizontal plates. *Phys. Fluids* **2023**, *35*, 017121. [CrossRef]
42. Zhao, B.B.; Wang, Z.; Duan, W.; Ertekin, R.C.; Hayatdavoodi, M.; Zhang, T. Experimental and numerical studies on internal solitary waves with a free surface. *J. Fluid Mech.* **2020**, *899*, A17. [CrossRef]
43. Seiffert, B.; Ertekin, R.C. Laboratory Generation of Solitary and Cnoidal Waves by the Green-Naghdi Equations. In Proceedings of the International Short Course and Conference on Applied Coastal Research, Florence, Italy, 28 September–2 October 2015.

MDPI AG
Grosspeteranlage 5
4052 Basel
Switzerland
Tel.: +41 61 683 77 34

MDPI Books Editorial Office
E-mail: books@mdpi.com
www.mdpi.com/books

www.ingramcontent.com/pod-product-compliance
Lightning Source LLC
Chambersburg PA
CBHW041017270926
51850CB00024B/790
9783725869947